Primate
Ethology

Primate
Ethology

Edited by
Desmond Morris
With a new introduction by Darryl Bruce

ALDINETRANSACTION
A Division of Transaction Publishers
New Brunswick (U.S.A.) and London (U.K.)

Third printing 2011
New material this edition 2006
Copyright © 1967 by Desmond Morris.

This book is printed on acid-free paper that meets the American National Standard for Permanence of Paper for Printed Library Materials.

Library of Congress Catalog Number: 2005052021
ISBN: 978-0-202-30826-5
Printed in the United States of America

Library of Congress Cataloging-in-Publication Data

Primate ethology / Desmond Morris, editor ; with a new introduction by Darryl Bruce.
    p. cm.
Originally published: London : Weidenfrld & Nicolson, 1967. With a new introduction.
Includes bibliographical references.
ISBN 0-202-30826-X (alk. paper)
    1. Primates—Behavior. I. Morris, Desmond.

QL737.P9P67246   2005
599.8'15—dc22   2005052021

Contents

Introduction to the AldineTransaction Edition   xi

1   Introduction: the Study of Primate Behaviour   1

    DESMOND MORRIS, *Zoological Society of London*

2   The Facial Displays of the Catarrhine Monkeys and Apes   7

    J. A. R. A. M. VAN HOOFF, *Laboratorium voor Vergelijkende Fysiologie, Utrecht*

3   Socio-sexual Signals and their Intra-specific Imitation among Primates   69

    WOLFGANG WICKLER, *Max-Planck-Institute für Verhaltensphysiologie, Seewiesen*

4   Allogrooming in Primates: a Review   148

    JOHN SPARKS, *Ethology Laboratory, Zoological Society of London*

5   Play Behaviour in Higher Primates: a Review   176

    CAROLINE LOIZOS

6   Variability in the Social Organization of Primates   219

    T. E. ROWELL, *Department of Zoology, Makerere University College, Kampala*

7   Comparative Aspects of Communication in New World Primates   236

    M. MOYNIHAN, *Canal Zone Biological Area, Smithsonian Institution*

8 The Effect of Social Companions on Mother-Infant
Relations in Rhesus Monkeys   267

  R. A. HINDE AND Y. SPENCER-BOOTH, *Sub-Department
  of Animal Behaviour, Madingley, Cambridge*

9 Mother-Offspring Relationships in Free-ranging
Chimpanzees   287

  JANE VAN LAWICK-GOODALL, *Sub-Department of Animal
  Behaviour, Madingley, Cambridge*

10 An Ethological Study of Some Aspects of Social
Behaviour of Children in Nursery School   347

  N. G. BLURTON JONES, *Department of Growth and Development, Institute of Child Health, University of London*

Index   369

List of Illustrations

Plates

Chapter 2 (*between pages 14 and 15*)

    I   Tense-mouth face of the male *Papio anubis*

   II   Silent bared-teeth face of *Mandrillus sphinx*

  III Silent bared-teeth face of chimpanzee

  IV Protruded-lips face of *Macaca nemestrina*

   V   Relaxed open-mouth face of chimpanzee

  VI Crab-eating monkeys with silent bared-teeth face

 VII Silent bared-teeth face of chimpanzee

VIII Pout-face of chimpanzee

  IX Temper tantrum response of chimpanzee

Chapter 3 (*facing pages 86 and 102*)

    I   *Papio anubis, C. aethiops, P. hamadryas* and Gelada

   II   *Cercopithecus aethiops* and Miamin

Chapter 8 (*facing page 278*)

    I   Rhesus monkey restraining her infant

Chapter 9 (*between pages 294 and 295*)

    I   Melissa with her one-day-old infant

   II   Mandy with her sleeping infant

III   Mandy on a branch
IV   Melissa with her infant
V   The infant Flint
VI   The infant Goblin
VII   Flo with her daughter
VIII   Circe and baby with a one-year-old infant
IX   Olly with her infant Gilka
X   Marina and her infant
XI   The mother Flo with her offspring
XII   Flo with her male infant
XIII   Flo with the infant Flint
XIV   The elder offspring of Flo

Figures

Chapter 2   page
1   Facial expressions in a Macaca type primate   64

Chapter 3
1   Female sexual swellings   75
2   Presentation of *Lemur catta* and *Pan troglodytes*   76
3   Social presentation of a female baboon   79
4a   Presentation of female rhesus   80
4b   Sexual presentation of Chacma baboon   80
4c   Submissive posture of Chacma baboon   80
5   Presentations of the Hamadryas baboon   81
6   'Protected threat' of a Hamadryas female   83
7   Posture of a high-ranking rhesus male   87
8   *Procolobus verus* genitalia   95
9   Female *Tarsius* genitalia   96
10a   Social presentation of young male lion   105
10b   Sexual presentation of female domestic cat   105
11   Genital display of young *Saimiri* male   111

12   Genital display in male 'sentinels'   117
13   *Papio hamadryas*, XVIII Dynasty bas-relief   118
14   Hottentot's skirt   125
15a, b Stone-age figurine from Laspugue, Haute-Garonne   126
15c Stone-age figurine from Savignano, Italy   126
15d Modern Kaffa figurine, Africa   126
15e Modern Caraja-Indian figurine, Brazil   126
15f Hottentot female from Bersaba   127
15g Bustle dress (1882)   127
16   Human male and female sitting postures   129
17   Pharaoh blessed by a deity   130
18   Hermes of Siphnos   131

Chapter 7

1   Tail-twining by *Callicebus moloch*   239
2   The arch posture of *Aotus trivirgatus*   248
3   Visual displays of *Callicebus moloch*   249
4   Ritualized facial expressions of different species   251
5   Ritualized facial expressions and pilo-erection patterns of
     different species   253
6   Pilo-erection patterns of *Cebuella pygmaea*   254

Chapter 8

1   Infants on the nipple with eyes closed   271
2   Infants on the nipple with eyes open   272
3   Infants on their mothers but not attached to the nipple   273
4   Infants off their mothers   274
5   Infants more than two feet from their mothers   275
6   Distance between infants and mothers increasing from
     less than two feet to more than two feet   276
7   Distance between infants and mothers decreasing from
     more than two feet to less than two feet   277
8   Infants on and off their mothers   278

9 Infants more than two feet from their mothers   279
10 Mothers grooming their infants   281

Chapter 9

1 The hunched gait of a mother   291
2 The mother Flo cradling her infant   293
3a Mother about to move off   297
3b Infant climbing up its mother   297
3c Infant and mother move off   297
4 Mean suckling frequency of the infant Flint   299
5 Mean length of suckling bouts of infants   301
6 Percentage of interference of mother Flo   307
7 Frequency of mother and offspring travelling about
        together   317
8 Frequency of maternal play   320
9 Frequency of Flo's play with her offspring   323
10 Distribution of grooming sessions of mothers and off-
        spring   326
11 Frequency of grooming sessions between mothers and
        offspring   328
12 Percentage of Flo's responses to the 'hoo' whimper   329

Chapter 10

1 Frequency of running and laughing in children   363

Introduction to the Aldine Transaction Edition

Darryl Bruce

MODERN ethology springs from biology and may be defined as "the comparative study of the natural behaviour of animal species" (Thorpe, 1979, p. viii). In practice, this has rarely meant the study of the behavior of primates.[1] *Primate Ethology*, edited by Desmond Morris, thus represents an exception in that it brings together reports by ethologists of their investigations of some of the behavior patterns of primates, both human and nonhuman. Modern ethology is actually a johnny-come-lately to this field, for there is a significant history of research on primate behavior that stretches back some 100 years prior to the present volume. Some of it was actually conducted by those who today would qualify as ethologists, but most of it was carried out by psychologists. In my introduction, I will give an overview of this history. My aim is to set a context for the research reported in *Primate Ethology* and to indicate the book's unifying significance to the science of animal behavior.

As Morris mentions in his opening chapter, the seeds of the scientific study of primate behavior, and animal behavior in general, lie in the publication in 1859 of Charles Darwin's book, *On the Origin of Species by Means of Natural Selection*. For one thing, the assumption that humans and mammals have the same ancestor—common descent as it is called—was now on the table, even though in 1859, at least, Darwin soft-pedaled the inclusion of man in the mammalian line of descent.

---

[1] Unless otherwise indicated, the term "primate" will be used to mean a nonhuman primate—a monkey, a chimpanzee, an orangutan, and so on.

But no one missed that implication. One need only consider Bishop Wilberforce's insolent question to Thomas Huxley, Darwin's defender, in the course of a debate over evolution and Darwinian theory in 1860 at the University of Oxford: "I should like to ask Professor Huxley . . . as to his belief in being descended from an ape. Is it on his grandfather's or his grandmother's side that the ape ancestry comes in?" (F. Darwin, 1892/1958, pp. 251-252).

A second important implication of Darwin's treatment of evolution was that just as bodily features have an evolutionary history, so too do behavior patterns. The critical agent in that history is natural selection, the most important component of Darwin's theory. The origins of many behaviors, like those of physical structures, can only be seen in the light of their service in meeting existing selection pressures. The evolutionary history of a behavior can be similar to that of a bodily characteristic in other ways. For instance, a behavior may become vestigial. Or it may acquire a different function altogether. For example, a specific courtship movement by a duck may have developed out of an action that initially had a preening function. In sum, it is the combination of common descent and the evolutionary history of behavior patterns that is the basis for the comparative study of animal behavior.

In view of Darwin's ideas on evolution, especially that of common descent, one might expect that scientists interested in animal behavior would turn their attention to man's closest relatives, primates. And so they did, but the initial steps were tentative, constrained certainly by the immense practical difficulties in doing research on primates. Again, Darwin (1872/1965) provides a starting point, specifically, his important psychological work, *The Expression of the Emotions in Man and Animals.* Among the data that Darwin presented in his monograph were his own observations, begun perhaps as early as 1838, of the emotional behaviors of primates—monkeys, baboons, chimpanzees, and orangutans—housed in the Zoological Gardens of London. In addition to merely observing their behavior, Darwin sometimes intervened to create conditions that might produce emotional experiences in the animals. For example, he recounts that he once placed a mirror on the floor before two young orangutans to see what expressive behaviors might occur.

Darwin's protégé, George Romanes, also studied the behavior of primates. Romanes' principal concern was the question of mental evolution (Romanes, 1882, 1884), a problem that Darwin had taken up earlier in *The Descent of Man* (1871). We need not enter into the wide-ranging case that Romanes mounted on this matter. Suffice it to say that he saw a steady increase in mental ability ranging from primitive animals to humans. Because the evidence concerning the mental abilities of primates was scanty, Romanes borrowed a monkey from the collection of the London Zoological Society. He housed it in the home of his sister and charged her with noting all points of interest connected with the monkey's intelligence. For roughly two and a half months, she recorded her observations. Romanes then published them verbatim in his 1882 book, *Animal Intelligence*. Despite such flimsy grounds, Romanes had little hesitation in concluding that the psychology of apes most closely approached that of humans.

Toward the end of the nineteenth century, Edward Thorndike began his influential research in North America on animal behavior. His investigation of cats and small dogs attempting to escape from puzzle boxes (Thorndike, 1898) was in sharp contrast to the work and ideas of Romanes. Where Romanes' methods were largely observational, qualitative, and often marked by subjective impressions, Thorndike's were experimental, quantitative, and objective. And where Romanes saw plenty of evidence for animal intelligence, Thorndike saw precious little. Instead, he maintained that animal problem solving was a process of trial and error followed by accidental success, repeated occurrences of which led to a gradual emergence of the correct solution behavior.

Thorndike (1901/1911) went on to examine the problem-solving ability of monkeys. Particularly relevant to the present discussion are the answers to three questions that Thorndike posed: How did the behavior of monkeys escaping from puzzle boxes and solving other mechanical puzzles compare with that of cats and dogs? Thorndike observed that though the learning was faster, it did not appear to be qualitatively different. Could monkeys learn the solution to a problem by watching the performance of their fellows who had already mastered the task? Or could they acquire it if Thorndike showed them how to solve the problem or physically guided them through the correct behavior? The answer to both questions was "no." Thorndike took these findings as providing no basis for the conclusion that monkeys possess the capacity to reason or that they have the ability to associate "ideas."

Was Thorndike suggesting that there was no mental continuity between humans and animals, a continuity that both Darwin and Romanes had argued for? Not at all. In fact, he explicitly opted for a continuity, but it was along an associative dimension that did not involve mental events. The continuity was a presumption of his laws of effect and exercise, which were intended as principles that governed learning and behavior in all species (Thorndike, 1911). Together, they set forth the conditions that strengthen a connection, not between ideas but between "situations" and "responses." Differences in intelligence among species concerned how fast such S-R associations are established, their number, delicacy, complexity, and durability. In short, species differences in intelligence were a matter of degree not kind.

Thorndike's research and his theoretical pronouncements stood at a fork in the road in the study of animal behavior. Four subsequent paths are clearly discernable, all of which reflect to varying degrees the direct or indirect influence of Thorndike. I will give capsule descriptions of each, paying particular attention to the primate behavioral research that occurred in each instance.

1. The question of whether animals reason.

Thorndike's position that animals possessed no ability to reason was controversial. Two individuals made prominent attempts to show otherwise and studied primates in the effort. The first was Leonard Hobhouse, a jack-of-all-intellectual-trades, one of which was biology and, specifically, mental evolution. Thorndike's 1898 monograph was certainly not the impetus for the research that Hobhouse reported in his 1901 book, *Mind in Evolution*, but the data and conclusions that Thorndike presented came in for detailed and, I might add, quintessentially articulate British criticism. For example, Hobhouse (1901) characterized Thorndike's explanation of animal learning as "repeated blundering efforts with fortuitous successes that are gradually selected" (p. 142) and then acidly remarked that the idea "has been applied as a sort of universal solvent by Mr. Thorndike" (p. 142). The subjects of Hobhouse's research were animals at the zoo in Manchester, the city where he lived. His initial studies tested nonprimates, but his later and more compelling work was done with a monkey and a chimpanzee. The upshot was that the primates learned to employ objects to solve problems, for example, using a stick to pull in food through the bars of their cages, food that was otherwise beyond their research. The nonprimate subjects were never able to demonstrate tool use to the same degree that the monkey and chimpanzee did. In clear opposition to Thorndike's views on the animal mind, Hobhouse concluded that apes were able to form and connect articulate ideas (e.g., of the relations of objects), connections that found expression in their problem-solving behavior.

Somewhat similar conclusions were reached by Wolfgang Köhler. Best remembered as one of the three German originators of Gestalt psychology, he served from 1913 to 1920 as director of a research station for the study of apes located on Tenerife in the Canary Islands. It was a productive period for Köhler, but arguably his most significant work during that time was his investigation of the problem-solving skills of chimpanzees. He described this research in a book, *The Mentality of Apes* (Köhler, 1917/1927)

Köhler's issue with Thorndike was that his puzzle-box tasks did not allow the animal to see all the elements of the problem-solving situation so that they could be combined into a solution. The formula for the problems that Köhler set his chimpanzees was to block the direct route to a solution but to leave open an indirect or roundabout path and allow the animal to see all the critical components of the whole situation. Köhler used two main variations on this theme. One was in the vein of Hobhouse. A chimp had to rake in food out of reach outside its cage by using a stick that was lying about. Sometimes two sticks had to be fitted together to make a rake long enough to reach the food. The second variation involved hanging a banana overhead that again was out of the animal's reach. To solve this problem, the subject had to position a nearby box suitably and then stand on it to reach the food. A more difficult version of the task required the chimp to pile a number of boxes on top of one another to seize the banana.

The apes varied in their success at these problems, but when a solution occurred, it often did so in a very characteristic fashion. First, there would be a pause in the animal's presolution activity. Then the correct behavior would run off in a continuous or nonstop sequence. Kohler's interpretation was that the chimps were perceptually restructuring the problem situation and then acting accordingly. In sum, he saw such behavior as both intelligent and insightful.

A Russian program of research on primate behavior that is not well known in the West also warrants mention. Ivan Pavlov's research center at Koltushi near Leningrad included an anthropoid research facility. Pavlov was critical of Kohler's conclusion that apes showed insight in problem solving. About 1933, he conducted his own tests of chimpanzees attempting to master the task of putting boxes on top of one another to reach a suspended bait. The animals were eventually able to do this, but only after a protracted period of fruitless effort. Pavlov's explanation of their behavior was couched within a conditioning framework and thus more in keeping with Thorndike's trial-and-error account of animal problem solving (Windholz, 1984, 1997).

After Pavlov's death, his colleague, E. G. Vatsuro, continued this line of investigation. His was a multi-year project and involved experiments with a single adult male chimpanzee named Rafael on a variety of problem-solving tasks (e.g., the stick problem). Rafael had also been tested in a number of Köhler-type situations over a four-year span prior to Vatsuro's investigation. A description of the animal's behavior in this program of research has been provided by Ladygina-Kots and Dembovskii (1969). The essential outcome was that Köhler's findings were corroborated, but the Russian camp interpreted them differently. Learning was accorded a greater role. As Ladygina-Kots and Dembovskii put it, Köhler "takes no account of the fact that the chimpanzee has his own past during which he has acquired many habits, and that he turns to them in the experimental situation" (p. 61). If that statement seems to have a Thorndikean flavor to it, the taste is even stronger in the following assessment of the ape's tool-use behavior:

> Higher apes are able to alter an unsuitable object in order to make it fit a specific task; they can even construct a tool from several parts, but unification of parts was done unintentionally, accidentally, during play. The associations which the ape establishes are space-time relationships and not cause-effect relationships. (Ladygina-Kots and Dembovskii, 1969, p. 69)

It may be noted that Ladygina-Kots was herself a prominent Russian primatologist. Among her accomplishments were establishing the zoo-psychological laboratory of the Darwin Museum in Moscow and raising a chimpanzee in her home along with her own child (Cole and Maltzman, 1969).

None of the primate research on reasoning had much of an impression on the early study of animal behavior in North America. The Russian work was undoubtedly even less known then than it is today. As for Hobhouse and Köhler's anti-Thorndike conclusions, there were likely a number of reasons why they had a lukewarm reception. One was surely that the methods Hobhouse and Köhler used did not have the rigor and the quantitative character of Thorndike's. As well, Köhler's book was not available in an English translation until 1925, which no doubt limited its North American audience. But the real resistance could well have been the kind of psychology that Thorndike's ideas may have encouraged, a matter that I now wish to consider.

2. The psychology of learning and general behavior theory.

For well over half a century—from approximately 1910 to 1970—the investigation of learning held center stage in North American experimental psychology. The objective was to uncover the basic principles of learning, principles that applied not just to learning in animals but in humans as well. Furthermore, the principles were assumed to account for behavior in general. Some have seen the hand of Thorndike in this tradition (Dewsbury, 1978; Jenkins, 1979). After all, hadn't Thorndike shown us that differences between the minds of different species of animals were a matter of degree not kind? Hadn't he demonstrated that there were no species-specific differences in learning? And hadn't he made a persuasive case for the basic continuity between the intelligence of animals and humans? Especially appealing was his use of the experimental method, a laboratory setting, and the objective measurement of behavior. But whatever the reasons, many psychologists, at least in North America, went down the trail that Thorndike had blazed.

The question of interest to the present discussion is how much the study of learning involved primate subjects. Suffice it to say that the short answer is, "a great deal." Warren (1965) provides a lengthier answer in his review of research up to the early 1960s of learning by primates and nonprimates in a variety of experimental tasks. Especially pertinent to the Thorndikean view that the principles of learning are common to all species is Warren's conclusion that "the differences in learning between primates and other mammals are not so great as they seemed to many primatologists a few years ago" (p. 275).

3. Comparative psychology.

Another North American research tradition in the study of animal behavior that developed after Thorndike was comparative psychology. The field defies any widely accepted definition (Dewsbury, 1984). One of its distinguishing features, however, is the investigation of a wide range of animal behavior patterns and problems, a range that does not ordinarily include learning. Sample areas of research interest to a comparative psychologist are reproductive behavior, aggression, dominance, exploratory behavior, feeding, and communication, to mention but a few.

My brief account of the history of comparative psychology during the first sixty years of the twentieth century will center almost entirely on Robert Yerkes, who with considerable justification may be called the father of North American comparative psychology and primatology. Yerkes received his Ph. D. degree from Harvard University in 1902. He was a comparative psychologist from the outset and pursued a variety of problems in a range of species. But it was the study of primates where Yerkes made his mark. A discussion of his many contributions to the field is beyond the scope of my remarks, but among the more prominent were his investigations (along the lines of those by Hobhouse and Köhler) of the problem-solving abilities of monkeys and an orangutan (Yerkes, 1916) and a book of comprehensive information about apes (Yerkes and Yerkes, 1929).

Perhaps his most enduring accomplishment was establishing a laboratory devoted exclusively to the psychology and biology of primates, an idea that occurred to him in 1900 (Dewsbury, 1996). Thirty years later—June, 1930, to be exact—it was a reality. With generous financial support from the Rockefeller Foundation, the Anthropoid Experiment Station of Yale University—where Yerkes was now a faculty member—opened for business in Orange Park, near Jacksonville, Florida. Yerkes served as director until his retirement in 1941, at which time the laboratory was renamed in his honor as the Yerkes Laboratories of Primate Biology. In 1965, the facility was relocated in Atlanta, Georgia, but it still bears Yerkes' name.

From its inception, the laboratory was a leading center for primate research. From a purely behavioral perspective, however, its halcyon days were undoubtedly the period from 1930 to 1960, when many of the most famous names in the history of comparative psychology were associated with the Yerkes laboratories. In addition to Yerkes, they included Karl Lashley, Henry Nissen, Donald Hebb, and Austin Riesen, all primatologists of the first rank. Henry Nissen bears singling out, for at the time of his death in 1958, he was described as the "Western world's leading authority on the biology and psychology of the chimpanzee" (Carmichael, 1965, p. 205). The evaluation is deserved. He was a pioneer in research on chimpanzees in their natural habitat, which he did for two months in Africa in 1930. Later, he conducted notable investigations of the behavioral and perceptual development of chimpanzees. In 1939, he was appointed assistant director of the Yerkes laboratories and in 1955, director, a position that he held until his death in 1958. His ability in caring for and otherwise managing the chimpanzee colony at Orange Park—by 1941, it numbered forty-five chimps, the world's largest captive collection at the time—was legendary and the success of the laboratories owed much to him.

In addition to Nissen's field research, Yerkes also stimulated the field studies of two other investigators. One was Walter Bingham, who conducted research on gorillas in their natural habitat in 1932 in what was then the Belgian Congo. The other was C. Ray Carpenter, whose fieldwork on primates can only be described as magisterial. Among his projects were an investigation of the behavior of howler monkeys in Panama in 1931 and a study of gibbons and orangutans in Southeast Asia in 1937. No aspect of the behavior of these animals—their territoriality, sexual activity, social behavior, and so forth—escaped the attention of this extraordinarily observant scientist.

The study of primates in the United States received a considerable boost in 1959 and 1960 when the U. S. Congress funded the Primate Research Center program. Yerkes became one of seven regional centers under the umbrella of that program. With this development, we may turn to the fourth of our post-Thorndikean paths in the study of animal behavior, that of ethology.

4. Ethology.

Gottlieb (1979) has claimed that "ethology came on the scene when most animal psychologists began to limit their study to only a few species under a narrow range of laboratory conditions, which were not designed with an eye toward the evolution or development of the species under study. . ." (p. 163). The veiled suggestion is that the emergence of ethology was a reaction to the North American tradition of learning and general behavior theory inspired by Thorndike's work. There is a surface plausibility to this view. Modern ethology certainly came into its own during the first sixty years or so of the twentieth century with the influential research of Karl von Frisch on the language of honey bees, Konrad Lorenz on imprinting in greylag geese, and Niko Tinbergen on nest finding in wasps, aggression in stickleback fish, and feeding activity of herring gulls. Moreover, it is true that in many ways, ethology and animal learning research in North America were a study in contrasts, contrasts that are spelled out by Morris in his introduction (see p. 5) and need not be repeated here.

The facts of the matter, however, are otherwise. In the first place, ethology has a long history. Its origins are to be found in France during the second half of the eighteenth century (Thorpe, 1979) and its path of development is clearly through Darwin and Romanes. As for the contrast between ethology and the study of animal behavior by North American psychologists, it is quite misleading to portray it as representing the differences between ethology on the one hand and comparative psychology (the third of my post-Thorndike paths) on the other, as Morris does. The reality is that ethology and comparative psychology have much in common, not the least of which is their strong biological emphasis. True, there has been dissent between the two fields (e.g., Lehrman, 1953), but the affinities have always far outweighed the differences (see Dewsbury, 1984, for an extended discussion of this issue). The commonality may also be seen in the many points of historical overlap between the two disciplines. For example, C. Lloyd Morgan has been called "one of the founding fathers of both comparative psychology and ethology" (Thorpe, 1979, p. 26); the comparative psychologist John B. Watson conducted early field studies of the nesting behavior of noddy and sooty terns that were as ethological as they come (e.g., 1908); and Karl Lashley, also a comparative psychologist, wrote a famous analysis of instinctive behavior (1938) that stated the ethologists' view of instinct even before they did. The list could go on.

The current situation is that the differences between ethology and comparative psychology have all but disappeared. Up to 1967 (the date of the present volume), however, there was still at least one area of separation between the two fields. As I emphasized in sketching the development of comparative psychology, the study of primate behavior has always been a major concern. Ethologists, by contrast, have historically directed little attention to this topic. The importance of *Primate Ethology*, then, is not simply that it stands as a significant contribution to primatology, but that it puts finis to that difference. Indeed, *Primate Ethology* may well mark the removal of the last significant gap between the two fields.

Darryl Bruce
Saint Mary's University

REFERENCES

CARMICHAEL, L. (1965). Henry Wieghorst Nissen, February 5, 1901-April 27, 1958. *Biographical Memoirs of the National Academy of Sciences*, 38, 204-222.

COLE, M. and MALTZMAN, I (Eds.). (1969). *A handbook of contemporary Soviet psychology*. New York: Basic Books.

DARWIN, C. (1859). *On the origin of species by means of natural selection, or the preservation of favoured races in the struggle for life*. London: John Murray.

DARWIN, C. (1871). *The descent of man and selection in relation to sex*. London: John Murray.

DARWIN, C. (1965). *The expression of the emotions in man and animals*. Chicago: University of Chicago Press. (Original work published 1872).

DARWIN, F. (Ed.). (1958). *The autobiography of Charles Darwin and selected letters*. New York: Dover. (Original work published 1892 as *Charles Darwin, his life told in an autobiographical chapter and in a selected series of his published letters*. New York: Appleton.)

DEWSBURY, D. A. (1978). *Comparative animal behavior*. New York: McGraw-Hill.

DEWSBURY, D. A. (1984). *Comparative psychology in the twentieth century*. Stroudsburg, PA: Hutchinson Ross.

DEWSBURY, D. A. (1996). Robert M. Yerkes: A psychobiologist with a plan. In G. A. Kimble, C. A. Boneau, and M. Wertheimer (Eds.), *Portraits of pioneers in psychology*, (Vol. II, pp. 86-105). Mahwah, NJ: Erlbaum.

GOTTLIEB, G. (1979). Comparative psychology and ethology. In E. Hearst (Ed.), *The first century of experimental psychology* (pp. 147-173). Hillsdale, NJ: Erlbaum.

HOBHOUSE, L. T. (1901). *Mind in evolution*. London: Macmillan.

JENKINS, H. M. (1979). Animal learning and behavior theory. In E. Hearst (Ed.), *The first century of experimental psychology* (pp. 177-228). Hillsdale, NJ: Erlbaum.

KÖHLER, W. (1927). *The mentality of apes* (Translation by E. Winter of the 2nd revised German edition). New York: Harcourt, Brace. (Original German edition, 1917).

LADYGINA-KOTS, N. N. and DEMBOVSKII, Y. N. (1969). The psychology of primates. In M. Cole and I. Maltzman (Eds.), *A handbook of contemporary Soviet psychology* (pp. 41-70). New York: Basic Books.

LASHLEY, K. S. (1938). Experimental analysis of instinctive behavior. *Psychological Review*, 45, 445-471.

LEHRMAN, D. S. (1953). A critique of Konrad Lorenz's theory of instinctive behavior. Quarterly Review of Biology, 28, 337-363.

ROMANES, G. J. (1882). *Animal intelligence*. London: Kegan, Paul, Trench.

ROMANES, G. J. (1884). *Mental evolution in animals*. New York: Appleton.

THORNDIKE, E. L. (1898). Animal intelligence: An experimental study of the associative processes in animals. *Psychological Review Monograph Supplements*, 2 (4, Whole No. 8), 1-109.

THORNDIKE, E. L. (1911). Laws and hypotheses for behavior. In E. L. Thorndike, *Animal intelligence: Experimental studies* (pp. 241-281). New York: Macmillan.

THORNDIKE, E. L. (1911). The mental life of the monkeys; an experimental study. In E. L. Thorndike, *Animal intelligence: Experimental studies* (pp. 172-240). New York: Macmillan. (Original work published 1901).

THORPE, W. H. (1979). *The origins and rise of ethology*. London: Heinemann Educational Books.

WARREN, J. M. (1965). Primate learning in comparative perspective. In A. M. Schrier, H. F. Harlow, and F. Stollnitz (Eds.), *Behavior of nonhuman primates: Modern research trends* (vol. I, pp. 249-281). New York: Academic Press.

WATSON, J. B. (1908). The behavior of noddy and sooty terns. *Carnegie Institution Publications*, 2(103), 187-255.

WINDHOLZ, G. (1984). Pavlov vs. Köhler: Pavlov's little-known primate research. *Pavlovian Journal of Biological Science*, 19, 23-31.

WINDHOLZ, G. (1997). Ivan P. Pavlov: An overview of his life and psychological work. *American Psychologist*, 52, 941-946.

Chapter One

Introduction: The Study of Primate Behaviour

DESMOND MORRIS

THE objective investigation of primate behaviour is probably the most demanding branch of zoological study. There are two reasons for this. Firstly, monkeys and apes are so intelligent, their brains so advanced, their social organizations so intricate and variable, that the student of animal behaviour can easily become dazzled by the complexity of the scene. All too often in the past he has stood at the edge of the problem, gesticulating airily with expansive generalizations, or puffing himself up with elaborate but fact-starved theories.

Alternatively, if he has been sufficiently subjected to experimentalist indoctrination, he has defended himself by plunging head-first into some minute facet of primate life, where the scale of the task is so limited that he can feel reasonably secure, and can quickly assemble fashionable quantifications.

Secondly, monkeys and apes are so closely related to the human animal that the dangers of anthropomorphism are constantly present. This 'humanizing' can be deliberate or accidental. In the former case, the primate is purposely used as a substitute for man in some rigorous test situation, with the specific intention of applying the results obtained from the animal to a human context. In particular, medical and psychiatric research has utilized primates extensively in this way, but this can hardly be classified as a true study of primates as primates; it is the study of primates as dummy-humans, and anything we learn about them as unique zoological specimens here is a purely secondary benefit.

Accidental 'humanizing' of primate activities can distort objectivity in a subtle but damaging way. If a zoologist sees that an animal such as a fish has, say, a 'fierce expression' he will be in little danger of drawing the conclusion that this generally indicates an aggressive motivation. He will be aware that the facial expression of the fish is entirely fortuitous, the result merely of a special configuration of jaw muscles and bones that give it an obviously accidental resemblance to human 'fierceness'. The anthropomorphism is so far-fetched that it is clearly recognized and discarded. But when an observer notes a 'fierce expression' on the face of a monkey or an ape, he is less likely to discard it. Monkeys and apes do, after all, possess aggressive facial expressions and they are, of course, closely related to the faces made by man. Closely related, yes, but identical, no; hence the real danger of anthropomorphic distortion. The student of primate behaviour must be constantly aware of this problem. He must work hard to keep a perfect balance when recording the similarities and the differences between man and his primate relatives.

In addition to these theoretical difficulties there are also formidable practical obstacles to be overcome, both in the laboratory and the field. In the laboratory it is possible to keep a vast army of rats or mice in the space required by a mere handful of monkeys or apes. A million fruit-flies can be housed in the area required for a dozen apes. Furthermore, primates are expensive to feed, difficult to keep clean, hard to handle, and often dangerous. If they are maintained in typical laboratory cages it is impossible to study any but their simplest behaviour patterns. If they are set up in a social group, then a large enclosure is required and it immediately becomes difficult to manipulate or control the individual members of the group for experimental purposes. A fish, a small bird, or a rat can be put into a social group or removed from it in a matter of seconds by means of a small hand-net, or some similar device. To perform the same operation with a monkey in a colony can easily become a major undertaking.

In the field, observations also provide special difficulties for the primatologist. Birds have fixed nesting-sites, display grounds, or feeding areas. Monkeys and apes tend to be more nomadic, the groups perpetually moving from place to place in an extensive home range. The good field observer must copy them, must roam with them, often in an unfriendly and hazardous environment. Even when face to face with them at close quarters there is the frustrating problem of the branch or clump of leaves behind which the monkey performers disappear at a vital moment in a behaviour sequence.

Given all these disadvantages and difficulties, how exactly has the study of primate behaviour fared? How did it begin and what advances have been made in recent years? There is little to record of any importance in pre-Darwinian times, and this is not surprising. Only after Darwin had clarified the nature of the true evolutionary relationship between man and the monkeys and apes, were these difficult objects of study considered to be worth the special trouble their serious investigation demanded. Previously they had been looked upon as some strange kind of 'counterfeit' of the human, interesting enough for morphological comparisons in order, as the nineteenth-century anatomist Richard Owen put it, to appreciate 'those modifications by which a material organism is especially adapted to become the seat and the instrument of a rational and responsible soul', but still 'brute beasts' unworthy of comparison with rational man when it came to the question of behaviour.

Early on in the post-Darwinian epoch a number of authorities began to realize that our new-found relatives demanded a closer scrutiny. It took an interestingly distorted form:

In effect, Darwin set off two trends (in the present context). One was to consider the behaviour of living animals, especially monkeys and apes, as worthy of detailed scientific study. As our near relations, their way of life became of more than passing interest. The other development, which accompanied it, was the 'uplifting' of our poor relatives to a level that, while keeping them in a suitably inferior status, nevertheless raised them to a plane that did not disgrace their richer, human cousins. ... The gap between monkey and man is narrowed – to satisfy Darwinism – while man remains on his pedestal – to satisfy the church. (From *Men and Apes* by Ramona and Desmond Morris, pp. 151–2.)

Observations in zoos and occasionally in the field were all heavily biased towards showing how startlingly complex and advanced were the social and other behavioural activities of our simian relations. It was over half a century later before the dust from the Darwinian explosion had settled sufficiently for the air to be clear enough to see what primates really did and did not do. In the 1920s and early 1930s a number of independent pioneering studies were undertaken that set off a new epoch of primate behaviour investigation. On the Canary Islands the German Wolfgang Kohler carried out his now classic tests of chimpanzee intelligence. In Moscow Nadie Kohts began an intensive comparison of the behaviour patterns of a young chimpanzee and a human infant. In the United States, Robert Yerkes set up a major ape station at Orange Park in Florida. In London, Solly Zuckerman made a detailed study of the social behaviour of a large colony of baboons at the Regent's Park Zoo. Harold Bingham, Henry Nissen and C. R. Carpenter all ventured into the field to investigate respectively, gorillas, chimpanzees and howler monkeys.

From these individual and separate endeavours came an exciting new picture of the primate way of life. The stage was set for a co-ordination of the new findings and an extensive attack on a broad front of the whole primate problem. Unhappily this did not follow. Instead, World War II concentrated attention strictly on to the aggressive behaviour of a single primate species.

After the war, comparative primate studies were slow to start again. The momentum had been lost. It was as if man's own behaviour had not endeared him to the group of which he was a member. A decade passed before serious work began again on any appreciable scale. Then, with increasing rapidity, field and captive studies were planned and initiated. Research workers from the United States, Switzerland and Britain took to the field again, concentrating their efforts largely in Africa. In Japan a flourishing Monkey Centre was established at Inuyama. In the last few years more and more laboratory behaviour studies on a wide variety of primate species have been set in motion, especially in the United States where, during the first half of the present decade, government grants of nearly twenty million dollars were made available to enable the establishment of seven major Regional Primate Research Centres. Primate investigations are no longer isolated individual projects, but concerted and co-ordinated programmes involving thousands of monkeys and apes and hundreds of research workers.

Much work of great value is being carried out, but there is a risk that, in the rush, certain aspects of the problem or, to be more precise, certain approaches to it, will be given insufficient attention. To understand why this danger should exist it is necessary to look back again at the inter-war period. At that time there were two major approaches to the study of animal behaviour in general: comparative psychology and comparative ethology. The former was well established, the latter a struggling newcomer. Comparative

4

psychologists were concentrating almost entirely on the white rat and on strictly experimental work in the laboratory. If they called themselves 'comparative', the comparison was a narrow one: the rat was compared with man. They were not trained as zoologists and lacked an evolutionary or functional approach – even scorned it. The comparative ethologists, under the leadership of Konrad Lorenz and Niko Tinbergen, were zoologists who felt that there was more to animal behaviour than rats running mazes or pressing levers. They worked in the field as well as in the laboratory and were basically Darwinian in approach. They stressed the importance of controlled experimentation but insisted that it must be relevant to the species under study. This relevance, they claimed, could not be established without an initial period of prolonged observation covering as much as possible of the whole behaviour repertoire of a species. Their basic unit of behaviour became the 'fixed motor pattern'. Their initial records in any investigation were 'ethograms' listing the nature and context of every classifiable action performed by the animal in question. Their further analyses attempted to unravel the inter-relationships between these actions, and their functions, causations and possible derivations. Whereas the comparative psychologists had reduced the stimuli and the responses to the most easily managed minimum and concentrated on the intervening variables, the ethologists attempted to take into their sphere of interest as many natural stimuli and responses as possible. Also, they were truly comparative, frequently investigating many related species alongside one another at the same time.

If comparative ethologists considered comparative psychologists too narrow and artificial in their approach to behaviour, it could, of course, be levelled at the ethologists that they were too broad, that they were little more than scientifically respectable 'natural historians'. It could be argued that by trying to encompass too much they could never hope to produce any really detailed analyses. The ethologists were fully aware of this difficulty and dealt with it to some extent by avoiding what they considered to be the more behaviourally complex forms of animal life. They restricted themselves almost entirely to insects, fish and birds and concentrated on the more rigidly fixed patterns of inborn behaviour. Mammals, especially primates, were avoided.

When ethology was gaining momentum after the war the fish

and birds still held the centre of the stage, but during the past decade certain of the younger ethologists from the ranks of those trained by Konrad Lorenz and Niko Tinbergen, or by their pupils, have been becoming increasingly interested in primate behaviour. They have been dissatisfied by certain aspects of existing primate research, especially by the lack of zoological and evolutionary thinking. Most primate behaviour research workers have, in fact, been drawn from the worlds of psychology and anthropology, and too little attention has been paid to detailed observation and motor pattern description.

By bringing together in this present volume, for the first time, the results of a variety of ethological attacks on primate behaviour problems, it is hoped that the great value of this type of approach will become evident. Whatever happens, it is clear that ethology is inevitably going to infiltrate this important area of scientific research, but I would like to think that by assembling these papers and presenting them as a volume at the present moment in the history of primatology, it will in some small way help this process along and that the infiltration will be more readily recognized and accepted.

Of the ten contributors, three (Morris, Moynihan and Blurton Jones) were trained by Tinbergen, and one (Hinde) came under his influence at an early stage. Three (van Hooff, Sparks and Loizos) studied under Morris, two (Rowell and Goodall) under Hinde and one (Wickler) under Lorenz. Wickler, like myself, made his earliest research studies on fish; Hinde, Moynihan, Blurton Jones and Sparks on birds. They have approached monkeys and apes from the humbler side of the evolutionary scale, looking up at them from simpler, less brainy species, rather than down from the dizzy behavioural heights of man. In so doing they have, I think, illuminated the subject in a new and exciting way.

Chapter Two

The Facial Displays of the Catarrhine Monkeys and Apes

J. A. R. A. M. VAN HOOFF

INTRODUCTION

IN the last few years a considerable number of publications have appeared dealing with the expression movements of primates, especially the facial expressions and calls. Five of these (Hinde and Rowell, 1962; van Hooff, 1962; Andrew, 1963a and c; Bolwig, 1964), pay tribute in their opening phrases to Darwin who in 1872 was the first to attempt a systematic, comparative description of the principal expression movements of man and a number of other mammals, especially primates. His interest was mainly focused on the complex mammalian displays, known as facial expressions. Although since that time much work has been done on the behaviour and sociology of primates (for a review see Carpenter, 1958), no comprehensive study of the above-mentioned specialized method of communication has been made, except for the classic work on the chimpanzee by Kohts (1935). As in other early studies that occasionally touch on the subject, Koht's and Darwin's interpretation of the expressions do not go much deeper than a rather anthropomorphic labelling of these in terms such as 'attention', 'astonishment', 'anger', 'timidity', and so on.

The rise of comparative ethology in recent decades has provided a great amount of data as well as a theoretical foundation in connection with, among other things, the kinds of behaviour known as displays (see, for instance: Tinbergen, 1948, 1959; Lorenz, 1951, 1960; Eibl-Eibesfeldt, 1957; Marler, 1959).

This development has obviously inspired a number of workers to turn their attention to the analysis of primate displays. These studies, all qualitative so far, have been started independently of each other. Hinde and Rowell (1962) have described a number of postures and facial expressions in the rhesus monkey (*Macaca mulatta*), which are likely to play a rôle in communication between individuals, and have made an attempt to interpret these. With the help of spectrographic analysis, they have been able to distinguish a considerable number of calls in this species (Rowell and Hinde, 1962). An account of the major categories of displays in which the facial structures take part has been given by van Hooff (1962) for the higher primates, especially the Old World monkeys and apes. A similar study has been made by Bolwig (1964). An extensive inventory, especially of calls of representatives of the major groups of lower and higher primates has been made by Andrew (1963a), who shows particular interest in the derivation and evolution of calls and facial expressions. The social behaviour patterns of the night monkey (*Aotus trivirgatus*) form the subject of a recent publication by Moynihan (1964).

The present paper closely follows van Hooff (1962). Unless otherwise stated, the observational data have either been taken from this publication or have been newly added.

CONDITIONS INFLUENCING THE APPEARANCE OF FACIAL DISPLAYS IN MAMMALS

Elsewhere (van Hooff, 1962) the conditions which have facilitated the development of a system of facial postures and movements which may serve as social signals have been considered extensively. These postures and movements suppose the presence of a more or less elaborate system of facial muscles.

In cold-blooded vertebrates and to a certain extent in birds, this system is still of a relatively simple nature. It is, throughout the vertebrates, served by the *nervus facialis*, which in the vertebrate ground plan innervates the part of the serially arranged *musculus constrictor superficialis*, belonging to the second segment of the hyoid arch. This muscle complex is attached to the branchial skeleton and serves the respiratory mechanism. In amphibians the afore-mentioned segment of this complex has become independent and has split up in two layers, of which the lower, in the facial region, gives rise to the *m. depressor mandibulae*. Reptiles show a similar situation; the superficial layer has formed the *m. sphincter colli*, which has a large extension in birds, especially in the long-necked species. In mammals this *m. sphincter colli* has split up again on two layers: the superficial *platysma* and the *deep sphincter colli*. Both extend their territory well into the facial region and give rise there to a more or less differentiated system of facial muscles (Huber, 1930, 1931).

Thus it appears that in the cold-blooded animals the musculature in the facial region is more or less restricted to muscles opening and closing the mouth, the eyes and the nose. In this group, moreover, the head skin is not movable; in reptiles it is often also plated. The facial expressions of these animals are, therefore, restricted to biting intention movements, which may have been ritualized and form part of the threat posture; see, for instance, Kitzler (1940), Eibl-Eibesfeldt (1955), Weber (1957). In birds a similar situation is found. Here, however, autonomic responses, for instance, pilomotoric, which in the facial region may be accentuated by the presence of tufts of feathers or crests, often have signal value (Morris, 1956).

The basic factor which, according to Gregory (1929) and Light-oller (1938), may have facilitated the appearance of facial musculature is the increase in metabolic activity in mammals (and birds), reflected by the change from poikilotherm to homoiotherm. This change made necessary a complete rearrangement of the layer that insulates the body from its environment; the main features emerging from this were the soft, pliable leathery skin and the growth of hairs and feathers together with the development of muscles to regulate their posture.

This development has taken place both in mammals and birds. Huber (1922–1923, 1930) has pointed out some factors which are more closely connected with the sudden appearance of the true facial musculature in mammals. Two mammalian inventions, e.g. the extensive chewing of food (note also the 'sudden' specialization of the teeth) and the suckling of the young are possible only with the help of such newly developed features as mobile lips and cheeks. New also are the mobile external ears, the movements of which may have acquired signal value. In primates this system is subject to deterioration. Furthermore, the nose, seat of the (for the average animal highly important) sense of smell was freed from its bony casing and, in principle, the muzzle became mobile (e.g. elephants and tapirs). Most important perhaps is that the probably

9

crepuscular, primitive mammal with its hidden way of life was compensated for its poor sense of vision by the development of tactile vibrissae, or whiskers. There are five pairs of groups of these in the mammalian ground plan, bilaterally inserted in the facial region. Their position can be controlled by groups of muscles attached to the layers of connective tissue in which they are implanted.

Although in practically all mammals the same pattern of facial musculature is present, a more or less elaborate system of facial displays has evolved, apart from the higher primates, only in some groups of a few orders, for instance, in the carnivores (Canidae–Schenkel, 1947; Felidae–Leyhausen, 1956a, b) and in the ungulates (notably in the Equidae–Antonius, 1937, 1939; Trumler, 1959; Zeeb, 1959, 1963). These groups have in common a more or less specialized and refined sense of vision. Especially in the tree-living primates and the hunting carnivores it is characterized by the development of a small zone of sharp binocular vision. This is brought about by the formation of an *area centralis* and a fovea, and by the move towards frontality, combined with an increased mobility of the eyes (Walls, 1942; Polyak, 1957).

Bolwig (1959a) has pointed out that it might be an advantage for animals, which have obtained a distinct area or fovea, to have their expression movements concentrated in a restricted zone (e.g. the face), so that they can be seen in one glance, from the distance at which most social communication will occur. Since, among other things, the fact that these animals are able to control visually the immediate neighbourhood in front of the head reduced the necessity for the possession of tactile vibrissae, the muscles attached to these structures came free to become the effectors of facial displays. This development may be seen in the Canidae, and especially, in the primates. Ruge (1887) and Huber (1930, 1931) have shown how the facial musculature in this last group has undergone a very rapid specialization, culminating in the highly complex organization found in man.

METHODS AND TERMINOLOGY

This report comprises the results of a qualitative study of those movements for which the structures of the face are the main effectors and which may be observed relatively frequently in social interactions between fellows. Although there is some variability in these movements and one movement may go over into another quite gradually sometimes, it is possible to distinguish a number of *compound expressions*. These are typical combinations of a number of *expression elements*, which are defined as the recognizable separate movements and postures of such facial elements as the eyes, the eyelids, the eyebrows and the upper head skin, the ears, the jaws, the mouth-corners and the lips; also, as the movements and postures of the body, the autonomic responses and the vocalizations. (For a description of these expression elements, see van Hooff, 1962.)

Some of the expression elements occur together more or less necessarily. A certain muscle, for instance, may affect more than one facial element at the same time (the independence of the expression elements apparently increases in the 'ascending scale' of the primates. Ruge (1887) and Huber (1930, 1931) have shown that compact muscles may split up in more or less well-defined units, able to act independently from their 'relatives'). One expression element may also directly influence another. Posture and movements of facial elements, such as the lips and the jaws, may determine the sound quality of vocalizations (Andrew, 1963a). The distinction of separate expression elements is, nevertheless, useful in descriptive analysis.

For the interpretation of expression movements it is necessary to establish in what context they occur. This can be done in different ways. Bolwig (1964), who admits that his approach is basically subjective, justifies this approach by saying that someone who is familiar with primates and accustomed to handling them daily, can draw conclusions about a certain condition or mood in an animal which shows a certain expression. These conclusions are responsible for his success in handling the animals. They are based on numerous small observations, which unconsciously have been sorted out and classified in his own brain. Although in many fields of human experience such intuitive understanding of relationships between phenomena may develop, it should be realized that for close investigation of such relationships it can only form the starting point (and a good one, if the intuition is good). It may lead the investigator to formulate a hypothesis.

It is not so much in this more general sense that Bolwig's approach is subjective, but more in the sense that it is rather anthropomorphic. Thus a list of conditions which, according to

Bolwig, are responsible for the occurrence of the facial expressions, consists of the following 'moods': joy, amusement, unhappiness, fear, sadness, anger, rage, love and affection.

In connection with primate displays, as with those of other animals, a number of questions can be asked. On the assumption that the display functions as a 'social releaser' (Tinbergen, 1951), it may be asked, what does it express, or synonymously: what is its motivation? In connection with this question one may relate the display to other activities of the performer; for instance, un-ambiguous social activities, such as flight, attack, approach to groom, to mate, etc. Thus, if in an animal the occurrence of a certain display is highly correlated with attack movements, this display can be said to express a high likelihood of attack. Saying, in agreement with common ethological usage, that the expression indicates a high tendency to attack, does not imply that both be-haviour patterns (attack and display), are acted upon by some unitary causal mechanism; it only means that they are dependent on (probably complex) systems of causal factors which have much in common. A second question may concern the nature and the interaction of the causal factors (internal and external, e.g. the releasing stimulus situation), which are responsible for the occur-rence of the respective displays. A third question may concern the function of the display. In the first place one wants to know whether it functions as a 'social releaser': whether the information which it carries concerning the motivation of its performer is being received by the fellows; furthermore, what its effect is on the behaviour of the fellows, what changes in motivation it provokes in these and what advantage this brings to the performer. Wide-ranging com-parative investigations may, finally, provide answers to questions concerning the derivation and evolution of displays. For a complete insight regarding their survival value for the species, studies on animals in the wild seem to be indispensable.

Although this exposé does not contain much that has not already been said, wholly or in part, in connection with other studies on displays (see, for instance, Tinbergen, 1960, 1963; Moynihan, 1964), it seemed desirable to discuss these points again in view of the differences in approach which exist in recent studies of primate expressions (Bolwig, 1964; Andrew, 1962, 1963a). This discussion may provide an answer to Andrew's recent criticism (*op. cit.*) of the use of motivational analysis as being very misleading since the motivational model it provides may lead to unjustified deductions concerning the causal mechanisms of displays (reflected, for instance, in the different meanings of the term 'drive').

Another criticism has been raised by Andrew (*op. cit.*). Suppose a certain display is found, for instance, to represent a high tendency to attack. A disadvantage of such a statement, according to Andrew, is that it unnecessarily suggests wide changes in the causation of the components of the display (for instance: raising the upper lip in a dog as part of the threat display), because these may be shown in the frame of quite different behaviour systems (for instance: while biting at food). Andrew concludes that 'if the movement is spoken of as a response which precedes biting, then its similarity of causation is emphasized'. In fact, no more has been done than making a kind of 'motivational statement' concerning the component of the display; however, on a lower level of integration. But it does not tell anything more about the causal mechanisms of the 'higher level' behaviour patterns in which it has been incorporated. Thus, in analysing the causation of the above-mentioned lifting of the lip, one may restrict oneself to establishing what the causes are for lifting the lip, if the animal is going to bite (low level of integration), but it does not give an insight in the reasons, which in such different circumstances make the animal bite (high level of integration). In this context see also the remarks of von Holst and von Saint Paul (1960) on 'niveau-adäquate Terminologie'. The same applies, of course, when considering such components as vocalizations.

In studying the causation of displays Andrew (1963a) proposes, following the advice of Kennedy (1954), to consider responses and the stimuli eliciting them.

It is suggested (Andrew, 1962, 1963a, 1963b, 1963c) that the amount of change in stimulation ('absence of change is neutral or unpleasant, small changes are pleasant and large changes unpleasant'; Andrew, 1963a, from McClelland *et al.*, 1953) is the factor which determines which calls or facial expressions occur. Examples given by him are 'the group of alert responses which serve to bring the major sense organs to bear on the stimulus' and are caused by moderate stimulus contrast, and a number of vocalizations and certain associated facial movements of which it is suggested that they 'had a common origin in a group of protective responses evoked by sudden intense stimulus contrast' (*op. cit.*). It is argued that certain patterns of stimulation which depart from normal or sought-after stimulation retain contrast very persistently and that certain patterns which have acquired 'valency' through conditioning or otherwise, may have very high perceptual contrast. It could also be argued that the fact that an animal does not always react identically in similar situations is the result of the fact that, as yet unknown, other factors influence the 'valency'; it is possible to think of: previous behaviour, endocrine balance, physical condition, etc. This is implicitly recognized by Andrew since he often mentions that a certain behaviour pattern is released by a certain stimulus situation in an animal which is, for instance, 'friendly (but somewhat frightened)', 'confident', 'unsure of itself', etc. (1963a).

Thus the hypothesis is proposed that primate vocalizations and to a large extent the facial expressions, the nature of which is determined to a considerable extent by the vocalization, remain 'evoked by stimulus contrast and become more intense as the contrast becomes more intense' (Andrew, *op. cit.*). On this Moynihan (1964), in his study on the night monkey, comments 'that it is undoubtedly true if "stimulus contrast" is defined broadly enough'. But platyrrhine signal patterns (Moynihan's study object) are certainly not produced by a single range of qualitatively similar stimulus contrasts of differing strengths. Certain stimuli usually ('normally') provoke only sexual patterns, others generally provoke only hostile patterns, still others usually provoke only parental responses, etc. Thus the causation of platyrrhine signal patterns can be described in terms of stimulus contrasts only if distinctions are made between qualitatively different types of stimulus contrast, e.g. between hostile stimulus contrasts and sexual stimulus contrasts.

This seems to be recognized by Andrew (1963a), since 'secondary complications may well be important. Thus it is always possible that there may be some attachment of specific calls to specific stimuli by conditioning.'

If Andrew (1963a) makes the criticism (first expressed by Hinde, 1959) 'that concepts such as "drive" are fundamentally unsound, in that they represent attempts to describe the effects of groups of independent causal factors in terms of a single variable, one might as well return this criticism and ask whether it is then justified to express the occurrence of responses in terms of a variable like

Plate I *The tense-mouth face* shown by a male of *Papio anubis*, with simultaneous lifting of the eyebrows. Note the tension in the region around the mouth (mouth-corners pulled forward) and in the region around the eye (straight under eyelid)

Plate II The peculiar *silent bared-teeth face* of *Mandrillus sphinx*, which is characterized by an opening between the lips in the form of a horizontal figure eight

Plate III *The silent bared-teeth face* performed by a chimpanzee. An identical expression is shown by the male in photograph a and b of Plate VII

Plate IV A male *Macaca nemestrina* shows *the protruded-lips face* while smelling at the genital region of a female in heat, which presents

Plate V *The relaxed open-mouth face* is shown by the chimpanzee at left during a playful interaction. Note the complete covering of the upper teeth

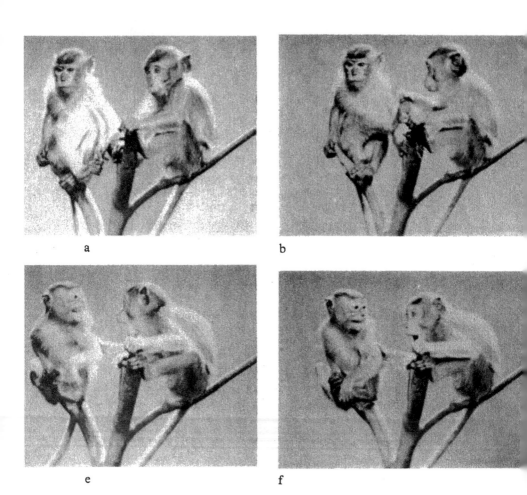

Plate VI Sequence in which a *silent bared-teeth face* is given in a context which may indicate an 'appeasing' function. Dominant crab-eating monkey (*Macaca irus*), male, at right; subordinate female at left (photograph a). The 'offence'; the female puts right leg on the branch occupied by the male (photograph b). The male is gazing fixedly at the female, showing the beginning of a *staring open-mouth face*, and lifts its hand for a hitting movement. The female, taking back its right leg, looks 'evasively' (i.e. out of corners of eyes) at the male (photograph c) The male, showing *staring open-mouth face* has hit the female, which withdraws the upper part of body. The female shows horizontal retraction of the lips (photograph d)

c

d

g

h

Intense *staring open-mouth face* on the part of the male; intense *silent bared-teeth face* on the part of the female (photograph e)

The male still showing *staring open-mouth face* (less intense) looks away from the female. *The silent bared-teeth face* of the female grows less intense (photograph f). Male shows relaxed face again; the display of the female fades away (photograph g)

The *status quo* restored (photograph h). The photographs are taken from a 16 mm cine-film (speed 24 fr/sec)

The distances between the successive photographs are, respectively 6, 7, 7, 14, 14, 8, and 41 frames

a   b

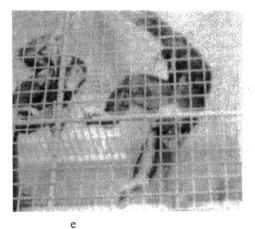

e   f

Plate VII Sequence in which a *silent bared-teeth face* is given by a chimpanzee in a context which may indicate a 'reassuring' function. The nearby adult male (right) approaches a younger female, which is sitting rather immobile and in a hunched posture, indicative of a slight fleeing tendency. The male shows a low intensity *silent bared-teeth face* (slight vertical and horizontal retraction, mouth not opened) (a and b).

The male settles in front of the female, who so far shows no reaction. *The silent bared-teeth face* becomes more intense; the mouth is opened

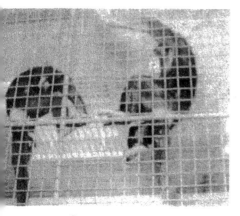

c

d

g

h

(photograph c). Now the female starts approaching the male cautiously, whereas the male maintains its, again more intense, facial posture (photographs d, e, f, g). Note the evasive looking (out of the corners of the eyes) on part of the female (photograph f). The female turns (photograph g) and allows the male to 'embrace' her from behind. Photographs taken from a 16 mm cine-film report of the introduction of a female chimpanzee to an established adult male.

Distance between the successive photographs respectively 36, 48, 40, 24, 16, 24, and 72 frames (speed 24 fr/sec)

Plate VIII *The pout-face* of the chimpanzee. The lips are parted in the middle and pressed together near the mouth-corners

Plate IX 'Temper tantrum' response of the chimpanzee, characterized by frantic body movements and full vertical and horizontal retraction of the teeth

"stimulus contrast", which clearly also reflects the effects of groups of independent causal factors'. At any rate, one may doubt with Moynihan (1964) whether a system like the one proposed by Andrew is any more convenient or useful than the employment of terms such as 'tendency'.

However, a more fundamental criticism may be made. Andrew (1962, 1963a) has arranged stimulus situations in a number of groups in each of which similar calls are evoked. These groups have been arranged in order of a progressive increase of intensity of the associated calls. If it is then stated (Andrew, 1962, 1963a) that the hypothesis (viz. the amount of stimulus contrast determines which calls or expressions are evoked) does fit all the facts, this means that it has been made plausible that in a series of stimulus situations which respectively release more intense calls the stimulus contrast increases accordingly. Apart, however, from the possible subjective appreciation of the fact that for the evocation of a strong response something powerful must happen, it might seem to be not an easy task to design a method of measuring such a complicated variable as stimulus contrast, which to a large extent is determined centrally as the outcome of interactions between the afferent input and other factors. Andrew, at least, has not indicated such a method. The danger of circular argumentation lurks when on the one hand the amount of stimulus contrast is derived intuitively from a more or less subjective appreciation of the stimulus situation, while on the other hand 'stimulus contrast has been defined in terms of alert responses' (Andrew, 1964, p. 64).

Explaining, for instance, the highly probable evolution of the human smile from the monkey grin (which Andrew considers to be a protective response originally, evoked by high stimulus contrast) he states (1963c): 'Progressive facilitation of a protective response such as the grin could thus reasonably be expected to lead to a condition in which the response could be evoked by changes in stimulation small enough to be judged pleasant (that is "interesting" or "amusing")'. I think it preferable, instead, to characterize evolutionary developments of displays in terms of shifts in balances between objectively definable and measurable qualities such as, for instance, the tendency to flee, the tendency to approach, etc.

In this report a number of compound expressions will be described which can be observed regularly in most species of catarrhine monkeys and apes. Detailed study of separate species may reveal that more subtle distinctions can be made, sometimes perhaps with different emphasis. The results of the present qualitative study provide a general survey. It contains some, still very crude, information as well as suggestions regarding the motivation, causation and function of the compound displays. It is hoped to report later on the results of quantitative studies which are now in progress.

The names of some of the compounds, described by van Hooff (1962), have been replaced by descriptive terms, so that all names are now short and schematic, descriptive qualifications.

SPECIES STUDIED

All the studied species have been observed in zoos (especially: London, Whipsnade, Arnhem and Rotterdam). Some species have been kept in an experimental set-up. Attention has been mainly focused on the Old World monkeys and apes.

Apes: *Pan troglodytes; Gorilla gorilla; Pongo pygmaeus.*

Catarrhine monkeys:
Colobus abyssinicus.
Erythrocebus patas.
Cercopithecus aethiops; C. petaurista; C. diana; C. neglectus; C. pygerythrus; C. hamlyni; C. ascanius; C. erythrotis; C. nictitans; C. mona; C. l'hoesti; C. albogularis; C. talapoin.
Cercocebus albigena; C. torquatus; C. atys; C. galeritus.
Macaca irus; M. sinica; M. mulatta; M. nemestrina; M. sylvana; M. maurus; M. silenus; M. speciosa.
Cynopithecus niger.
Papio cynocephalus; P. papio; P. anubis; P. hamadryas.
Theropithecus gelada;
Mandrillus leucophaeus; M. sphinx.

Platyrrhine monkeys:
Cebus apella; C. capucinus; C. albifrons.
Lagothrix cana.

THE COMPOUND EXPRESSIONS

The Relaxed Face

Common to all monkeys, this facial posture is seen when the animals are not engaged in any particular activity, but are just lying, sitting or walking around. Though it does not accompany particular activities or indicate a tendency to perform particular activities, this negative condition as such is expressive. The posture is mentioned here because it is the one with which all the others are compared; all facial elements are in their 'neutral position'. The expression shown during sleeping may be regarded as an extreme form of it.

The Alert Face

This expression is common to all monkeys. It differs in only a few measurable features from the relaxed face and there is a gradual change between the two postures. The most obvious difference is to be seen in the eyes. Whereas in the relaxed expression the upper eyelid is not lifted completely, the iris being only partly exposed, a true alert posture is characterized by a fully opened eye. Rather difficult to describe, but evident for a close observer, is the more abrupt character of the body movements in some cases, and the higher tension in the facial skin, especially in the region around the mouth, indicating a higher muscle tonus.

The alert face occurs in many different situations. It may accompany a number of social behaviour patterns, such as grooming, mounting, smooth directed approach (as contrasted with brusque approach which is usually associated with attack). The expression, therefore, cannot be said to be indicative of a particular tendency or balance of tendencies. It may be observed, for instance, when the animal is in 'pure', unobstructed flight.

The Tense-mouth Face

This face is closely related to *the alert face*; the facial elements behave as follows:

The eyes may be opened rather widely and are staring fixedly towards a partner.

The eyebrows are normal or lowered in a frown. The presence of a frown may be difficult to establish, when the eyes are opened more or less widely. It certainly is not very conspicuous in the catarrhine monkeys. Bolwig (1964) mentions that the *m. corrugator supercilii*, which is responsible for the frowning movement, is much less well developed in the lower catarrhines than in apes and man. It is possible that in lower catarrhines contraction of this muscle, independent of contraction of the *orbicularis oculi*, is not possible. In a photograph of an attacking run by a rhesus monkey, presented by Hinde and Rowell (1962), a lowering of the brows is said to be perceptible.

The mouth is kept in a tightly closed, or almost closed position; there is obviously a great tension in the jaw muscles. According to Hinde and Rowell (1962) the lower jaw is often thrust a little forward in the rhesus monkey (*Macaca mulatta*). This may be the case in other species as well.

The lips are closed or almost closed and the lips may be drawn inwards (especially in *Papio* and *Theropithecus*).

The mouth-corners are brought forward. As a result the mouth often looks like a narrow slit.

The expression may be seen during a fast attack run or, sometimes, during a rigid sneaking approach, which results in attack. Whenever the expression is shown without these movements an attack is likely to follow. The attack run is most easily seen in animals living under conditions in which they have plenty of room at their disposal. In most species no vocalizations are heard. *Pan troglodytes*, *Theropithecus gelada* and the studied *Papio* species have been noticed to utter a rather low-pitched bark during the run. For *Papio ursinus* this has been described by Hall (1962). In *Theropithecus gelada* this bark is often given at the moment the partner is reached.

The compound expression is usually performed by the dominant animal or animals in a group (the dominance relationships can be expressed in terms of who has priority over whom in the access to food and other needs; the most dominant member of the group is almost without exception the adult male or, when several males are present, the biggest of the adult males). The behaviour pattern can be released in a number of different situations. Although it is not always quite clear what actually elicits the behaviour (it seems to occur rather spontaneously sometimes), it has been possible to establish that in a number of cases physical obstruction by a partner or the presence of a partner near a food source or a favoured resting place might have been the cause. A more systematic analysis is obviously needed here. In most cases the partner flees. If it does not succeed in overtaking the other, the performer may stop. If it does succeed, there may be a short struggle in which the performer may deliver a fierce bite or blow. The partner often performs the compound expression which will be described under the heading of *'frowning bared-teeth scream face'*.

It is likely that the performer is largely, if not completely, motivated by attack. The expression movement has been observed convincingly in most of the species studied. Notable exceptions are *Gorilla gorilla* and *Pongo pygmaeus*. An almost identical description has been given for *Macaca mulatta* by Hinde and Rowell (1962) under the heading 'attacking run'. Bolwig (1964) has described it as the expression of 'anger'. Van Hooff (1962) termed it: 'the aggressive threat face'.

In *Theropithecus* and the studied *Papio* species a facial expression can be frequently seen in which the posture of the mouth, the lips, the mouth-corners and the eyes is as described, perhaps in a less marked way. *The eyebrows*, and *the upper-head skin*, however, are retracted. Seen from a distance the lifting of the brows is the most conspicuous feature. The expression can be performed when the animal is in a sitting position. It goes over into attack less readily and the reactions of the fellows are less vehement. Kummer (1957) has described this expression ('Brauenziehen'), as a low intensity aggressive threat.

The Staring Open-mouth Face

The eyes are staring fixedly at the partner.

The eyelids are fully apart, so that the iris is completely exposed.

The eyebrows. The behaviour of these facial elements varies through the range of monkeys in which *the staring open-mouth face* has been noticed. It is possible to discern three categories:

1. To the first category belong the eight species of *Macaca* which have been studied, *Cynopithecus niger, Theropithecus gelada,* the studied species of *Papio, Mandrillus* and *Cercocebus,* as well as the *Cebus* species. The animals of this group are characterized by the fact that during the performance of this expression movement the eyebrows are lifted in a very marked way. This lifting is emphasized in most species by the fact that the skin of the upper eyelid and the region immediately above it, which is of a much

brighter shade than the skin around it, is exposed. Most conspicuous is the movement in *Cercocebus torquatus*, where the region of
the upper eyelid bears a blueish-white spot, sharply marked off
from the surrounding black skin. Simultaneously the upper head
skin is strongly retracted. In *Macaca sinica* and *Cynopithecus niger*
the presence of a crest emphasizes the movement (these species,
by the way, do not posses a contrasting upper eyelid).

2. A second category comprises the studied *Cercopithecus* species,
Erythrocebus patas and *Colobus abyssinicus*. When comparing the
animals of this group with those of the former group, one is struck
by the relative immobility and lack of expression in their faces. It is
almost impossible to detect any change in the position of the eyebrows. Only in a few instances was it established with certainty
that there was a slight lifting of the eyebrows, the animals in these
cases being particularly close to the observer with the face turned
more or less in his direction when the action was performed. The
species were: *Erythrocebus patas*, *Cercopithecus nictitans*, *C. petaurista*, *C. albogularis* and *C. aethiops* (the latter having been kept in a
special observation set-up for the purpose of introduction tests).
So it is likely that the animals belonging to this group (comprising
all the guenons) perform a mild lifting of the eyebrows in this
context. Typically, however, it is below the threshold of perception
(at least for the human observer).

3. The only member of the great apes in which *the staring open-
mouth face* has been observed satisfactorily is the chimpanzee. In
this animal, forming the third category, the eyebrows may be
lowered in a frown. Schaller (1963) reports the same for *Gorilla
gorilla beringei*.

Not only in connection with the behaviour of this element, but
also in other respects, these three major categories can be discerned.
Henceforth, therefore, reference will be made simply to the
monkeys of the first, second and third category.

The ears. Linked with the movements of the eyebrows and the
upper head skin are the movements of the ears. In the monkeys of
the first category the ears are retracted and flattened against the
head more or less simultaneously with the retraction of the upper
head skin. In the monkeys of the second category, movements of
the ears are much more difficult to detect and have with certainty
only been seen in *Cercopithecus aethiops*. In chimpanzees and other
apes the ears seem to be rather immobile.

The mouth may be slightly or rather fully opened with all degrees of openness in between. As we will see, this compound facial expression indicates the presence of both the tendency to attack and the tendency to flee. From qualitative data available it is not possible to decide whether the degree of opening of the mouth is related to the intensity of the motivation (the absolute level of both conflicting tendencies) or to the qualitative character of the motivation (the relative strength of the conflicting tendencies in comparison with each other) or, perhaps, both.

The mouth-corners are brought forward and *the lips* are fully tensed and pressed against the teeth. These are not visible except when the mouth is very widely opened; in this case particularly the canines may be visible. Especially in the monkeys of the first category (except *Macaca irus* and *M. sinica* which in this respect resemble the guenons) strong contraction of the *orbicularis oris* may pull the mouth-corners forward to such an extent, that, with moderate opening of the mouth, the aperture appears as a circular hole. This has never been seen to such an extent in the guenons and in *Macaca irus* and *M. sinica*. Here the angle between the lips at the mouth-corners does not get lost (compare the mouth-posture of Fig. 1b with that shown in Plate VI, photographs d and e). If the mouth is widely opened the mouth-posture has a great resemblance with the one shown in the display which will be described (see below) under the name *relaxed open-mouth face*. This expression, however, is in other respects different from the present one. In *Macaca sylvana* two distinct mouth-postures may be noticed. In the first one the mouth is rather widely opened and so resembles the guenon-type posture. In the second one the mouth-corners are brought well forward and, simultaneously, the lips are protruded. The display in which this latter posture is incorporated has been described by Darwin (1872) as the expression of anger of this species. The second posture greatly resembles the one shown by many monkeys and the apes in the compound expression which will be described as *the pout face* (see below). Here, again, other expression elements are different.

An aberrant mouth-posture which is rather distinct from that of the other monkeys is shown by *Cercopithecus diana*. Whereas in the other species the mouth is opened more or less suddenly to be kept open for a second or more, this species opens it in a jerky manner quite widely, but immediately afterwards closes it again.

The closing phase is somewhat slower than the opening phase. The whole movement lasts about a second; it may be repeated several times in quick succession. The other expression elements are the same as in other guenons.

In the chimpanzee (*Pan troglodytes*) the lips are not pressed against the teeth as in the majority of catarrhines, but protruded more or less. The teeth may become partly visible. This protrusion is usually quite different from the one which is shown as part of *the pout face*. In the present case the lips are fully parted, whereas in *the pout face* they are only parted over a small region in the middle, but joined together on both sides, thus producing a kind of funnel.

The vocalizations which accompany this compound expression can best be described as rough, rather staccato, usually rather low-pitched grunts or loud barks. Andrew (1963a) has been able to give detailed descriptions of calls which may be heard in threat: e.g. grunts, hoarse sounds and barks in *Pan troglodytes* (chimpanzee); sharp calls in *Cercopithecus* and 'arrr' bouts and sharp calls in *Macaca*. Hinde and Rowell (1962) distinguish a number of threat calls in *Macaca mulatta*: bark (Andrew's sharp call?), pant-threat, roar and growl. (The last three perhaps being Andrew's 'arrr' call.) Both authors made these distinctions on the basis of sound spectograms.

The body posture shows some variability. The animals may sit or stand. In the latter case the fore-legs may be bent and the head may be kept lowered in what clearly is an intention movement of jumping. Forward directed leaps as well as short retreat movements may be seen. A detailed description of the possible body postures in *Macaca mulatta* (rhesus monkey) is given by Hinde and Rowell (1962). The lowering of the head may be repeated frequently in an exaggerated manner in some species (e.g. *Macaca mulatta* – Hinde and Rowell, 1962; *Papio ursinus* – Hall, 1962). In the mandrill and the drill (*Mandrillus* sp.) the movement has become so exaggerated that it is only possible to recognize its original function when comparable movements of other species are known. Here it is a sudden and abrupt movement of the head consisting of a short upward jerk, a long downward jerk and finally a long upward jerk. The jerks follow each other directly and are performed with the same high speed. The whole movement takes much less than a second. It strongly contrasts with the vehement horizontal head shaking, which forms part of *the silent bared-teeth face* of this species (see below). This contrast nicely illustrates Darwin's (1872) principle of antithesis. In some species (e.g. baboons, macaques) typical hitting or rubbing movements with one of the fore-legs on the ground or the branch can be observed (Kummer, 1957; Bolwig, 1959b). These may be considered as ritualized intention movements of hitting or pushing down the fellow (Bolwig, 1959b).

An autonomic, pilomotoric response is the erection of hairs, mainly on shoulders and back, which makes the performer look much more sturdy. It is clearly to be seen in baboons. The adult males of *Papio hamadryas* (sacred baboon) and *Theropithecus gelada* (gelada baboon) possess impressive collars of long hairs of a colour contrasting with the rest of the body. These greatly increase the effect of the response. In chimpanzees the erection of the long hairs on the under-arm is also obvious.

The Circumstances

This compound expression is often, though not always, performed by the dominant animal involved in the sequence. In the performer it usually gives rise to attack or to nothing (it is the end of a sequence). Occasionally the performer will change over into flight (defined as rapid retreat) or avoidance (defined as slow, more 'cautious' retreat). The opponent may react with different expressions, but frequently seen are fleeing or avoiding. This leads to the conclusion that this compound on the average expresses a probably dominant tendency to attack, thwarted by the tendency to flee. For this reason the compound was termed *aggressive threat face* by van Hooff (1962). This does not mean that other tendencies may not be activated. Thus chimpanzees may show this display directed to a partner, while jumping on the spot on all fours; hands and feet beat the floor with force. This particular display, which has been observed especially during a number of introduction tests where it was directed towards the newcomer, might alternate with different behaviour patterns: for instance, a fast run past the fellow, during which this animal is hit or kicked with force; flight or avoidance; smooth approach resulting in investigation, grooming, embrace, etc.; gnaw-wrestling accompanied by *the relaxed open-mouth face* (see below). Here, apart from the tendency to attack and to flee, a tendency to approach (smoothly), resulting in positive social behaviour patterns such as grooming, huddling, mating, or simply being close together (for short, termed '*social attraction*') may express itself. It would be interesting to know whether any significance must be attached to the fact that this display occurred in four of those five (out of nine) introduction tests in which the newcomer was not much younger than the partner (i.e. in which the future hierarchical relation could not be foretold at the first glance).

Although the impression exists that *the staring open-mouth face* is released in situations in which some activity or some tendency of the actor is thwarted by a fellow or its 'rights' are violated in some way or another, it is not at all easy to point out what the reason is in an individual case. Complicated interactions of more than two individuals occur, in which redirection and 'gesichertes Drohen' (Kummer, 1957) may take place. There may be factors which influence the level of aggression (the probability that agonistic encounters will occur) in a group or certain members of a group and which are not given by the eliciting stimulus situation in a more narrow sense. Thus, for instance, an increase of the mating activity of the dominant male, in contrast with that of the other males, has been found to increase the number of aggressive episodes between females in a group of *Papio ursinus* (Hall, 1962).

The present compound expression can be seen frequently in animals of the first category. In animals of the second category, on the average, it is observed comparatively less frequently. Further analysis may show that more agonistic encounters are decided by actual attack or fleeing, or that the level of aggression in this group is on the whole lower.

The compound has been described in a detailed manner by Hinde and Rowell (1962) under the heading 'threat' and 'backing threat' for *Macaca mulatta*. A general description has been given by Andrew (1963a) for the genera *Macaca, Cercopithecus* and *Pan*, and by Bolwig (1964). *Gorilla gorilla beringei* shows it when 'annoyed', in slightly disturbing situations (Schaller, 1963).

The Staring Bared-Teeth Scream Face

The ears. These are folded back in the monkeys of category one. Movement is barely perceptible in the representatives of the second category, and in the chimpanzee.

The eyes in most cases are staring fixedly towards the opponent.

The eyelids are fully apart, so that the eyes are widely open.

The eyebrows. Here the response is the same as in *the staring open-mouth face.* In the species grouped in category one, the eyebrows are fully lifted and the upper head skin is retracted. In the species of category two, it is extremely difficult to see any change in the posture of the eyebrows. In the chimpanzee they are clearly lifted, although much less obvious than with the monkeys of category one. The chimpanzee lacks the emphasizing markings possessed by them; on the other hand, the forehead immediately above the eyebrows contains a number of clearly visible wrinkles, which become more obvious when the eyebrows are raised.

The mouth is opened, usually widely.

The mouth-corners are retracted completely.

The lips are, usually, completely retracted vertically. The action of the last two elements results in a complete baring of the teeth, including the molars. The gums may even be exposed partly. The appearance of wrinkles curving around the mouth-corners may emphasize this expression. In *Theropithecus gelada* the upper lip is not only lifted but turned upwards with the inside out, thus exposing a large patch of bright pink skin, strongly contrasting against the dark face.

The vocalizations which accompany this display vary from a high-pitched, staccato barking (shrill bark, Hinde and Rowell, 1962; sharp call, Andrew, 1963a) to a more prolonged high-pitched screaming. The sound is well qualified as 'rrah', 'aach' or 'eech'.

The body posture which is shown together with this expression may vary considerably. It may, for instance, show strong forward tendencies and so resemble the posture which accompanies *the staring open-mouth face*: the fore-legs are bent; the head kept low and sham-attacks towards the opponent may be noticed. It may also contain strong backward tendencies, sometimes giving rise to actual backward movements. It is not uncommon to see the animal present its side to the opponent, obviously the result of a conflict between simultaneous urges to go forward and to go backward. The movements (e.g. the locomotion) of the animal may be brusque and rapid.

An autonomic response which can be observed is the raising of the hairs on shoulders, back and arms (see: *Staring Open-mouth Face*).

The Circumstances

This compound expression is most commonly seen in response to an attack or a *staring open-mouth* display by a more dominant member of the group. The expression may cause the fellow to attack or to show *the staring open-mouth face*; these displays may also be stopped in response to this expression movement; possible reactions of the partner are also fleeing and avoiding. On the withdrawal of the dominant animal the display usually fades away. There are, however, notable exceptions to this. Thus the subordinate animal may continue its display and pursue the other animal which is retreating; this may sometimes even lead to an attack. In many cases the display changes over into flight or into the compound expression which will be described as *the frowning bared-teeth scream face*. It sometimes develops out of *the staring open-mouth face*.

The data suggest that this expression is the result of the competition of the two incompatible tendencies to attack and to flee, which both may be rather high. It is probable that in general the tendency to flee dominates. In the chimpanzee it is usual that this display accompanies an attack or a fight; it seems to be shown equally readily by dominant and subordinate individuals. Thus it is likely that in this species the display covers a similar range of qualitative balances of the competing tendencies as *the staring open-mouth face*, the difference being that the absolute level of the competing tendencies is much higher in *the staring bared-teeth scream* display. A similar situation seems to exist in *Gorilla gorilla* (Schaller, 1963; own obs.).

The compound has been shown frequently by the monkeys of the first category and by the apes (sufficient observations have only been obtained from the chimpanzee and the gorilla). It is, however, rarely seen, and then only for a very short time, in the monkeys of category two. Convincing observations have only been obtained from *Cercopithecus aethiops* and *C. diana*. Short salvoes of rather high-pitched, staccato sounds were uttered. The display has been described by Darwin (1872), Andrew (1963a) and Bolwig (1964); these authors talk about the expression of 'rage'.

A peculiar property of this posture (the same applies to the compound called: *frowning bared-teeth scream face*) is the fact that, when it is performed by an animal living in a group, other animals, not engaged in the encounter that caused the display, run to the scene and take part in the encounter.

Zuckerman (1932) has described this phenomenon: 'the characteristic squeal of a terrified animal, as it threatens another, usually provokes an immediate response of an aggressive protection from any neighbouring dominant animal'.

Bolwig (1959) records the same. Although the subordinate animal often benefits from the intervention, in that the attention of the opponent is diverted, this is not always the case.

The newcomers may well direct their aggressive behaviour towards the subordinate, if as it seems, it is the first animal they meet.

According to Kummer (1957), screaming may even release attack or threat from fellows.

Darwin describes this display as 'terror'. Bolwig (1964) records that it may either represent high intensity 'anger' or 'horror' (i.e. high intensity 'fear'). This compound expression can be seen not only in direct social interactions between fellows, but also in more complicated situations, instances of which will be given below.

In this context the display may be rather variable: in addition to *the staring bared-teeth scream face,* also *the frowning bared-teeth scream face* and intermediate forms occur. The body movements may become quite frantic and the animals may occasionally throw themselves against the ground or whirl around and dash themselves at objects. This intense form of the display is mainly seen in chimpanzees, particularly in young ones. It is described by several authors, e.g. Darwin (1872), Köhler (1926), Yerkes and Yerkes (1929), Kohts (1935), Hooton (1942), Yerkes (1943), and is often referred to as the 'temper tantrum response'.

The display is shown, for instance:

When an animal, especially if it is a pet (Andrew, 1962), is left in a situation in which it is not possible for it to follow the keeper (e.g. when it is put in a cage). Andrew suggests that this represents an abnormal persistence of infantile behaviour.

When an animal is shown something attractive, but not allowed to take it. In these circumstances the expression movement is often preceded by the display which will be described as *the pout face* (see below). A comparable situation may exist, when at feeding time the keeper with food is in sight. The reaction may become more intense when there are 'unexpected' unusual delays in the feeding routine.

When, as is usual in zoos, animals can see that their neighbours in another cage are getting food, they may show the display, alternating with patterns such as *the staring open-mouth face*. It may also alternate with biting own hands and wrists (Andrew, 1962; own obs.).

The display was shown on numerous occasions by a female *Papio anubis* which was living next to a pair of *Macaca nemestrina* with an infant, a couple of months old. The female baboon used to be regularly at the partition separating the cages, fervently showing *the lip-smacking face* towards the infant, which would often react by approaching the baboon at the partition. This animal then tried to finger the infant. Frequently, however, the female *Macaca nemestrina* would react to the lip-smacking on part of the baboon by taking up the baby and carrying it away. In these cases the display was shown, alternating with *the staring open-mouth display*, *the lip-smacking display* (both directed to mother and young) and ferocious biting of own legs and hands.

Andrew (1962) gives yet another example: Finch (quoted in Lawson and Marx, 1958) has found that difficult discriminations offered during a test can evoke the display in the chimpanzee. They were more intense the longer the preceding time of deprivation of food, the greater the number of previous rewarded trials and the more difficult the discrimination.

In these cases the tendency to flee is not necessarily present. Andrew (*op. cit.*) concludes this from the fact that in the instances where the display is shown as a reaction to the offering of food to neighbours, animals were involved which were relatively dominant in their own society and never had had an opportunity of losing a fight with their neighbours. Moreover, the partition is an effective territorial boundary, behind which the animal's confidence may have increased. The same is probably true for the given instance of the female baboon and the infant pig-tailed monkey.

A certain tendency to attack may be present in a number of cases. The ferocious self-biting may form an indication for this. Andrew (*op. cit.*) has drawn attention to the fact that spoilt children may show intense aggressive responses (kicking, striking, biting) towards their opponent during a tantrum. It is likely that in all instances the presence of a fellow (e.g. keeper, experimenter) is of great influence. Andrew, therefore, concludes that the great intensity of the vocalization is probably due to summation between the effect of an opponent in producing vocalization and the effect of a 'frustration' situation.

Expressed in terms of motivation, the display may probably be said to occur when a certain tendency (for instance, to feed or to approach) which appears to be strongly activated is thwarted by some direct activity of the partner (e.g. a dominant fellow in the normal social relation, or a keeper or experimenter in the situation typical for captivity). In the cases in which the display was directed towards neighbours, these could withdraw, thanks to the barrier, not by active avoidance, but simply by ignoring the partner, from the adequate encounter which normally would have occurred. These instances may well appear to be an artefact of captivity.

In nature the display might permit the performer to maintain a highly prized advantage which is being disputed by a dominant fellow. This can be the result of an increase of the tendency to flee of the dominant or of other mechanisms. In chimpanzees, for instance, the display often evokes embrace by the partner. Thus it has been observed on many occasions (Arnhem Zoo – one dominant male and two females) that one of the females, in the possession of a delicacy, performed the temper tantrum response in reaction to the approach of the dominant male when it showed the obvious intention of stealing. In a number of cases (though not always) the male then embraced the performer after showing *the silent bared-teeth face* (see below). Afterwards he remained watching how his mate consumed the tit-bit (if sometimes only for a short while, before 'cautiously' trying to steal again).

The Frowning Bared-teeth Scream Face
This facial expression bears great resemblance to *the staring bared-teeth scream face*, as it is characterized by the same obvious mouth posture. The differences in form between the two compound expressions are given by the elements of the upper half of the face and by the body posture.

The expression elements:
The eyes are closed or opened only to a small degree. When not closed the eyes are never directed straight towards the opponent; the animal looks away, and often moreover 'faces away'.
The eyebrows are lowered in a frown.
The ears are retracted.
The mouth is opened.

The lips are vertically retracted and *the mouth-corners* are fully retracted, thus baring all the teeth.

The body posture is mostly hunched or crouched and turned away from the partner. Often the animal presses its belly against the ground and does not move. Kummer (1957), who has given an accurate description of this compound expression for *Papio hamadryas*, for this reason terms it 'Pressen'.

The vocalization is usually a loud high-pitched screaming, sounding like 'aach' or 'eech' and has been termed *shriek* (Andrew, 1963a), *squeak* (Rowell and Hinde, 1962) or 'Kreischen' (Kummer, 1957).

Urination and *defecation* may also be observed often (Kummer, 1957).

The Circumstances

This compound facial expression (cf. Kummer, 1957) has always been observed in agonistic situations and usually marks the end of such a situation (unless more than two animals take part). It is shown, without exception, by the subordinate animal involved in the encounter. It occurs as a reaction to an imminent or actual attack and, in the performer, is frequently alternating with flight. The attacker will stop its action fairly soon in most cases and turn away. It now depends on the beaten animal, whether the agonistic encounter continues. If the next movements of the subordinate are very conspicuous (for instance, if it starts running away immediately after the dominant has turned its back towards it) this may release a new attack or pursuit. The loser was never seen to resume the fight. Rowell and Hinde (1962) have observed in *Macaca mulatta* that the display of the subordinate changes after the attack. It may sit and its *squeaks* may go over into screeches. These are much longer and do not reach such a high pitch as the squeaks. In screeching the mouth is opened widely, but the lips are not drawn back to show the teeth. The postures of eyes and eyebrows have not been described. According to the authors, 'screeching seems to occur when fright is gradually lessening'.

The data indicate that a high fleeing tendency is present. The animal does not actually flee, because it is inhibited by its 'strategic' position. The dominant may have cornered it or be immediately above it, or the subordinate may, perhaps, be too tired to flee any further. Here the tendency to flee is blocked by a *physical obstruction* or *physical incapacity*.

This compound expression, named the 'crouch face' by van Hooff (1962), has been observed in most monkeys (very infrequently in the guenons) and in the chimpanzee. It is basically the same in all observed species. The same facial expression, sometimes directed, however, to the mother, may be shown by young infants, while venturing to walk or climb alone (observed in *Macaca sylvana*, *M. nemestrina*, *Pan troglodytes*). It is not always clear what releases the response; in some cases the approach of humans or other animals might have been the reason. It releases immediate recovering by the mother.

Silent Bared-teeth Face

The eyelids: The degree of opening of the eyes varies; it may be maximum, or normal to slightly opened.

The eyes may either be staring at the opponent in a fixed way or be evasive (i.e. it may throw short glances towards the partner out of the corners of its eyes; sometimes the face is turned away sideward or upward).

The eyebrows and the upper head skin may be relaxed or in a lifted position.

The body posture may be erect and contain strong forward tendencies, occasionally giving rise to forward movements (always of a smooth and not a brusque nature). It may also contain strong backward tendencies; a somewhat hunched posture or a posture in which the head – or even the upper part of the body – is turned away, may be observed. It results occasionally in actual avoidance or in actual fleeing.

The mouth may either be more or less closed, so that there is no opening between the teeth, or opened more or less widely.

The mouth-corners are retracted.

The lips are usually vertically retracted, thus baring the teeth, and, with strong retraction, part of the gums.

Urination and defecation may sometimes be observed.

Vocalizations are absent.

The postures of the facial elements in the region of the mouth show a great similarity with those shown in *the staring* and *frowning bared-teeth scream faces*. The first four elements show considerable variability. Fully opened staring eyes may be associated with a

31

forward body posture; looking away, sometimes with moderately opened eyes, may go together with a backward body posture.

So we might speak of a *staring silent bared-teeth face* and a *looking-away silent bared-teeth face*, which are likely to represent different motivational states. The two forms may grade into each other quite smoothly.

The Circumstances

In the performer itself it never gave rise to attack; it may alternate with or occur during fleeing, avoiding, approach, or with such displays as, for instance, *the staring bared-teeth scream face*. If it occurs or alternates with smooth approach (this has only been observed in a limited number of species – see below) it may go over into such activities as embracing, huddling, play or, simply, being in each other's neighbourhood. In the other animal it was rarely seen to elicit attack. It may give rise to the same behaviour as in the performer. This full array of contextual possibilities is found in *Pan troglodytes*, in the genus *Mandrillus*, in *Theropithecus gelada*, and perhaps in some members of the genus *Macaca* (e.g. *Macaca maurus*), *Cynopithecus niger*, and *Papio ursinus* (Bolwig, 1959; Hall, 1962).

In the majority of monkeys, e.g. *Macaca irus, sinica, mulatta* (cf. Hinde and Rowell, 1962), *nemestrina* and probably in the other species the tendency to flee, though not being strong, seems to dominate and the tendency to attack does not seem to be present. Actual fleeing may be inhibited, for instance by different kinds of site-attachment (i.e. the tendency to stay at a certain spot, for example, a feeding or resting place). The compound expression may be shown in reaction to the approach of a dominant fellow. The latter does not necessarily show any overt signs of aggression; it may 'just happen' to come in this particular direction. The effect of the expression on the approaching animal can be rather negative, in the sense that the dominant animal does not pay any attention to the subordinate. This, however, may be the exact function of the response, since it permits the performing animal to stay where it is.

On a number of occasions it was noticed that aggressive behaviour (e.g. *staring open-mouth face*) on the part of the dominant monkey stopped, after the subordinate started *the silent bared-teeth display* (see photograph series VI). If this is the result of this display, it can be said to have an appeasing function. On the basis of the given evidence this more cautious formulation should be preferred to the one given by van Hooff (1962).

Macaca mulatta has been mentioned as one of the species in which *the silent bared-teeth face* seems to be motivated mainly by flight. An exception in this respect may be formed by the 'silly grin' display which has been observed by Hansen (1962) on a few occasions in a very particular context. In the experimental set-up, used by this author, four cages in each of which a mother with her infant lived were situated around a playpen. The passage between each of these four cages and the playpen was of such dimensions that only the infant could get through and the mother was confined to the living cage. When the infant was out of reach of the mother in the playpen, the mother might show the display. It incorporated a wide retraction of the lips interspersed with lip-smacking, and sometimes a specific type of vocalization. It brought about an immediate return of the infant to its mother.

In the listed species in which the full array of contextual possibilities is found, the display may be shown by a dominant animal to a subordinate or an infant; a few times the subordinate clearly expressed a tendency to flee or, as has been observed in the chimpanzee, performed the temper tantrum response. The subordinate became quiet and an approach followed (cf. the photograph series VII). In these species the compound expression can clearly express *social attraction* (as defined above). The tendency to approach may perhaps be thwarted by a low tendency to flee in the performer or by the experience of the performer that actual approach only increases the tendency to flee in the fellow and so may result in an actual increase of the distance between the partners.

If the advertisement of a non-hostile motivational state in the performer results in a lowering of the tendency to flee in the partner, approach may become possible. Van Hooff (1962) proposed to talk of a *reassuring function* in this context.

The possibility for the emancipation, in a number of species, of *the silent bared-teeth* response from a display expressing the tendency to flee into one facilitating mutual approach, may have been given by the fact that, already in its original form, it tended to lower both agonistic tendencies. Andrew's (1963a) observations show that a similar development has taken place in the platyrrhine monkeys. It is clear that the suggestions, given above, need further verification by detailed investigation.

The silent bared-teeth face can be regarded as a separate display, in the genera *Macaca, Cynopithecus, Theropithecus, Mandrillus* and *Pan*. In the genera *Cercopithecus* (except for *Ceropithecus talapoin*) and *Papio* it seems to be more an initial and final stage – and thus a low intensity stage – of vocalized bared-teeth displays. Hall (1962), however, has observed the response occasionally in a dominant *Papio ursinus* male while mating.

In both the drill and the mandrill the mouth posture is somewhat different from the one described above.

The mouth-corners are pulled back.

The mouth is closed.

The lips, are vertically retracted near the mouth-corners, but in the middle they still more or less meet. So only the canines and, partly, the molars are bared. The mouth-opening has the shape of a horizontal figure eight (see Plate II).

Marked *horizontal head-shakes*, strongly contrasting with the vertical head-nods given during threat (see: *staring open-mouth face*), may accompany this display. The exaggerated snarling and the intense head-shaking, 'as though tearing at an opponent' (Andrew, 1963a), create a ferocious impression. They may have led Andrew (*op. cit.*) to interpret the display as a threat. The expression movement has been observed both in dominant and subordinate individuals. It alternated with simple staring, avoidance, approach (sometimes followed by grooming or mating) or gnaw-wrestling. In dominant animals it occasionally alternated with *the tense-mouth face*. As yet it has never been noticed to be followed by attack. In the partner it may evoke similar responses. It is likely therefore that it indicates strong social attraction, at times mixed with a tendency to attack or to flee.

The head-shaking may perhaps be interpreted as a ritualized 'looking-away' movement (looking away has been observed to occur in other species in those cases where *the silent bared-teeth face* appears to be flight motivated). Schloeth (1956) has observed the movement in his introduction tests with *Mandrillus sphinx* and noted that it might be followed by play. Bolwig (1964) has described it as the expression of 'joy' of this species.

An identical mouth-posture has been noticed, so far only sixteen times, in *Cercopithecus talapoin*. With this feature it distinguishes itself greatly from the other members of the genus. The expression was shown during approach and was followed by a typical embrace, during which the animals stood on their hind-legs in front of each other. It alternated a couple of times with *the teeth-chattering face* (see below).

In *Theropithecus gelada* the compound expression is accompanied by the characteristic inside-out turning of the upper lip which also forms part of the other bared-teeth displays of this species (see: *staring bared-teeth scream face*). The movement may be made in a jerky manner, lasting for part of a second; at a distance a pink flash is noted then. The posture may also be maintained for some time (a second or more). The display frequently alternates with *the lip-smacking face*. Like the latter expression it is often performed during, or followed by, smooth approach. Unlike the latter it also alternates with *the tense-mouth face* or leads to play. It is shown by dominant and subordinate animals. It is often answered with the same expression. The data suggest that 'social attraction' plays a certain rôle and that a tendency to attack may be present as well.

In *Pan troglodytes* two extreme forms of *the silent bared-teeth face* can be distinguished: one in which the teeth are completely bared by vertical retraction of both lips and another in which the upper lip is not retracted, so that only the lower teeth are bared. Whether these forms possess a different motivation is not known at present.

Bared-teeth Gecker Face

The expression elements are all identical to those shown as part of *the silent bared-teeth face* except for the fact that *the lips* may not be completely retracted vertically and that *vocalization* is present.

In the genera in which this display has been distinguished (*Macaca, Papio, Theropithecus*), this vocalization consisted of a number of usually not loud, moderately to highly pitched, short, staccato sounds. They are given singly or, more often, in short rapid bursts.

The vocalization has been described as a 'gecker' by Hinde and Rowell (1962) for *Macaca mulatta*. These authors regard it principally as a baby noise and note that it occurred when the mother tried to get up while the offspring was sucking and often when it appeared to have lost the teat. The movement was also found to be associated with violent, spasmodic movements of the limbs and the body and to be followed by clinging tightly. The accompanying facial expression was not observed. Andrew (1963a) quotes Lashley and Watson (1913) who heard a baby 'chatter' when handled roughly by its mother and considers this call to correspond with his 'arrr'-bout or titter.

Infants of *Macaca mulatta, M. irus, M. sylvana, M. nemestrina* and *Papio anubis* have been observed to perform the display in the situation described above. A baby *Macaca nemestrina* would also gecker frequently, without vertical retraction of the lips, some time after it had left the mother on a private excursion. The display then often alternated with *the pout face* and was performed during, or followed by, climbing back to the mother, who reacted much more hesitantly to it than to shrieking (see: *frowning bared-teeth scream face*). Andrew (1963a) recorded titters, given without vertical retraction of the lips, and shrieks, given with vertical retraction in infants of *Macaca mulatta* when seeking their parent-substitute.

A display, resembling the infant geckering, with vertical retraction of the lips, has been observed in subordinate adult members of the species *Papio anubis* and *Theropithecus gelada*. Detailed information is available from the latter. Here it alternates with the *lip-smacking face*, avoiding movements and hesitant approach. It may follow *the staring bared-teeth scream face*. This suggests that here both a mild tendency to flee and to approach is present. According to Kummer (1957), *Papio hamadryas* expresses 'mild fear' by 'Keckern'.

A similar display, in all probability expressing the same motivation, can be observed in the chimpanzee. In this species, it often alternates, for example, with *the pout face* and with *the temper tantrum response*.

An intermediate form between the display and *the pout face* may be distinguished, in which the mouth-corners are withdrawn, *the underlip* is vertically retracted to some extent and *the upperlip* is protruded. The vocalization is intermediate too, and is colloquially known as 'whimpering'. This facial expression is, therefore, called *the stretched pout face*. In the adults of *Papio, Theropithecus* and *Pan, the bared-teeth gecker face* more or less occupies the place which in most of the *macaques* is taken up by *the teeth-chattering face*. The latter compound expression has never been observed as a true separate display in the above-mentioned species (see below). It is not unlikely that the gecker face in adult baboons and chimpanzees is an emancipation of the juvenile response described at the start, from which it may derive an appeasing and attracting effect. It may even have become an expression of submissive attachment in *Papio ursinus*, where, according to Bolwig (1959), an intermittent, almost coughing sound is uttered, with full vertical and horizontal retraction of the lips, when the animal is 'amused or full of joy over the keeper's presence'.

The gecker display has not been distinguished by van Hooff (1962).

The Lip-smacking Face

Monkeys which are grooming each other occasionally pick up and eat a small particle from the skin of the partner. Directly they have brought it to their mouth, they perform a few clearly audible smacks. The motor pattern of such a smack consists of movements of the underjaw, the tongue and the lips. The underjaw goes down and up, but the teeth do not meet; simultaneously the lips open slightly and close again; the tongue is brought forward between the teeth. The whole process gives the impression that the animal is tasting the particle. When one animal starts grooming another, the first groom-movements may be accompanied by a smacking that is not preceded by the act of bringing a particle into the mouth. It differs from the functional smacking in that it is performed with a higher, 'typical intensity' (Morris, 1957). Usually this initial vacuum smacking soon stops and during the rest of the grooming period the functional smacking may be observed. In a number of species the movement is more loosely connected with the grooming act and may be shown in other circumstances as well. The facial elements then behave as follows:

The eyelids are usually fully apart, so the eyes are opened completely.

The eyes are usually staring in a fixed way at the partner. The head may also, however, be turned sidewards a little, so that the animal is looking out of the corners of its eyes.

The eyebrows and the upper head skin are often lifted in the monkeys which have been classified in category one (see p. 20). As far as the monkeys of category two are concerned, the participation of these elements is obscure.

The ears may be retracted.

In the functional smacking seen during grooming, the eyes, eyelids, eyebrows and the ears are in the neutral position.

37

The mouth is rapidly opening and closing (almost or completely).
The mouth-corners are brought forward.

The lips. In different species the movements of these elements
may vary slightly. Mostly resembling the original functional
smacking is the pattern in which, together with the opening of the
mouth and parting of the lips, the tongue is protruded between the
teeth and lips. During the closing phase the tongue is withdrawn
again. This form is present in almost all species in which the
display has been observed. These are: *Erythrocebus patas, Theropithecus gelada, Cynopithecus niger,* and the studied species of
Mandrillus, Papio, Macaca and *Cerocebus.* The movement has
rarely been observed in a few *Ceropithecus* spp. (e.g. *C. talapoin,
nictitans, mona,* and *aethiops*). It is the only form, present in
Theropithecus gelada, where the tongue protrusion element has
been exaggerated. We might term the display of which it forms
part *the tongue-smacking face.*

In another movement pattern, the tongue protrusion element
has been reduced greatly. Thus the pattern almost solely consists of
a rapid opening and closing of the lips, which is performed with a
higher speed than *the tongue-smacking face* of the species concerned. Often the head is tilted upwards and the performer peers
over its nose to the partner. The display of which it forms part
may be termed *the true lip-smacking face.* It has been observed in
Macaca (except *M. nemestrina* and *M. sylvana*), *Papio,* and
Cynopithecus.

A different pattern, again, is found in *Macaca sylvana.* While
the underjaw is moving up and down rapidly (the molars can be
heard to meet), the lips remain pressed together and slightly
rolled inwards. This type of smacking has been termed *the chew-
smacking face.* In this species *the tongue-smacking face* has been
observed rarely. Like the other smacking faces it is performed
with a 'typical intensity'.

The same is true for the fourth pattern which is typical for the
studied *Cercocebus* species. Besides the original lip-smacking, a
form can be observed in which the upper lip is kept in a lifted
position, thus baring the upper incisors and canines. The tongue
is protruded rhythmically. The mouth-corners are pulled forward.
The compound which is characterized by this typical pattern
might be called *the snarl-smacking face.*

The body posture may vary greatly. The animal can be standing, sitting or walking. Forward as well as backward intention movements occur. The movements are mostly smooth.

Vocalizations. Soft, to moderately pitched, short grunting sounds may be uttered.

The Circumstances

In most species, the movement is not frequently seen between animals which are associated in an established relationship with one another (except, perhaps, for the *Papio* spp.). It may be shown, however, quite frequently during a short time when the animals meet for the first time, or after a period of separation.

In the performing animal it rarely gives rise to aggression; equally, it rarely releases aggression in the other animal. In the performer it may be followed by fleeing or by approaching. The approach is usually more or less 'cautious'; this means it is done without the brusque and rigid movements which characterize the agonistic approach. The other animal may react mainly by approach, lip-smacking, teeth-chattering (see below) or fleeing. An approach is followed by mutual grooming, huddling together, play, mating or simply being in each other's neighbourhood. The display may be shown by dominants to subordinates (irrespective of whether it concerns males or females) or conversely. In a number of cases it has been observed that the compound expression was directed towards very subordinate individuals, which showed a strong fleeting tendency, or by a mother towards her young which performed *the bared-teeth gecker display*; then, often, the subordinates or the baby became quieter again. It is mostly responded to in the same way. As a rule the posture facilitates non-hostile approach between the individuals and this may be achieved by an appeasing or reassuring effect on the partner (see also: *the silent bared-teeth face*). The display may reflect a conflict between the tendency to flee and the tendency to approach. The latter may frequently be dominating, since the impression exists that the display is given most often by a dominant animal which makes approach movements. In the latter case the tendency to flee may even be negligible.

These conclusions of van Hooff (1962) are shared by Hinde and Rowell (1962). These authors present a list of situations in which lip-smacking in *Macaca mulatta* occurs, that is rather similar to the one presented above. 'Since lip-smacking is more often

P.E.—4

39

given by an established monkey to a newcomer than vice versa, since it is sometimes associated with threat postures, and since it has been seen to intergrade with chewing (see above), it is possible that it may also involve an element of slight aggression' (*op. cit.*).

Although more detailed investigations will be needed to verify this, the impression exists that *the true lip-smacking face, the chew-smacking face* in *M. sylvana* and *the snarl-smacking face* in *Cercocebus* are the result of a higher tendency to approach than the less ritualized *tongue-smacking face* in the same species. The form of particularly the two latter displays may indeed suggest that here, also, a slight tendency to attack is present.

Besides its possible appeasing function (lowering the tendency to attack) and reassuring function (lowering the tendency to flee) it may also have an *attracting function* (raising the tendency to approach). This supposition is supported by the fact that an animal (e.g. a subordinate female) may direct the display to a partner (e.g. a dominant male) who is lying or sitting somewhere completely relaxed. The partner may then rise (and approach) and indulge in some social activity (e.g. mating).

Zuckerman (1932) mentions the posture in *Papio hamadryas* and states that animals greet each other and youngsters in this way when they meet.

Bolwig (1959) has reported it for *Papio ursinus*. In this species it is said to precede copulation and last during this act; it is, therefore, interpreted as a 'friendly gesture, probably with a slight sexual motivation and connected with grooming'. In other species it is certainly also shown a great deal outside any sexual context. This has been confirmed for *Papio ursinus* as well by the observations and experiments of Hall (1962) on a group of *Papio ursinus* in the wild.

Van Hooff (1962) reported to have seen only the functional smacking (i.e. accompanying grooming) in *Macaca nemestrina*. Later observations on other groups convincingly showed that also *the tongue-smacking face,* used as a signal from a distance, is present in this species. It frequently alternates with the (for this species characteristic) *protruded lips face* (see below).

In *Colobus abyssinicus, the lip-smacking face* is slightly aberrant. Whereas in the other monkeys the successive smacks in a bout follow each other immediately, there is a considerable pause between them in this species, resulting in an average of one to two smacks per second. In the other monkeys the rate varies around seven smacks per second.

In comparison with the other monkeys, *the lip-smacking face* is most readily released in *Papio* spp. (Bolwig, 1964). In the related genus *Mandrillus*, the display has been observed rather infrequently. Here it was only noticed in close connection with grooming behaviour. In other circumstances, which in *Papio* usually evoke lip-smacking, *Mandrillas* may perform its characteristic form of *the silent bared-teeth face* or a mixture of this expression and *the teeth-chattering face*.

In none of the great apes does this facial expression play a rôle of importance. In *Pan troglodytes* a kind of lip-smacking movement can be observed, sometimes, during grooming or autogrooming. It seems to have lost its original functional connection with the grooming act. It has been observed frequently without any particle from the skin having been taken into the mouth. It does not resemble the original functional smacking in that the protrusion of the tongue is barely noticeable. It almost solely consists of a comparatively slow (average frequency two to four smacks per second) opening and closing of the mouth and the lips. The mouth-corners remain in a neutral position and the rest of the face is completely relaxed. These data may indicate that it is a relic of a former ritualized display which may have been much more important in the ancestors of the present African apes.

Especially in baboons, lip-smacking may be accompanied by or alternating with *grunts*. These grunts may be produced also in a number of other situations (Andrew, 1962, 1963a and b). This will be discussed more fully below.

The Teeth-chattering Face
This compound is closely related to *the lip-smacking face*. It is likewise characterized by a rapid opening and closing of the mouth. Here, the teeth meet when the mouth is fully closed. The lips, on the other hand, do not meet, since they are fully vertically retracted.

The expression elements:
The eyelids may be fully apart or in the normal position.
The eyes are either staring at the partner in a fixed way or making evasive movements.
The eyebrows and the upper head skin may be retracted.

The ears may be retracted as well.

The mouth is opening and closing rapidly and the teeth chatter audibly.

The mouth-corners are retracted.

The lips are vertically retracted, thus baring all the teeth.

The body posture may contain backward as well as forward tendencies. It may be slightly hunched or the broadside of the body may be presented.

The Circumstances

This display is in its ritualized form typical for the genus *Macaca* (except perhaps *M. nemestrina*). Here it is well distinct from *the lip-smacking face*. In the species which have been studied, it alternates quite often with lip-smacking. The transition from the one display to the other, however, is not at all smooth. Thus the changing of the position of mouth-corners and lips is abrupt and no smacks are given while these elements are in an intermediate position. Thus when relatively tame individuals of *Macaca irus* were approached from a distance by the experimenter, in an experiment room where they had plenty of opportunity to avoid him, lip-smacking was often released initially. At a certain moment single teeth-chatters appeared between the lip-smacks. The number of teeth-chatters now increased rapidly, while the number of lip-smacks decreased. Thereafter only teeth-chatters were seen. The appearance of teeth-chatters coincided with the appearance of backward glances. Finally the animal might jump away. As in lip-smacking, the frequency and amplitude of the teeth-chattering movements are fairly constant.

In *Macaca irus, sinica, maurus, silenus* and *speciosa* the display was shown mostly by subordinate individuals. It often alternated with *the lip-smacking face*. It was frequently preceded or followed by avoidance and fleeing or intention movements of fleeing, but it might also alternate with approach or intention movements of approach. In the performer it was never preceded or followed by *the staring open-mouth face* or attack. Usually it was shown in response to activities of the partner, such as simple staring, approach, *the lip-smacking face*, or *the staring open-mouth face*. It did not evoke the same response in the partner, in contrast to *the lip-smacking face*. In a number of cases (though not always) aggressive displays on part of the partner stopped when *the teeth-chattering face* was shown. Quantitative investigations may show that the teeth-chattering indeed produces a decrease in the tendency to attack, and that the display can be said, therefore, to have an appeasing function.

The data suggest the tendency to flee is predominantly present in the performer and that it is thwarted by the tendency to approach.

One of the studied species of *Macaca*, namely *M. sylvana* clearly was an exception to this. Here *the teeth-chattering face* was equally shown by dominant and subordinate animals and was frequently followed by approach, grooming, huddling, etc., and retreat movements were not as manifest as in the other species. It was also shown during embracing (in the other macaques as well as in the baboons, lip-smacking typically accompanied this activity). The partner might perform *the teeth-chattering face* as well. Often the display was initiated with a few *chew-smacking* movements. The data suggest that in this species a tendency to approach is present which may be stronger than the tendency to flee on the average. Perhaps the display covers approximately the same range as *the chew-smacking face*, reflecting a stronger motivation. The impression exists that *M. speciosa* takes an intermediate position in this respect between *M. sylvana* and the other mentioned species; quantitative investigations are clearly needed, however, to establish these points. *The teeth-chattering face* has not yet been observed in *Macaca mulatta* and *M. nemestrina*.

Although teeth-chattering movements have been observed in the genus *Papio*, they occurred rarely and only for a very short time. They seem to be not so much a separate display as a stage of transition between *the lip-smacking face* and *the bared-teeth gecker face*.

On a few occasions a dominant male *Mandrillus sphinx* performed *the teeth-chattering face* during mounting and mating. Here the posture of the lips was slightly aberrant. The lips were vertically retracted at the mouth-corners, but they covered the incisors, with the result that they met more or less in the middle of the end of the closing phase of the movement (cf. *the silent bared-teeth face* of this species).

The teeth-chattering face has not been observed in the guenons, except for *Cercopithecus talapoin*, where it occurred in alternation with the typical *silent bared-teeth face* of this species.

Among the platyrrhines the display appears to be common in

the genus *Cebus* where it may be shown in 'friendly greeting' (Andrew, 1963a).

The only member of the apes in which I have observed this expression is *Hylobates lar*.

The Protruded-lips Face

When a female pig-tailed monkey (*Macaca nemestrina*) is in heat, a male which has access to this female may frequently show a most peculiar response. During this period the male may repeatedly smell at the genital region of the female, which bears large swellings. It then shows a facial posture which is mainly characterized by a protrusion of the lips. The upper lip moreover is slightly curled upwards and the lower lip is pressed against it tightly. The smelling may last a few seconds: after the male lifts its head and with the face directed slightly upwards and the eyes gazing up in an undirected way, it maintains the facial posture for a short time. In a number of cases copulation follows. Schneider (1931, 1932a and b, 1934) has described a behaviour pattern in many mammals, especially the ungulates, which appears basically similar (apart from the fact that the lower lip is not pressed against the upper lip). This posture, which in German is known as 'Flehmen', can be observed, whenever these animals perceive certain natural or artificial scents. Especially the males may show the response when they smell at genitalia of females or urine, particularly of a female in heat. After smelling, the animals may tilt their head upwards and show the facial posture while gazing in the air. Schneider thought of this as a careful 'tasting' of the smell. Because of the resemblance of the present display with 'Flehmen', van Hooff (1962) called this display *the flehmen face*. Since this term suggests a homology, which is not yet certain, this name is replaced, presently, by one which is purely descriptive.

In *Macaca nemestrina* this facial expression is not only seen in the context described above, but also in a number of situations where it is directed to a partner which reacts to the posture. It is thus a social signal.

The expression elements:

The eyelids may be either fully apart or in the normal position.

The eyes may either be staring at a partner in a fixed way or making evasive movements.

The eyebrows and the upper head skin are lifted completely.

The ears are folded back against the head.

The mouth is closed.

The lips are protruded. The upper lip is curled upward more or less; the lower lip is pressed against it tightly.

The mouth-corners are brought forwards.

Vocalizations are usually not emitted. Rarely grunts have been heard.

The body posture may contain strong forward elements, actually giving rise to forward movements, or it may contain backwards elements, giving rise to backward movements. The head may be tilted upwards, so that the performer is peering down over its nose towards the partner.

The Circumstances

The male *M. nemestrina* may show this posture during mating. He may also direct it to the female from a distance; this will often induce the female to 'present'. The facial expression has been observed, however, in a great number of cases outside any obvious sexual context between animals of both the same and the different sex, and between animals of roughly all ages (for example, adults, especially the mother, towards infants). It often alternates with *the lip-smacking face*, which is present in this species (see above), and seems to have much the same motivation, expressing mainly of the tendency to approach and the tendency to flee, the first probably often dominating.

The display has been observed in two other species of *Macaca*. A male *M. silenus* frequently performed the expression movement towards a female before or during walking up to her to mate. If the female did not present, but walked on, he went in front of her and performed again. Usually the female (stopped and) presented. *The protruded lips face* was never observed in another context.

Only once was the display observed in *Macaca mulatta*. An adult male walked in front of a female and performed the expression movement. The female did not react and the movement was not followed by any other behaviour pattern.

The Pout Face

This compound expression consists of the following expression elements:

The eyelids are either in the normal position or farther apart.

45

The eyes may be directed towards a partner.

The eyebrows may be lifted, as well as the upper head skin, especially when the expression movement is directed towards a partner. Simultaneous contraction of the *m. depressor supercilii* and perhaps the *m. corrugator supercilii*, antagonistic to the *m. frontalis*, responsible for the lifting, may result in a curious slanting position of the eyebrows (Bolwig, 1964).

The mouth is slightly opened.

The mouth-corners are pulled forwards.

The lips are protruded; typically they stay pressed together near the mouth-corners, but are lifted in the middle region, thus creating a small round opening.

The sound which accompanies the expression is a clear, portato, not very loud, moderately pitched 'ooo'. Sometimes the sounds are given in bouts; the separate sounds then tend to be more staccato.

The body posture may be rather relaxed or it may be directed forward.

The Circumstances

In its typical form (i.e. with the described mouth-posture) this compound is a baby expression. It has frequently been seen in infants of a number of species, when they are away from the mother, for instance, after they have started on a private excursion. The duration of such an excursion may vary. During the final stage (this may be some minutes before either the mother comes to collect it or the infant reaches its mother on its own account), the display may be performed a great number of times. It may either be directed to the mother or given in an undirected way. The mother usually takes up the baby, but this does not happen always immediately. Sometimes the mother deliberately avoids the infant. *The pout face* may alternate with *the gecker face* and may eventually lead to shrieking (see above). The data suggest that it concerns a response which expresses the need to be given body comfort or the opportunity to feed. In this context *the pout face* has been observed extensively in representatives of different genera, namely *Pan troglodytes*, *Pongo pygmaeus*, *Papio anubis*, *Macaca nemestrina* and *M. mulatta*. This characteristic expression has been described for the latter species by Hinde and Rowell (1962). Bolwig (1963) mentions that a baby *Erythrocebus patas* during the first days after its arrival in captivity would frequently show the expression, especially when a human being was in sight. On withdrawal of the human being the animal used to shriek.

Schaller (1963) reports that youngsters of *Gorilla gorilla beringei* may utter a whining sound with slightly protruded lips, when the mother is out of reach. The mother usually reacts. The display may go over into screeching.

In the chimpanzee (*Pan troglodytes*) *the pout face* can be observed frequently both in young animals and in adults. Here, the expression movement was most often seen in relation to a person showing some attractive object like food. As soon as this is acquired by the animal the response fades away. In other cases the display stops when in response to it the animal is being cuddled. If the animal does not attain these 'goals', the display may be followed by the temper tantrum response. The expression was also observed in circumstances which are more distantly related to the 'original' context of 'begging' for food or body comfort. For instance, in the tests done by Crawford (1937; from Hooton, 1942) in which two chimpanzees had to co-operate in order to reach a certain goal, one animal might 'ask' the other to co-operate in this way. The display may be observed in interactions, in which lip-smacking might be shown in most of the catarrhine monkeys.

The typical 'clear call' (Hinde and Rowell, 1962) which forms part of the display may also be heard in a number of other situations, when it can be produced by adults. Although contraction of the *orbicularis oris*, resulting in a slight protrusion of the lips, is usually present, the protrusion is normally not so strong as in the exaggerated 'funnel'-mouth which may be seen in infants. Hinde and Rowell (1962) have given an extensive account of the different kinds of clear calls and have shown that they may differ both in different individuals and in the same individual in different circumstances. Thus clear calls are given in response to the sight or sign of food. They become shorter, the nearer the moment comes that the food can be eaten. When food is actually being taken, short harsh calls (grunts) are given. Clear calls were heard when the group was shut out in the outside pens. Frequent clear calls are given by animals separated from their cage mates. Animals in the group may emit them for no external reason apparent to the observer, and they are often answered by a partner (contact-calls). Furthermore, the appearance of other animals or humans is always remarked on with, in both cases different, clear calls. Although the function of these different clear calls is still unknown, the range of situations in which they are given suggests a motivational relationship with *the pout face*.

The clear calls of Hinde and Rowell (1962) certainly correspond with the 'woo', described for *Macaca mulatta* by Andrew (1963a). The 'woo' has been heard when a human acquaintance leaves the room; it is also given with 'arr' bouts in response to the approach of a 'frightening human'. Corresponding with the *Macaca* 'woo' is the *Papio* and *Cercopithecus* 'moan' which is reported to occur in similar situations and the chimpanzee 'oo' call.

In the chimpanzee, facial displays which are intermediate between *the pout face*, described here, and *the staring open-mouth face* (see above) occur. Vocalizations also intermediate between the clear 'ooo' characteristic for *the pout face* and harsh grunting·or barking noises occur. As far as their motivation is concerned they may well be rather complex. A muted 'boo' was given by a young animal to strange objects (Andrew, 1963a, under 'warning and excitement'). Such calls are accompanied by mild to pronounced protrusion of the lips, which for the greater part may be separated, and the gaze is fixed on the object (for example, a centipede). It immediately brings companions to the spot, which look in the same direction. The behaviour pattern may thus have a 'pointing' function.

Köhler (1926) and Yerkes and Yerkes (1929) mention the occurrence of spells of 'excitement' (described by Andrew, 1963a, under 'crying'). They can be frequently watched in most adult zoo chimpanzees, where they take the following form. Rather spontaneously (however, only when public is standing in front of the cage) the animal may, slightly rocking, start to give soft clear 'ooo' sounds, which accompany *the pout face*, as described above. These sounds become increasingly louder; the lips become separated more and more and the typical funnel form gets lost. The hairs bristle and the animal may start to jump on all fours or to drum with its hind-legs on some heavily resounding metal slide. Its hooting develops into loud shouting, during which the mouth is widely opened, whereas the still protruded lips do not any longer cover the teeth. With a loud roar it may finally charge into the direction of the public. The motivation of this behaviour is not fully clear, although aggression obviously plays a rôle.

The Relaxed Open-mouth Face

The mouth-posture of this display may superficially resemble the one shown in the version of *the staring open-mouth face* in which the mouth is rather widely opened (see above). Closer examination shows nevertheless that there are many subtle differences in the separate expression elements between the two displays. The expression elements are as follows.

The eyelids are usually in the relaxed position or slightly drawn together (Bolwig, 1964). According to Bolwig (*op. cit.*) the outer corners of the eyes are always slightly lifted due to a pressure from the zygomatic muscle, which draws the mouth-corners backwards and upwards. The same would be true also in the other displays in which the zygomaticus contracts (bared-teeth displays), but may be less visible there because of the tension which is often present in the musculature controlling the degree of opening of the eye.

The eyes may occasionally be directed to the partner. The gaze, however, is less fixed than in the *staring open-mouth face* (Bolwig, 1964).

The eyebrows and the upper head skin are usually in the normal position. They may, however, be lifted when the display is directed to a partner from a distance.

The ears are usually in the normal position.

The mouth is rather widely opened.

The mouth-corners may be more or less in their normal position or drawn backwards and upwards (Bolwig, 1964). In *the staring-open-mouth face* they are usually more or less drawn forward.

The lips are not retracted in the majority of species; the upper lip is usually rather tight, and covers the upper teeth completely; the lower teeth may partly be visible. When the mouth is opened rather widely the upper lip may slide up and bare the upper teeth. This seems to occur especially easily in species with large pointed snouts (cf. Bolwig, 1964).

The body posture may be very variable. Positions change fast and in an unpredictable way. Forward as well as backward movements are frequent. The movements are often quick and supple.

Vocalizations have not been heard in most species. Often there may be fast rhythmic staccato breathing, which is clearly audible. In some species the breathing is accompanied by low-pitched soft noise.

49

The Circumstances

This facial display is frequently shown when animals are playing together. In monkeys and apes social play is characterized by a number of movements which also may be seen in agonistic and sexual behaviour. They are performed in a somewhat exaggerated way and in an 'inconsequent' order. The gnaw-wrestling shown during play is also less tense than the struggling which may occur during an agonistic encounter. Hansen (1962) distinguishes four major types of social play: rough-and-tumble play (contact), approach-withdrawal play (non-contact), mixed play (rapid oscillation between contact and non-contact patterns), aggressive play (vigorous contact patterns accompanied by barking, pilo-erection and yawning – if there is unilateral participation, the non-participant may make frantic escape movements). Play is predominantly a juvenile activity, but not exclusively (Kummer, 1957).

The relaxed open-mouth face is obviously an intention movement of the gentle biting and skin-gnawing movements which occur during wrestling (see also Bolwig, 1964), but it is also shown during the approach-avoidance play. Occasionally the play may stop for a while. Suddenly one of the partners may adopt some kind of 'jumping intention posture' and perform *the relaxed open-mouth face* in the direction of the other who will often immediately react by resuming the play. So it may evoke the same response in the partner.

The display has been observed in many representatives of all the studied catarrhine genera. Recently Bolwig (1964) has given a detailed description of this display which he considers to be the expression of 'happiness and joy'. Andrew (1963a) mentions that the play face of macaques and baboons much resembles the face shown during 'aggressive threat'.

In the genus *Mandrillus* the expression movement only occurs during the actual gnaw-wrestling; in this species the upper lip may often slide back from the upper teeth. In most play contexts the typical *silent bared-teeth face* of this species can be frequently observed. In *Theropithecus gelada* the *relaxed open-mouth face* often also alternates with the typical inside-out turning of the upper lip which has been described as a part of the bared-teeth displays.

In the chimpanzee the fast rhythmic breathing is often accompanied by bouts of short staccato vocalizations, which sound like a soft, low-pitched, noisy 'ah, ah, ah'. This facial expression which is often referred to as 'laughing' has been described or mentioned by many authors, for instance, by Darwin (1872) who notes that the main feature in which this chimpanzee 'laughing' differs from our own laughter is the fact that the teeth in the upper jaw are not exposed. It may not only be observed during social play, but also, occasionally, during autoplay. Köhler (1926) remarks that sometimes a slight retraction of the mouth-corners may be noticed 'during the leisurely contemplation of any objects which give particular pleasure' (for example, little human children). The facial expression shows some resemblance with the human smile. A similar observation is reported by Kohts (1935).

The relaxed open-mouth face can also be evoked by tickling a chimpanzee. Andrew (1963a) mentions that with increased intensity of tickling, vertical retraction of the lips, frowning and closing of the eyes may occur. It probably indicates that the tickling is becoming unbearable. During normal social play this extreme form has been observed only very rarely by the present author.

Schaller (1963) reports that playing mountain gorillas partially open their mouths with the corners drawn far back into a smile, but without exposing the gums and teeth. Occasionally the display was accompanied by a very soft panting chuckle, sounding like 'a-a-a-a'.

The facial posture shown during play by lowland gorillas is identical. While playing, the animals emitted a continuous soft, low-pitched purring sound, occasionally going over into chuckles.

GENERAL COMMENTS

It is possible to distinguish in the catarrhine monkeys and the apes a number of compound expressions (i.e. typical combinations of certain expression elements). Some of these can be observed frequently in most catarrhine monkeys and apes. Others belong to the repertoire only of certain groups or species. Instances of these latter ones are *the silent bared-teeth face, the teeth-chattering face* and *the protruded-lips face*.

Thus *the silent bared-teeth face* can hardly be distinguished as a separate display in the genus *Cercopithecus* (with the notable exception of *C. talapoin*, which appears to be aberrant in other respects as well). Although the bared-teeth display may occur silently, it forms almost always an initial or final stage for one of the vocalized bared-teeth displays (cf. Andrew, 1963a and c). Andrew has shown that the situation found in this genus is not yet very different from the one found in more primitive primates, e.g. the Lemuroidea. The retraction of the lips is still almost completely connected with the occurrence of loud vocalizations. In the other catarrhines this low intensity stage has started to lead an individual life, not only in that it has become ritualized to such an extent that it gets a quite distinct form (e.g. *Mandrillus*), but also in that it may represent a qualitatively different motivational state (e.g. *Pan troglodytes*).

In the same way *the teeth-chattering face*, which can be regarded as a transitional stage between a bared-teeth display and *the lip-smacking face*, and which can be observed as such, for example, in *Papio*, becomes ritualized in that it gets a typical intensity and frequency and occupies a distinct place in the repertoire of only one group of catarrhine monkeys, namely, the genus *Macaca*. The range of motivational states which it represents can, as we saw, still differ in the different species of this genus.

As another example *the protruded-lips face* has been presented. So far this compound expression has been observed convincingly only in two species, *Macaca nemestrina* and *M. silenus*. It may perhaps occur occasionally also in other species; this is probable, for example, for *M. mulatta*.

The other displays can be seen in almost all monkeys and apes, though some may play a more important rôle in certain groups than others. Thus one gets the impression that *the silent bared-teeth face* and *the lip-smacking face* may both be chosen as a means to facilitate mutual approach. Among the closely related genera *Papio* and *Mandrillus*, *the lip-smacking face* has obviously been given preference by the first genus, whereas the second genus has given preference to *the silent bared-teeth face*. In *Theropithecus gelada* the situation is intermediate. In the chimpanzee (*Pan troglodytes*), *the lip-smacking face* has disappeared almost completely and its place has been taken by *the silent bared-teeth face* and *the pout face*, which, in its typical form, is especially important as a baby-expression in probably most of the other species.

In a number of cases more than one compound expression seems to represent a certain range of motivational states, at least on the basis of crude qualitative data like those presented here. Thus both *the pout face* and *the silent bared-teeth face* may, in the chimpanzee, represent a tendency to approach and to engage in positive social activities, which may be blocked in some way or another. More detailed analysis of the situational context may reveal in what aspects these compounds do differ. Such investigations may confirm the impression that *the pout face* is shown in those circumstances in which the tendency to approach and to engage in positive social interactions is blocked by a lack of 'willingness' to co-operate (i.e. a lack of such tendency) on part of the companion, whereas *the silent bared-teeth face* may be shown when this tendency is blocked by the fact that the companion withdraws, because of the presence of a tendency to flee. Such a situational difference would be in accord with the history of both compound expressions.

In the same way explanations must be sought for the apparent overlapping in motivational range of *the protruded-lips face* and *the lip-smacking face* in *Macaca nemestrina*.

Most of the compound expressions discussed in this paper appear to cover a broad range of motivational states. In each instance a suggestion has been given as to the nature of this range. *The lip-smacking face*, for example, reflects in many species the presence of a tendency to approach and a tendency to flee. The tendency to approach seems to dominate on the average. The reader may have noticed that the compound expressions have mainly been distinguished on basis of the mouth-posture and, to some extent, the nature of the vocalizations. Clearly, the postures and movements of such elements as the eyes (e.g. staring fixedly, turned away or intermediates), the eyelids, the body (forward or backward inclinations), etc., may still vary greatly. Further consideration of the separate combinations within a certain compound expression will probably justify more subtle distinctions of compound expressions, each with a much more specific meaning. Especially in the case of *the staring bared-teeth scream face*, it is clear that the situational context in which it occurs to a large extent determines its meaning.

The derivation of the different components of compound expressions is clear in a few instances. It has already been pointed out that *the lip-smacking face* may be considered as a ritualized intention movement of one of the components of the grooming behaviour, namely the tasting and swallowing of particles picked from the skin (Hinde and Rowell, 1962; van Hooff, 1962; Andrew, 1963a and b). In other mammals similar intention movements may come to serve as expressions indicating the 'willingness' to engage in positive social activities (Eibl-Eibesfeldt, 1957). Recently, Zeeb (1959, 1963) has shown that the horse possesses an 'Unterlegenheitsgebärde', performed especially by subordinate individuals, which is characterized by up and down movements of the underjaw; the mouth does not close completely and the teeth are fully covered by the lips. The display frequently alternates with grooming. Andrew (1963a and c) has, moreover, suggested that the sucking of the infant may have been a source which has contributed to the shaping of the display.

Andrew (1963a and c) has related the contraction of the *M. orbicularis oris*, which is characteristic for *the staring open-mouth face*, with the respiration reflex which can be observed in primitive primates; the contraction of this muscle, which accompanies strong respiration, may have become an intrinsic part of displays which are often accompanied by calls given with strong expiration, but without the mouth-corner withdrawal which develops with high glottal tension (see below). It remains questionable, however, whether this can explain the typical, exaggerated funnel form of the lips, shown in *the pout face*, especially since it is always associated with low intensity smooth vocalization. Van Hooff (1962) suggested that it can be regarded as an intention movement of sucking, since the same mouth posture can be observed when a baby is seeking the nipple. Moreover, an infant which shows *the pout face* will, immediately after reaching the mother, if only for a few seconds, take the nipple in its mouth. It is, therefore, highly likely that the reaching for the nipple has, at least, contributed to the shaping of this response.

The main source from which, according to Andrew (1963a and c), many of the expression elements are derived is formed by the so-called 'protective responses'. Originally they were given following the perception of a noxious smell or taste, and they can still be observed in this context. The head is shaken from side to side and the mouth-corners and the lips are drawn back, so that anything in the mouth tends to fall out of it. This is further aided by tongue protrusion. The eyes are closed, the eyebrows lowered and the ears flattened against the head for protection. The glottis is closed and then suddenly opened with a vigorous expiration which may help to clear the mouth and respiratory tract from noxious substances. The same reaction, though usually less intense, may be shown to other strong stimuli. The same components appear to form part of the responses of many mammals given in reaction to threat by a dominant fellow or predator. Here, the activity of the glottis has led to the development of usually loud calls. Many of these protective responses have also been incorporated in the facial display of the primates.

This hypothesis is valuable particularly because it offers a new and interesting possibility for the explanation of the derivation of the intense calls and the baring of the teeth given by animals which are subject to some form of threat. Favourable evidence is provided by the fact that in the lower primates baring of the teeth is almost always accompanied by sudden intense vocalizations, like clicks, crackles, etc. (Andrew, *op. cit.*).

Yet it is not so certain whether in other species the same explanation concerning the derivation of the baring of the teeth is valid. As evidence from other species, supporting his hypothesis, Andrew (1963a) gives the situation found in the ungulates, especially the Equidae. The author considers the baring of the teeth which is shown by the Equidae during the rut face, the greeting face, and the threat face, as a protective response, related to the mouth posture shown during the 'flehmen' face which is interpreted as a protective response against strong scents. In the rut face and the greeting face, which have been shown to be modifications of the defensive threat face, and in the defensive threat face itself the teeth are bared by vertical and horizontal retraction of the lips. The situational context in which the rut face and especially the threat face are shown clearly indicates that this posture is related to the act of biting and can be regarded as an intention movement of it (Antonius, 1937, 1939; Trumler, 1959; Zeeb, 1959). The mouth posture which characterizes the 'flehmen' face, however, greatly differs from the one characteristic for the threat face. The mouth-corners are pulled forwards and upwards rather than backwards, and the upper lip is simultaneously lifted and protruded (curled upwards), thus baring the incisors only (cf. photograph given by Zeeb, 1959). Besides, it seems questionable whether the 'flehmen' face should be regarded as a protective response at all. It is certainly anthropomorphism to assume that the smell produced by a female ungulate in heat is unpleasant to a male partner. The frequency and the intensity with which males remain smelling may even suggest the opposite.

It is thus clear that in some cases the classic explanation of the derivation of the baring of the teeth as a preparation to bite, which has given rise to a ritualized showing of the weapons (e.g. Eibl-Eibesfeldt, 1957), should be preferred. Andrew indeed envisages this possibility also in connection with the derivation of the baring of the teeth in primates – these do withdraw their lips immediately before biting at food or at an opponent – and rightly concludes that both possibilities are not mutually exclusive.

It has been pointed out already that the mouth posture in *the protruded-lips face* (particularly the posture of the upper lip) bears great resemblance to the one shown during the 'flehmen' display of many mammals. It may, therefore, be homologous with the 'flehmen' face, in that the lip posture is primarily related to the act of smelling. An almost identical expression, though much less intense, is shown by human beings, when smelling intently or sniffing. However, the mouth posture could perhaps also be thought of as the closed phase of *the true lip-smacking face* with the strongly protruded lips, which has been emancipated into a separate display. In other macacques this form of lip-smacking can often be observed during sexual encounters.

Horizontal head-shaking is one of the protective responses (Andrew, 1963a) and may have been incorporated in displays. In some instances a more simple explanation might be preferred. The horizontal head-shaking which often accompanies *the silent bared-teeth face* in *Mandrillus*, for instance, is not seen in related genera in this context. However, in macacques this display regularly alternates with looking-away movements. This head-shaking may thus well be a ritualized looking-away movement.

Ear-flattening is undoubtedly a protective response. The scalp retraction found in *Papio, Cercocebus, Macaca, Theropithecus, Cynopithecus* and *Mandrillus* is probably connected with it (Huber, 1931; Andrew, 1963a).

Frowning has often a protective function as well (Andrew, 1963a and c), for example, in *the frowning bared-teeth scream face*. When bringing the face close to an object, contraction of the muscles around the eye may also help to achieve convergence of the optical axes and better accommodation (Andrew, *op. cit.*). When trying to distinguish an object in the distance it may exclude disturbing light from above. In these circumstances man will often put his hands horizontally over the eyes (Darwin, 1872). Moreover, the simultaneous contraction of the muscles controlling the degree of opening of the eyes, with which the muscle responsible for frowning is connected, helps to steady the eye during fixation, reducing its trembling movements. Widely opened eyes, as occur, for instance, in the human expressions of astonishment and fear, tremble considerably; the advantage of this situation, however, is that their mobility is greatly increased (Frijda, 1956). In the catarrhine monkeys the eyes are usually farther open in *the tense-mouth face* and *the staring open-mouth face* than in the relaxed face. The nature of this openness may yet be different in comparison with the one shown during displays which reflect a tendency to flee. In the latter displays both eyelids are far apart and the eye opening has a more or less round form. In *the tense-mouth face* and sometimes in *the staring open-mouth face*, however, the upper eyelid is indeed lifted, but the lower eyelid is drawn upwards slightly as well and forms a more or less straight line. There is obviously a great tension in the region surrounding the eye, due to a simultaneous contraction of the muscles controlling the degree of openness of the eye. The difference can be seen when one compares Plate I, showing *the tense-mouth face*, with tensely opened eye in a male *Papio anubis*, with Plate II, showing *the silent bared-teeth face* with normally opened eye in *Mandrillus sphinx*. The same tensed eye posture can be noticed in man when he is likely to attack a rival. Here it goes together with staring eyes, a slightly hunched posture, a slightly lowered and forward thrust head, clenched fists, forward-thrust under-jaw and tensely closed lips. It is also characteristic of an expression which probably reflects a much lower tendency to attack, namely 'the mean look'. The work of Frijda (1956) on the understanding of human facial expressions shows that in man the eyes and the region around the eyes contributes greatly to the expressivity of the face, probably more than in monkeys and apes. The development of a white region around the iris is perhaps an adaptation, selected for during evolution to accentuate the expressivity of the eyes.

Surveying the catarrhine monkeys and apes one notes that *the tense-mouth face, the staring open-mouth face, the staring bared-teeth scream face* and *the frowning bared-teeth scream face* are present, in basically the same form, in all the studied representatives. They are the displays of which it can be assumed that they have mainly a distance increasing or maintaining function. The other displays, of which probably the main function is to decrease the distance between individuals, show a much greater variability. This has been discussed already above.

Especially, the evolution of *the silent bared-teeth face* is highly interesting. Andrew (1963a and c) has shown that the expression element of baring the teeth is almost always linked with intense vocalization in primitive primates. The guenons (*Cercopithecus* spp.) show a similar situation. It is in the macaques that the element of baring the teeth has become independent of vocalization. The display in which it is incorporated mainly reflects the tendency to flee. In a number of species (*Theropithecus, Mandrillus, Pan*) a motivational shift occurs, in that the display may come to express social attraction as well and to some extent (completely in *Pan*) replaces *the lip-smacking face*. Both in its form and its context the human *smiling response* resembles to some extent *the silent bared-teeth face* found in the latter category of monkeys and in the chimpanzee. So it is not unlikely that they may be regarded as homologous. In connection with the subject at issue, it should be remarked that Lorenz's (1963) supposition, that in the apes no functional equivalent of the macaque lip-smacking and the human smiling exists, is certainly not correct, as far as the chimpanzee is concerned.

It is customary to think of smiling and laughing as displays which are expressions of different intensities of the same motivational state. This point of view is explicitly stated by Lorenz ('Lächeln und Lachen entsprechen sicher verschiedenen Intensitätsgraden derselben Verhaltensweisen, d.h. sie sprechen mit verschiedenen Schwellen auf dieselbe Qualität aktivitäts-spezifischer Erregung an.' – 1963, p. 276). The same is implied by Andrew (1963c) who regards laughing as a response evoked by a greater 'amount of stimulus contrast' than smiling. The 'amount of stimulus contrast' evoking laughter is assumed to be near the verge between pleasant and unpleasant; this would be in accord with the fact that smiling and laughter are evoked by jokes, the essence of a joke being the right degree of discrepancy between the real ending and that anticipated. A similar point of view is obviously taken by Bolwig (1964) who draws attention to the fact that the initial and final stage of *the relaxed open-mouth face*, which sometimes may be observed (see above), 'very much resembles the smile as known in humans and should undoubtedly be classified as such' (p. 181). That this is the popular opinion is also reflected by the names these responses have in some languages, e.g. 'Lächeln' and 'Lachen', 'Sourire' and 'Rire'.

Smiling and laughing frequently alternate with each other. They both occur in a great variety of situational contexts. Yet a tentative attempt may be made to qualify a typical smiling situation and a typical laughing situation.

It is possible to distinguish a number of situations in which smiling is the appropriate response and laughing is both quite out of place and not at all easily evoked. For example: when a person introduces himself to his future boss; when a mother reassures or comforts a child; when someone expresses sympathy to someone else, etc. True cordial laughter can be observed in children during play, or in adults, when they have an established relationship with one another. It then marks a playful attitude, if mostly only in a spiritual sense. It is in this situation that jokes ('intellectual social play') are responded to by hearty laughter. Responding by smiling may even not be appreciated. A laughing encounter greatly strengthens the bond of comradeship. The ability of man to produce, within certain limits, expression movements 'at will' gives him the possibility to break the ice by laughter, for instance in a group of people who have no established relationship, in the above sense, with one another, but have at the most a 'smiling relationship'. Thus the wanted comradely relationship may be 'feigned' (the 'reception laugh').

In previous theories concerned with laughter (for a review, see Andrew, 1963a) the fact that this display is in its essence a social response has not always been fully appreciated and many a theory has been mainly concerned with the explanation of humour. Andrew (*op. cit.*) recognizes that not any 'stimulus contrast' which is on the verge between pleasant and unpleasant will evoke the response, but that stimuli from fellows are involved. 'Few laugh heartily when reading a joke alone' (p. 89).

This tentative attempt to qualify the difference between smiling and laughter may thus lead to the supposition that typical smiling and laughter may be regarded as responses which reflect a qualitatively different motivation, notwithstanding the fact that their motivational ranges obviously overlap considerably. These tentative considerations do not pretend to be much more than speculations; they illustrate how badly needed is an ethological analysis of the human smiling and laughing response.

The comparative data show that there is a similarity in form between the smiling response and *the silent bared-teeth face*. Apart from the similarity of the facial expression elements, both movements are often characterized by a 'cautious', perhaps 'inhibited', nature of the body movements. On the other hand, the laughing response shows a great similarity in form with *the relaxed open-mouth face*, especially in connection with the rhythmic, low-pitched staccato vocalizations and the boisterous body movements. A great difference is that, in man, laughing is always accompanied by both horizontal and vertical retraction of the lips, whereas vertical retraction of the lips does not normally occur in *the relaxed open-mouth face* of catarrhine monkeys and apes.

If the above distinction between typical human laughing and smiling has value, the similarity is not even restricted to the form, but also applies to the situational context. This would suggest that the human smiling response has evolved from *the silent bared-teeth face* and is thus homologous with it, and that the human laughing response has evolved from *the relaxed open-mouth face* and is thus homologous with it.

In contrast with man, these displays do not much overlap in the chimpanzee. This is clearly demonstrated when two chimpanzees are introduced to one another for the first time. In three introduction tests where the established chimpanzee was nearly adult, whereas the partner was much smaller and showed avoidance tendencies, *the silent bared-teeth face* was, on one occasion frequently, shown by the potentially dominant animal during the first minutes of the introduction. It was accompanied by 'cautious, inhibited' approach movements which led to embraces, etc. (see Plate VII). After this period the two animals began to move more freely; *the silent bared-teeth face* disappeared and *the relaxed open-mouth face* appeared. The behaviour developed into boisterous play.

The incorporation of the bared-teeth element in the human *relaxed open-mouth face* may be due to the broadening and consequent overlapping of motivational range of the two displays, resulting also in an increase of the formal similarity.

A similar overlapping may be observed in *Mandrillus*. Thus the peculiar *silent bared-teeth face* of this genus frequently occurs in play situations. Only during the actual bite-wrestling may a form of *the relaxed open-mouth face* be seen as well.

In the platyrrhines *the relaxed open-mouth face* sometimes resembles our human laughter much more closely (Andrew, 1963a). Full retraction of the lips accompanies it in *Lagothrix cana* (own obs.). Andrew (*op. cit.*) has shown that this 'facilitation of the grin' (i.e. the fact that it becomes progressively more attached to greeting and play displays) is quite general in this group, even more so than in the catarrhines. It is interesting to note that *the lip-smacking face* has never acquired the important place here that it occupies in the catarrhines.

Apart from the facial displays which have been described, there are certainly postures and movements which act as social releases, but which are not necessarily associated with a particular facial posture. Presenting and mounting are well-known examples. In the apes there are a number of distinct gestures (Köhler, 1926). Characteristic is, for example, the chest-beating display of the gorilla (Schaller, 1963).

Similarly there are a number of calls which do not necessarily form part of a particular facial display, although in a number of cases changes in the mouth-posture may be visible during the uttering of the call. Examples can be found for *Gorilla gorilla beringei* in Schaller (1963) and for *Macaca mulatta* in Hinde and Rowell (1962). The 'clear calls' of the latter species have been mentioned already. Besides these there are still a number of low to moderately pitched grunting sounds which have been called 'friendly noises' (Hinde and Rowell, 1962). Especially in baboons such grunting sounds occur frequently (Andrew, 1962, 1963a and b).

In all these species they may be given when a 'desired object' (for example: food) is perceived and also in a number of 'relaxed' social situations which have in common that approach to a partner or more specific positive social interactions with a partner may be involved. They can thus be said to reflect a predominantly positive (though not specific) attitude. A characteristic property of these grunts (and probably the reason for which they have been prized during evolution) is that they may attract and direct the attention of fellows and thus cause the same response or motivational attitude in these. The typical 'ooo'-calls given by chimpanzees and small children to novel objects have obviously the function to direct the attention of partners to these (Andrew, 1963b). They are clearly ritualized versions, since the protrusion of the lips which accompanies them may last considerably longer than the call (see also: *the pout face*).

There are a number of situations in which the frequency of grunting is rather high. In macaques and baboons this is the case when the animals hug each other (this occurs, for instance, after temporary separation, but the reasons are not always clear. This is, furthermore, the case during the first period of feeding after a temporary absence of food, and during mating, especially in the final stage. Hall (1962) reported that a high rate of grunting on part of the female *Papio ursinus* marks the final stage of copulation during which the male dismounts. The same has been observed in *Theropithecus gelada*, *Macaca irus* and *M. nemestrina*. Here, one even gets the impression that the grunting is performed with a typical intensity and frequency. (Has it been selected as a signal indicating successful mating and thus raising the attractiveness of the female concerned?)

SUMMARY

This paper presents the results of a qualitative study concerning those displays of the catarrhine monkeys and apes which are mainly characterized by the participation of the muscles and the structures of the facial region. It closely follows van Hooff (1962). New data have been added and the results are compared with those of a number of studies on the same subject which have appeared during the last few years.

The introduction gives a survey of the conditions which during evolution have favoured the development of a facial musculature which eventually could become the 'organ' of facial expressions. The possibilities which the presence of a facial musculature offered, in this respect, have been used most extensively by, especially, the social-living representatives of a few orders, namely the primates, the carnivores and the ungulates. This development appears to be correlated with a typical specialization of the sense of vision in these groups.

The facial displays in the primates show a great variability. They consist of a combination of postures or movements of such facial elements as the jaws, the lips, the tongue, the ears, the eyebrows and the upper-head skin, the eyes and the eyelids. Apart from these, there are the postures and the movements of the body or parts of the body and the vocalizations. Notwithstanding the great variability, it is possible, in some cases more easily than in others, to distinguish a number of compound expressions, i.e. typical combinations of expression elements. These are defined as the distinct postures and movements of the above listed elements.

The displays have been related to such basic and objectively definable social behaviour patterns as attack, approach (followed by positive social interactions or simply being together) and flight. Thus an insight can be acquired concerning the motivation of the display. Suggestions have also been given concerning the changes in motivation which occur in the partner when a particular compound expression is performed. Furthermore it has been attempted to establish in what situation the display is evoked.

A tendency to attack is present in animals performing *the tense-mouth face* and *the open-mouth face*. This is often also the case in an animal showing *the staring bared-teeth scream face*. The latter two compound expressions often also indicate the presence of a tendency to flee. The last display occurs in a number of social situations, which seem to have in common that some strong tendency of the performer (not necessarily one of the three social tendencies) is thwarted. The partner seems to play a rôle in that his activity is the cause of the thwarting or can perhaps remove the block. A strong fleeing tendency which is blocked is indicated by *the frowning bared-teeth scream face*. Both displays seem to have as a result that they may cause the partner to 'give in'.

In the macaques, *the silent bared-teeth face* expresses a probably low tendency to flee which may be thwarted, for instance, by site-attachment. Especially in *Pan*, *Mandrillus* and *Theropithecus* it may also indicate a tendency to approach. *The bared-teeth gecker face* can be observed in the infants of most species when 'uncomfortable' while clinging at the mother or when trying to get back to the mother. In adults, especially of baboons and of the chimpanzee, it may both indicate a low tendency to approach and to flee. *The lip-smacking face* represents a tendency to approach, which usually seems to be thwarted by a lower tendency to flee. This conflicting tendency to flee may be more notable in *the teeth-chattering face* which is a characteristic macaque expression. *The protruded-lips face* has been observed in only a couple of macaque species and represents a tendency to approach (often more particularly a tendency to mate). According to whether the motivation represents

63

predominantly a fleeing tendency or approach tendency, the function of these displays is likely to be appeasing, reassuring, or attracting respectively.

SCHEMATIC REPRESENTATION OF THE MAIN COMPOUND
FACIAL EXPRESSIONS IN A MACACA TYPE PRIMATE

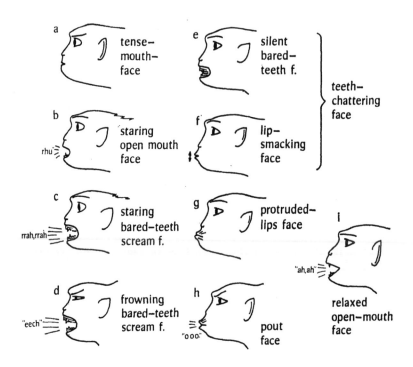

The pout face is in most species a baby expression, indicating the 'wish' to be with the mother. In the chimpanzee this compound expression can also be performed by adult individuals and may lead to positive social interactions.

The relaxed open-mouth face often occurs during social play or leads to play. It is present in most species.

The derivation of the expression elements is discussed as well as the evolution of the displays. It is suggested that the human smiling response may be partly homologous with *the silent bared-teeth face* and that human laughing may be homologous with *the relaxed open-mouth face*.

Addendum

After the completion of this manuscript an article by Altmann (1962) came to my attention which contains data relevant to the present subject. It gives a catalogue of behaviour patterns which may be socially significant in the case of *Macaca mulatta*. The data concerning the facial displays are in agreement with those presented here. The author furthermore pays attention to the way these patterns are incorporated in social interactions and the way they may function to determine the social structure of a group. In the 'tranquilizing' or 'pacificatory' patterns, for instance, he makes a distinction between those gestures appeasing aggression and those assuaging submission; this corresponds with the concepts of 'appeasement' and 'reassurement' as discussed here, the first of which has received ample attention in ethological studies, in contrast to the latter. For further details the reader is referred to the original publication.

Also the results of a number of field studies have become public, some of which contain information relevant to this comparative study. In particular, the reader is referred to the papers assembled by DeVore (1965), one of which, by Marler, gives a review of the work on primate displays.

REFERENCES

ALTMANN, S. A. (1962). 'A field study of the sociobiology of rhesus monkeys (*Macaca mulatta*)', *Ann. N.Y. Acad. Sci.* **102** (2), pp. 338–435.

ANDREW, R. J. (1962). 'The situations that evoke vocalization in primates', *Ann. N.Y. Acad. Sci.* **102**, pp. 296–315.

—— (1963a). 'The origin and evolution of the calls and facial expressions of the primates', *Behaviour* **20**, pp. 1–109.

—— (1963b). 'Trends apparent in the evolution of vocalization in the old world monkeys and apes', *Symp. Zool. Soc.* **10**, pp. 89–101.

—— (1963c). 'Evolution of facial expression', *Science* **142**, pp. 1034–41.

—— (1964). 'Vocalization in chicks, and the concept of "stimulus contrast" ', *J. Anim. Behav.* **12**, pp. 64–76.

ANTONIUS, O. (1937). 'Über Herdenbildung und Paarungs-eigentümlichkeiten bei Einhufer', *Z. Tierpsychol.* **1**, pp. 259–89.

—— (1939). 'Über Symbolhandlungen und Verwandtes bei Säugetieren', *Z. Tierpsychol.* **3**, pp. 263–78.

BOLWIG, N. (1959a). 'Observations and thoughts on the evolution of facial mimic', *Koedoe* 2, pp. 60–9.

—— (1959b). 'A study on the behaviour of the chacma baboon', *Behaviour* 14, pp. 136–63.

—— (1963). 'Bringing up a young monkey (*Erythrocebus patas*)', *Behaviour* 21, pp. 300–30.

—— (1964). 'Facial expression in primates with remarks on a parallel development in certain carnivores (a preliminary report on work in progress)', *Behaviour* 22, pp. 167–93.

CARPENTER, C. R. (1958). 'Soziologie und Verhalten freilebender nicht-menschlicher Primaten', *Handb. Zool. Berl.* 8, (18)–10, (11).

CRAWFORD, M. P. (1937). 'The co-operative solving of problems by young chimpanzees', *Comp. Psychol. Monogr.* 14, p. 2.

DARWIN, C. (1872). *The expression of emotions in man and animals.* London.

DEVORE, I. (Ed.) (1965). *Primate Behaviour: Field studies of monkeys and apes.* Holt, Rinehart and Winston, New York.

EIBL-EIBESFELDT, I. (1955). 'Der Kommentkampf der Meerechse (*Amblyrrhynchus cristatus*) nebst einigen Notizen zur Biologie dieser Art', *Z. Tierpsychol.* 12, pp. 49–62.

—— (1957). 'Ausdrucksformen der Säugetiere', *Handb. Zool. Berl.* 8 (8) 10, 6, pp. 1–26.

FRIJDA, N. H. (1956). *De betekenis van de gelaatsexpressie*, Amsterdam.

GREGORY, W. K. (1929). *Our face from fish to man*, New York.

HALL, K. R. L. (1962). 'The sexual, agonistic and derived social behaviour patterns of the wild Chacma Baboon (*Papio-ursinus*)', *Proc. Zool. Soc. Lond.* 2, pp. 283–328.

HANSEN, E. W. (1962). 'The development of maternal and infant behavior in the rhesus monkey', Thesis, Univ. Wisconsin.

HINDE, R. A. (1959). 'Unitary drives', *J. Anim. Behav.* 7, pp. 130–41.

HINDE, R. A. and T. E. ROWELL (1962). 'Communication by postures, and facial expressions in the rhesus monkey (*Macaca mulatta*)', *Proc. Zool. Soc. Lond.* 138, pp. 1–21.

HOOFF, J. A. R. A. M. VAN (1962). 'Facial expressions in higher primates', *Symp. Zool. Soc. Lond.* 8, pp. 97–125.

HOOTON, E. (1942). *Man's poor relations*, New York, Doubleday.

HUBER, E. (1922–3). 'Über das Muskelgebiet des Nervus facialis beim Hund', *Morph. Jb.* 52, pp. 353–414.

—— (1930). 'Evolution of facial musculature and cutaneous field of trigeminus', *Quart. Rev. Biol.* 5, pp. 133–88 and 389–437.

—— (1931). *Evolution of facial musculature and facial expression*, O.U.P.

KENNEDY, J. S. (1954). 'Is modern ethology objective?', *J. Anim. Behav.* **2**, pp. 12–19.

KITZLER, G. (1940–1). 'Die Paarungsbiologie einiger Eidechsen', *Z. Tierspsychol.* **4**, pp. 353–402.

KÖHLER, W. (1926). *The mentality of apes*, London.

KOHTS, N. (1935). ['Infant ape and human child'], *Sci. Mem. Mus. Darwinianum Moscow* **3**.

KUMMER, H. (1957). 'Soziales Verhalten einer Mantelpaviangruppe', *Beih. Schweiz. Z. Psychol.* **33**, pp. 1–91.

LASHLEY, K. S. and J. B. WATSON (1913). 'Notes on the development of a young monkey', *J. Anim. Behav.* **3**, pp. 114–44.

LEYHAUSEN, P. (1956a). 'Verhaltensstudien an Katzen', *B. Z. Tierpsych.* **2**.

—— (1956b). 'Das Verhalten der Katzen (Felidae)', *Hand. Zool. Berl.* **10**, 21, pp. 1–34.

LIGHTOLLER, G. (1938). 'Probable homologues: A study of the comparative anatomy of the mandibular and hyoid arches and their musculature', *Trans. Zool. Soc. Lond.* **24**, pp. 349–444.

LORENZ, K. (1951). 'Ausdrucksbewegungen höherer Tiere', *Naturwissenschaften* **38**, pp. 113–16.

—— (1960). 'Prinzipien der vergleichenden Verhaltensforschung', *Fortschr. Zool.* **12**, pp. 265–94.

—— (1963). *Das Sogenannte Böse*, Wien, Borotha-Schoeler.

MARLER, P. (1959). 'Developments in the studie of animal communication', *Darwin's biological work: Some aspects reconsidered*, pp. 150–206, Cambridge Univ. Press.

MCCLELLAND, D. C., J. W. ATKINSON, R. A. CLARK and E. A. LOWELL (1953). *The Archievement Motive*, New York.

MORRIS, D. (1956). 'The feather postures of birds and the problem of the origin of social signals', *Behaviour* **9**, pp. 75–113.

—— (1957). ' "Typical intensity" and its relation to the problem of ritualisation', *Behaviour* **11**, pp. 1–12.

MOYNIHAN, M. (1964). 'Some behavior patterns of platyrrhine monkeys. I. The night monkey (*Aotus trivigatus*)', *Smithsonian miscellaneous Coll.* **146**, (5), pp. 1–84.

POLYAK, ST. (1957). *The vertebrate visual system*, Univ. of Chicago Press.

ROWELL, T. E. S. and R. A. HINDE (1962). 'Vocal communication by the rhesus monkey', *Proc. Zool. Soc. Lond.* **2**, pp. 279–94.

RUGE, G. (1887). *Untersuchungen über die Gesichtsmuskulatur der Primaten*, Leipzig, Engelmann.

SCHALLER, G. B. (1963). *The mountain gorilla,* Chicago Univ. Press.

SCHENKEL, R. (1947). 'Ausdrucksstudien an Wolfen', *Behaviour* **1**, pp. 81–129.

SCHLOETH, R. (1956). 'Zur Psychologie der Begegnung zwischen Tieren', *Behaviour* **10**, pp. 1–80.

SCHNEIDER, K. M. (1931). 'Das Flehmen II', *Zool Gart., Lpz.* **4**, pp. 349–64.

—— (1932a). 'Das Flehmen III', *Zool. Gart., Lpz.* **5**, pp. 200–26.

—— (1932b). 'Das Flehmen IV', *Zool. Gart., Lpz.* **5**, pp. 287–97.

—— (1934). 'Das Flehmen V', *Zool. Gart. Lpz.* **7**, pp. 182–201.

TINBERGEN, N. (1948). 'Social releasers and the experimental method required for their study', *Wilson Bull.* **60**, pp. 6–51.

—— (1951). *The study of instinct,* Oxford Univ. Press.

—— (1959). 'Comparative studies of the behaviour of gulls (Laridae): a progress report', *Behaviour* **15**, pp. 1–70.

—— (1963). 'On aims and methods of ethology', *Z. Tierpsychol.* **20**, pp. 410–33.

TRUMLER, E. (1959). 'Das "Rossigkeitsgesicht" und ähnliches Ausdruckverhalten bei Einhufern', *Z. Tierpsychol.* **16**, pp. 478–88.

WALLS, G. L. (1942). *The vertebrate eye,* The Cranbook Inst. of Science.

WEBER, H. (1957). 'Vergleichende Untersuchung des Verhaltens von Smaragdeidechsen, Mauereidechsen, und Perleidechsen', *Z. Tierpsychol.* **14**, pp. 448–72.

YERKES, R. M. (1943). *Chimpanzees, a laboratory colony,* Yale Univ. Press.

YERKES, R. M. and A. YERKES (1929). *The great apes,* Yale Univ. Press.

ZEEB, K. (1959). 'Die "Unterlegenheitsgebärde" des noch nicht ausgewachsenen Pferdes', *Z. Tierpsychol.* **16**, pp. 489–96.

—— (1963). 'Equus caballus (*Equidae*) Ausdrucksbewegungen', *Encyclopaedia Cinematographica,* E505.

ZUCKERMAN, S. (1932). *The social life of monkeys and apes,* London.

Chapter Three

Socio-sexual Signals and their Intra-specific Imitation among Primates

WOLFGANG WICKLER

ON account of their close relationship with man, the so-called non-human primates are of particular interest to zoologists, anthropologists and sociologists. Systematic comparative studies of behaviour under field conditions have, however, been initiated at a remarkably late stage. This is partially due to the fact that large animals range over large areas. The observer must be able to identify individually the members of a group in order to be able to investigate the social behaviour, and he must not disturb the animals concerned. This latter requirement becomes increasingly difficult, and demands longer periods of habituation, the nearer the observer is to the animals. For this very reason, the first observations and the great majority of accurate observations to date have been carried out on ground-living primates, which can be observed from a fair distance on an open landscape. Such groups can also be followed to some extent, whilst an ape colony in the forest repeatedly disappears from view in the treetops. In fact, this shift of emphasis on to ground-living primates, which largely belong to the Old World monkeys (Catarrhines) is not an unlucky factor in the comparison with man. Firstly, the Old World monkeys are more closely related to man than the New World monkeys, and secondly it is assumed that primitive man occupied a ground-living niche analogous to that of present ground-living monkeys; i.e. the forms closest to man for good reasons are the forms which are easiest to observe.

One general difficulty in this field is inexact identification, even of those primates used in important experiments (Washburn, 1950; Herschkovitz, 1965); generalizations are therefore particularly

69

unreliable. In this respect, it is not the unsettled problems concerning grouping of larger systematic units which are crucial,
but above all the numerous outstanding questions in fine systematics and in the separation of species. This is because many races
are distinguished, (e.g. in *Cercopithecus aethiops*, after Schwarz,
1926 and Dandelot, 1959) or because previously separate species
are lumped together as one species (as in the genus *Papio*, for
example, for which Freedman (1963) postulates a geocline extending from West to East across Northern Rhodesia (*P. cynocephalus*),
then on down into Southern Africa (*P. ursinus*) and also up into
North-Eastern Africa (*P. doguera*)). It is my impression that
behaviour patterns and social signals may be important characters
in the definition of species, sub-species, etc., provided that the
data gained from different populations are not prematurely
lumped together. Apart from this, the intraspecific range of variation in social behaviour of the primates is presumably greater than
was previously supposed and the amount of interest taken in the
primates to suit various purposes far outweighs the amount of
available data of a reliable kind.

In spite of confident statements in the animal sociology literature, it is
not known whether baboon males are promiscuous, having a kind of
rotating mateship, whether each large male retains for long periods,
more or less for his exclusive sexual use, the same group of females,
whether there is a coming and going between 'harems', and so on. It has,
of course, been assumed, on the analogy of 'rut' animals like deer or
seals, that the male baboon is very alert in keeping his females from
straying, but, as there is no 'rut' in Primates, he has been assumed to
maintain his harem more or less permanently. That all such assumptions
are at present completely unjustified will be appreciated by any careful
observer who has watched wild baboons or other monkeys (Hall, 1960).

Kaufmann and Kaufmann (1963) emphasize the differences in
behaviour between captive and free-living primates and the
reluctance of many zoologists to take to the field, and zoologists
are also guilty of neglecting even the biological material available
for scientific purposes in many zoos. Social signals are in fact
more intensive and more conspicuous in many cases where captive
animals are compared with those in the wild, and certain signals
(and their significance) are rendered much more obvious when seen
at the highest level of intensity. Over and above this, details
necessary in comparison can be better observed in captive animals,

since close approach is possible. This is not to deny that field studies are indispensable for the formation of a biologically correct picture of the significance of these behaviour patterns, as appearing in the normal existence of the animals concerned. Kummer (1965), however, found from a comparison of captive and free-living hamadryas baboons only slight differences in their social behaviour (i.e. contrary to the statements of Freedman, 1954).

In the present paper, an attempt is made to interpret some morphological and behavioural peculiarities among Old World primates with respect to the characteristics of group-life. The term 'group' is intended to designate a relatively stable association of a limited number of conspecifics which stay together without the existence of a common extra-social goal and without (as far as we know) possessing a nest- or group-odour. The animals differentiate between group members and foreign conspecifics; a rather stable rank order within the group further shows that the group members recognize each other individually. An exchange of members between groups does not normally occur, even if the opportunity is frequently present.

As far as we know, all such groups consist of families whose members stay together beyond the actual period of pairing and broad care, in some cases including several generations. A rational expectation would be that the inter-relationships of the group members can be derived from the inter-relationships of family members; since family life is predominantly based on brood-care and mate-relationships, these latter functional patterns might be expected to play a rôle in wider intra-group relationships.

Living in such groups surely represents a special evolutionary level, a 'grade' that has been independently attained by separate phyletic lines of vertebrates, mammals and possibly even of primates. This is important, since common features in such convergencies can be employed in the derivation and retrospective evaluation of the biological prerequisites which must be fulfilled for the evolution of a vertebrate towards the ability to exhibit a group-existence.

The concepts presented here are based on: (1) a ten-years' comparative study of cichlid behaviour in our laboratory aquaria (large enough to closely simulate natural conditions for at least some of the species studied); (2) sporadic observations on primate behaviour in zoological gardens over the course of many years;

(3) observations on free-living baboons, vervet-monkeys and spotted hyaenas during three trips to East Africa (1964–6); and (4) an extensive survey of the literature concerning primate and other vertebrate behaviour.

Much information needed to test these concepts further has not been found in the literature, and it would seem that only with these ideas in mind will observers recognize special behavioural characters supporting or disproving the concepts. This is largely the justification for presenting the conceptual framework now, following Dobzhansky's statement that 'the primary function of a working hypothesis is to arrange facts into suggestive patterns capable of guiding observations and experience into meaningful channels'.

ACKNOWLEDGMENTS

The author is deeply indebted to Professor Dr K. Lorenz for his continuous interest and many stimulating discussions; to the Max-Planck-Gesellschaft, which generously provided a grant for the field studies in Africa; to the Director of the Tanganyika National Parks for his hospitality; to Dr W. Kühme for his guidance in the East African Plains and to Mrs L. Kühme for her assistance during the field studies and subsequently in the sifting of the observations. My thanks are also due to H. Kacher for the excellent illustrations and to R. D. Martin for carrying out the translation.

THE HAMADRYAS' HIND-QUARTERS PROBLEM

A very conspicuous character of the hamadryas baboon, which is familiar to all zoo visitors, is the 'luminous' red backside of the male. The biological significance of this feature has, however, remained uncertain; even in recent behavioural studies of this species the selective value of this character has been vainly sought after. A large area of naked skin in the anal region is present in a number of ground-dwelling monkeys and it has been suggested that this might be an adaptation to frequent sitting in wet grass and/or aiding heat-regulation, a function which has been recently demonstrated for the naked tail of the rat (Rand, Buston and Ing, 1965). However, in baboons and macaques this area is sometimes black and sometimes bright red; it is this colour difference which will concern us here. Whatever may be the original or the main function of this coloured area, there must be at least one additional, signal function of this very conspicuous colour of the hind-quarters, which has some social meaning. If so, one may ask: (1) what does this signal mean?; (2) in what situation(s) is it displayed?; (3) what agent is the normal signal-receiver and how does it respond to the signal?

Red hind-quarters in baboons and other Old World monkeys usually signifies a female in oestrous. Besides the red coloration, the females may develop during oestrous a sexual swelling, which in some species reaches such proportions that it appears to the uninitiated visitor as pathological, as is exemplified by the following quotation from Murie (1872): in a female *Macaca cyclopis* in the London Zoo at maturity 'the region generally at the root of the tail became extremely developed. Indeed, at certain times, the buttocks acquired such hideous proportions that it was necessary, for decency's sake, to remove her from the public gaze.'

In the earlier literature – e.g. in Elliot's review of the primates (1913) – no clear distinction is made between ischial callosities and sexual skin. The callosities or sitting pads are a distinguishing characteristic of the Cercopithecidae; they are not found in the Platyrrhines. They are referred to as ischial callosities because they overlie the tuberosities of the ischia, to which they are firmly attached in the Old World monkeys and gibbons; in chimpanzees they are movable upon the tuberosities. The callosities are cornified, thickened, hairless, almost avascular areas, usually greyish in colour and sharply outlined from the adjacent skin (for details see Miller, 1945). Some of the Celebes macaques, however, seem to have flesh-coloured or even red callosities throughout their lives. Washburn (1957) considers the callosities as adaptations for the peculiar sleeping position of these species, sitting in branches with the feet off the floor and hands next to feet, so that the full weight of the animal rests upon the ischial tuberosities.

The sexual skin in females is a specialized area of skin which is contiguous with the external genitalia; it is flesh-coloured to bright red in coloration and characterized by a thick dermis, a marked oedema of the subcutaneous connective tissue and an unusually rich blood supply. It shows variations in general conformation and colour in phase with the menstrual cycle. Periodic catamenial swellings, brought about by the deposition of water in the inter-cellular spaces and by imbibition of water by connective tissue elements, are associated with the follicular phase and so roughly coincide with the fertile phase of the cycle (Nesturkh, 1946; Bartelmez, 1953; Gilbert and Gillman, 1955; Eckstein and Zuckerman, 1956). The region that is mostly involved in the gross cyclical changes during oestrous in the adult female baboon is a triangular-shaped area extending from the root of the tail (above) to the junction of the lower part of the abdominal wall with the front of the thighs (below). This region is referred to anatomically as the perineum. Above, the perineum is expanded into an oval-shaped area lying between and above the callosities, whilst below it tapers gradually and ends abruptly between the legs at the lower margin of the symphysis (Gillman, 1935). One may distinguish, therefore, between an anterior, pre-vaginal part (with the clitoris and its sheath) and a post-vaginal part of the sexual skin (posterior part of vaginal opening plus circum-anal skin). The latter is usually more affected by the swelling, rapidly overlaps its basal attachment (hiding the ischial callosities) and distorts the anal orifice (Hill, 1958; Eckstein and Zuckerman in Parkes, 1960). The enlargement differs among genera (Fig. 1), from species to species and even between individuals, but always takes the form of a very conspicuous, hyperaemic, oedematous formation, generally brilliant red in colour. Among baboons it is most pronounced in *Papio porcarius*, a detailed description of which is given by Pocock (1925) and Gillman (1935).

In which context is this signal displayed? This, too, is well known: females in heat present the genital region as a mating invitation to males (Fig. 2). A female so presenting approaches the male, turns her hind-quarters towards him and often looks at him over her shoulder simultaneously. In intensive presentation, a female may lower the fore-quarters and reach backwards with one hand (Fig. 5).

A male confronted by presentation of this kind can respond variously: the male may give no visible response; it may simply gaze at the female; it may gaze and manipulate the genital region of the female for a varying length of time, simultaneously exhibiting rapid tongue and lip movements, using the fingers and/or the mouth; finally the male may mount briefly, with or without erection of the penis, and perform a feigned or actual coitus with friction-movements. All of these actions correspond to different levels of intensity in the male mating-behaviour. Which action is

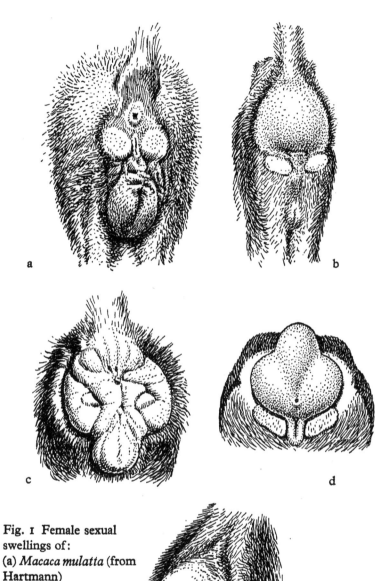

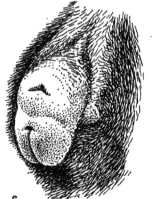

Fig. 1 Female sexual
swellings of:
(a) *Macaca mulatta* (from
Hartmann)
(b) *M. irus* (from
Spiegel)
(c) *Miopithecus talapoin*
(from Tomilin)
(d) *Macaca maura* (from
Büttikofer)
(e) *Pan troglodytes*
(from Yerkes)

75

involved depends upon the arousal state of the male and also on various other factors (e.g. intimidation by higher-rank conspecifics, etc.).

Fig. 2 Presentation of (a) *Lemur catta*, (b) *Pan troglodytes*

Presentation in the sexual context as the invitational stage of posturing for copulation is widely distributed among Old World monkeys, even in those species lacking the sexual swelling (e.g. in *Colobus polykomos*, Ullrich, 1961). Sexual swelling and the coloration of the sexual skin underline and magnify the effect of this well-defined social signal. In general, females with the most conspicuous heat-symptoms are preferred (Zuckerman, 1932; Spiegel, 1954); this has been individually investigated for *Macaca irus mordax* by Spiegel (1930) and Harms (1956), for the hamadryas baboon by Kummer (1956), and for the chimpanzee by Yerkes and Elder (1936a). The females in heat present most intensely and these conspicuous heat-symptoms must have a high stimulus-value to the male. In this connection, it is not necessary that recognition of the signal be innately determined; the signal could also be learnt by a process of conditioning. 'It is difficult to see how such a coloured excrescence can fail to convey the said information in the case of animals so intelligent and keensighted as Monkeys' (Pocock, 1906).

THE INTERFERENCE WITH RANK-ORDER

The predominantly ground-living primates exhibit a more or less defined rank-order within groups. In such rank-orders, the adult males are regularly at the top; an effect which increases with increasing sexual dimorphism, although it is still a matter of dispute how the rank-order should be assessed and how far the rank-differences (set out according to aggressiveness, vigilance behaviour, priorities to food and to females in heat) vary within one and the same group. The females can be incorporated into the group rank-order or may develop a separate rank-order within the group. In general, females in heat appear to become more aggressive; a factor which is brought directly to bear on like- and lower-ranking conspecifics. Apart from this, high-ranking females accompany mating-approaches with intensive presentation. In *Papio ursinus*, *P. anubis* and *Macaca mulatta*, the highest-ranking males copulate exclusively and very frequently with fully oestrous females. The more advanced the oestrous condition of the female, the more exclusively is copulation confined to high-ranking males (Hall, 1962). This is confirmed by Washburn and DeVore. In baboons, the female in early oestrous mates with juvenile males and less dominant adults, which she actively approaches. In the later part of oestrous, the female forms a temporary consortship with a dominant adult male. In *Macaca mulatta* (Bernstein and Mason, 1963) and *Presbytis entellus* (Jay, 1965), the females in oestrous themselves seek out the highest-ranking males. Conaway and

Koford (1964) confirm that in *Macaca mulatta* the highest-ranking males are most active in breeding.

In the rhesus monkey, oestrous females often cross group boundaries, according to Carpenter (1942); 'estrus only makes it possible for the female to penetrate and to be tolerated within a foreign group'. Washburn and DeVore (1962) state that oestrous in baboons disrupts all other social relations; the female leaves her preference group and child (if present).

Most authors agree that the rank of the female is altered for the heat period; whether the females then rise or fall in the rank-order may possibly vary from species to species, but probably depends upon the rank-criteria employed. Bopp (1962) is of the opinion that baboon females sink in the rank-order during the oestrous phase, whilst the female chimpanzee during oestrous becomes dominant over the male and over other females, as measured by placing a morsel of food between pairs of animals and measuring dominance by the number of food morsels taken (Yerkes, 1940; Nowlis, 1942; Birch and Clark, 1950). The female in oestrous even takes food morsels from the hand of the male.

That females in oestrous leave their preference group and follow the highest-ranking male is an indication that they are very 'self-confident' during this phase. In some cases, this has been more closely investigated: rhesus monkey females in oestrous are particularly aggressive (Carpenter, 1942); they become aggressive towards other females and at the same time begin to approach males (Rowell, 1963), they even threaten their prospective male consorts (Chance, 1956). The same applies to the chimpanzee (Reynolds and Reynolds, 1965). Receptive female chimpanzees present without any sign of fear; at other times they present in a timid manner, and this at least indicates that fear is compensated when the female is in heat.

It is known for some species that the males are at first aggressive to oestrous females (*Macaca fuscata, M. mulatta*), but it is not certain how far this aggression is a response to the changing rank of the female. In spite of this increasing aggression on the part of both partners, dangerous encounters are avoided and this is very often achieved by presentation, which inhibits aggression as well as functioning as a mating invitation: the readiness to fight in the threatening animal can be suppressed through sexual interest (Huxley, 1939).

PRESENTING AS A 'GREETING CEREMONY'

Presentation is not only exhibited by females in heat, but is also shown by females of all ages and at all stages of the oestrous cycle and even by males and very young animals. Regularly, if not exclusively, such presentation is directed towards a higher-rank partner and signifies general submissiveness (Fig. 3), as can be seen from the analysis of individual situations (Kummer, 1956). 'Presenting is the submissive form of approach to a dominant animal as a sort of greeting' (Knottnerus-Meyer, 1901). This has already been mentioned by Fisher in 1876 for the drill, mandrill, hamadryas,

Fig. 3 Social presentation of a female baboon towards a mother with young (from DeVore)

baboon, rhesus, and *Cynopithecus niger* and is discussed by Darwin (1876), who adds: 'The habit of turning the hinder ends as a greeting to an old friend or new acquaintance, which seems to us so odd, is not really more so than the habits of many savages, for instance that of rubbing their bellies with their hand, or rubbing noses together'. The social function of presentation, according to Huxley (1939) and Hediger and Zweifel (1962), has been secondarily derived from the sexual function.

Presentation as a submissive gesture in males and females, in response to higher-rank individuals, has been described for:

Papio hamadryas	by Kummer, 1956, 1957
Papio papio	Bopp, 1953
Papio ursinus	Bolwig, 1959

Papio anubis	Hall and DeVore, 1965
Papio sphinx	Schloeth, 1956–7
Macaca mulatta	Carpenter, 1942; Chance, 1956
Macaca radiata	Simonds, 1965
Macaca silenus	Inhelder, 1955
Macaca irus	Schloeth, 1956–7
Cerocebus torquatus	Galt, 1947
Cercopithecus aethiops	Bolwig, 1959a
Cercopithecus sp.	Hamilton (see Blin and Favreau, 1964)
Presbytis entellus	Jay, 1965
Pan troglodytes	Crawford, 1940, 1942a; Goodall, 1965

This list is, of course, incomplete, but it does show the broad distribution of this submissive gesture amongst the Old World monkeys. According to Hall, Boelkins and Goswell (1965), the patas monkey (*Erythrocebus patas*) provides an exception in that females do not present even as an invitation to copulation. That

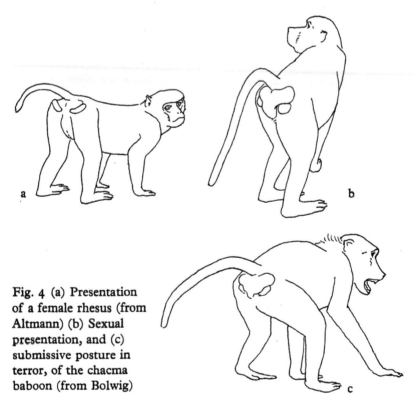

Fig. 4 (a) Presentation of a female rhesus (from Altmann) (b) Sexual presentation, and (c) submissive posture in terror, of the chacma baboon (from Bolwig)

this typical female sexual gesture does not occur at all in this species still needs further confirmation, however. It should not be forgotten that there are very different levels of intensity in presentation and that the finer forms can vary according to the situation concerned (Figs. 4 and 5). The more conspicuous the coloration of the hind-quarters, the more difficult it becomes to divert the attention of conspecifics from the signal; or, conversely, the smaller the effort required to display the signal. In captivity,

Fig. 5 Variations in presentation of the hamadryas baboon (from Kummer); the partner presented to is always thought to be on the left (behind the presenting animal)

whether in a cage or in a rock enclosure, situations requiring intensive presentation arise much more frequently than in natural conditions. A typical occurrence in captivity is the social conflict situation, where a lower-ranking individual is offered food, which it cannot take without provoking a higher-ranking individual. In just this context, presentation can be clearly seen as a largely preventive measure and is often inadequate to prevent an already initiated attack. In the hamadryas baboon society, each animal must follow very specific social rules; when these rules are broken, the individual can be attacked and bitten in the neck (= punished) by the dominant male, despite presentation (Schenkel, 1959; Kummer and Kurt, 1963). It can also be observed that, in at least some species, the female in oestrous takes up a rank differing from that normally occupied (particularly obvious in captive chimpanzees) (Yerkes and Elder, 1936a and b; Yerkes, 1940; Young and Orbison, 1944). In fact, this has only been described until now for species with pronounced sexual skin activity, but this may be due to the fact that the phases of the oestrous cycle are difficult to determine in other species.

It is known for various species that presentation in a social context occurs far more frequently than in purely sexual situations. Reynolds (1961) states that presentation in adults of the rhesus monkey colony at Whipsnade Zoo is solely an expression of submission, exhibited equally by males and females; presentation, sexually and as invitation to copulation, has been seen only for young females, never for adult females.

How does the presentation signal operate on the recipient? It emerges from all observations that presentation reduces aggressive acts from the recipient. It does not, however, reduce aggression in general, but only aggression towards the presenting animal. This is clearly shown by the fact that, at least in *Macaca mulatta* (Chance, 1956; Reynolds, 1961; Altmann, 1962) and *Papio hamadryas* (Kummer, 1957), presentation may be combined with units of the threat complex towards a third monkey ('Gesicherte Drohung' = 'protected threat'). An animal presenting towards a high-ranking male and at the same time threatening another monkey may orient an attack by the high-ranking male towards the third animal (Fig. 6).

The simplest assumption arising from this is that the presenting animal is protected from the recipient because of the sexual 'meaning' of presentation, that is by some kind of sexual response (which may remain sub-threshold) which it elicits within the animal towards which it is directed. Presenting 'appears to diminish threat especially when combined with facial gestures directed towards a more dominant animal, and may in this way provoke mounting by the dominant animal in place of an aggressive approach' (Chance, 1956).

The selective advantage in the development of sexual swellings which at first look extremely unadaptive is not yet apparent. Rowell (1963) found, amongst the animals of her rhesus monkey colony, a tendency to mate *after* the fertile part of the female's cycle. She suggests that this – if not an artefact of captivity – might direct selection towards better synchronized matings by developing sexual swellings as a more conspicuous recognition mark of the fertile part of the cycle. This, however, necessitates the presupposition that the only purpose of copulation is the provision of succeeding generations and that this goal should be attained with the minimum possible number of copulations. If, on the other hand, copulations are important for social adhesion, no such rigid synchronization would be expected. Another possible explanation would be that sexual swelling might act as an additional stimulus underlining the appeasing function of the presentation, since it seems especially well-developed in ground-living species, which tend to develop a more pronounced sexual dimorphism leading to larger and more aggressive males as an adaptation to the

Fig. 6 'Protected threat' of a hamadryas female (from Kummer)

increasing exposure to predators in a habitat lacking in trees, to which flight from danger might otherwise be possible (DeVore and Washburn, 1962). However, in the present context it is not particularly important which explanation best fits the development of these conspicuous structures.

In connection with the behavioural features involved, I think that the following statements have been sufficiently substantiated:

1. Presentation occurs as the female invitation for copulation in the sexual context ('sexual presentation').

2. Presentation occurs as an appeasement or 'greeting' gesture, which reduces aggression on the part of group members ('social presentation'); in the social context it is shown by both males and females and occurs in young animals as well as in adults.

3. The sexual swelling is most conspicuously displayed towards a conspecific during presentation; it probably reinforces the appeasement function of presentation.

4. The social or appeasement function of presentation in both sexes is biologically just as important to the species as is the sexual function of presentation by the females.

5. Social presentation is of equal importance to males and to females.

If this is correct, one may ask what additional stimulus the males may use to compensate for their lack of sexual swellings in those species in which the females exhibit swelling. If the question is thus phrased, one will immediately think of the bright red hind-quarters of the male hamadryas baboon, which are as conspicuous as those of the females. They, too, are presented towards higher-ranking group members and it is almost certain that they act upon the signal receiver in much the same way as do the sexual skins and sexual swellings of the female. This strongly suggests the hypothesis that the male's hind-quarters imitate the female's sexual skin, in order that the male's presentation be socially as effective as that of the females (Wickler, 1963). This hypothesis has one disadvantage: the testing of the hypothesis is more-or-less confined to experiments with models, and these animals are particularly unsuitable for this form of experimentation. It is, however, possible to assemble supporting arguments and to test the hypothesis by a comparison with other species.

COMPARISONS WITH OTHER CATARRHINES

The hamadryas baboon is not the only Old World monkey in which the male has a conspicuous anogenital region, but it is very difficult to compile a comprehensive survey indicating which species exhibit this feature and how far correspondence with the female heat-symptoms is present. Many details important to this study are not mentioned in the literature. For some species, there are exhaustive descriptions of the female heat-swellings, but for other species it is not clear whether these swellings are conspicuously coloured or whether they occur at all. In textbooks, the concept is very often confined to 'true swellings'; consequently 'no sexual swelling' is stated for species whose females nevertheless show conspicuous cyclic colour changes of the sexual skin. The descriptions of the male anogenital region are even more sparse. Those details which I have been able to collect are evaluated in the following account. As a general principle, it is assumed that, within the different systematic categories, ground-living and distinctive sexual dimorphism (in body-size, weight, coat characteristics, dentition) are derived characters and phylogenetically younger than the arboreal mode of existence and slight sexual dimorphism.

Genus *Papio*. The Mandrill (*P. sphinx*) and the drill (*P. leucophaeus*) are regarded by modern authors as baboons which have secondarily returned to a forest habitat. Little is known about these two forms, but I shall be dealing with the mandrill later in a different context. In all baboon species of the *cynocephalus*-group, the female oestrous swelling is bright red, though the size varies. The hindquarters of the adult male are hairless in *P. doguera*, *P. anubis* and *P. ursinus*, but there are differences in colour. The young baby's perineal skin is red, later permanently becoming black. The ischial callosities in adult males are horny and black and they may show a reddish or bluish sheen, particularly when the sun is low over the horizon. In subadults, the ischial callosities are flesh-coloured (rather like the palmar skin of the hand of a European). With age, the callosities darken progressivley from the periphery to the centre. In the centre, where the callosities fuse in males, the light coloration is retained longest. According to my observations, the baboons of different regions differ in the coloration of the male callosities, though the age of the observed animals could not be determined exactly and had to be estimated from body-size and mane-development. In the Nairobi National Park (Kenya), the fused callosities are entirely black; at Lake Manyara (Tanganyika), the central zone remains light for a long period, if not throughout the life of the individual; in the neighbourhood of Voi and in the Tsavo National Park (Kenya), I have seen adult males with juvenile coloration of the callosities, e.g. resembling the coloured plates in Cuvier's *Mammifères*. The Guinean *Papio* males have particularly pink callosity-areas. In accordance with our experience with other vertebrates, especially birds and fish, I would suggest that emphasis in the fine systematics should be laid upon these structures, which are possibly regionally variant intraspecific signals.

The hamadryas baboon is considered to be the more specialized species. The male's hind-quarters are bright red, and this is very probably yet another specialized character, which has evolved within the genus *Papio* (Plate I).

Genus *Macaca*. Within this large genus, oestrous swellings seem to be less developed in the long-tailed species and they are very differently developed in the others. Data concerning the coloration of the male's anogenital region are extremely scarce.

Macaca mulatta: The rhesus macaque's sexual skin is somewhat unusual in that a sexual skin swelling occurs only during maturation; in the adult stage, the oedema disappears, but a reddening of the skin (at perineum, buttocks, hips, base of tail, upper two-thirds of thighs) remains (Hartman, 1928). Further, this reddening is due not to pigmentation but to vascular engorgement (Collings, 1926). There is an increase in colour in mid-cycle; however other behavioural factors influence the coloration as well, so that in some females it changes rather erratically and rapidly. The sexual skin coloration of isolated females fades, but colour rises rapidly if a male is introduced (Rowell, 1963). Rhesus males have a red perineum. According to the author's observations in the Frankfurt Zoological Garden, only the highest-ranking males show the red colour, the others being as pale as frightened animals under natural conditions.

The rhesus macaque is sociologically the best studied monkey, and we have a large bulk of information on its dominance and submissive behaviour (Carpenter, 1942; Chance, 1956; Reynolds, 1963; Altmann, 1962). Presenting is a social greeting ceremony, shown by both sexes and usually towards higher-ranking animals, regardless of their sex. Besides the difference in colour-intensity of the perineal skin, the high-ranking males carry the tail in a raised position (Fig. 7) and show dominance mountings; the low-ranking males carry the tail in a lowered position and show repeated presentation (Altmann, 1962). 'Mounting of a female by a male, even with intromissions, may be a response to a true sex drive or may occur as a greeting response, as an expression of an affinitive relationship or as a substitute act to foil an attack. Mounting is not specific for oestrous nor, indeed, for the male-female relation because males mount males, and females mount females' (Carpenter, 1942). Females may be seen mounting males, and 'it should be pointed out that a monkey did not necessarily always assume the same rôle, even in its relations with a particular individual in a particular sequence of interactions' (Altmann, 1962).

Macaca radiata: According to Pocock (1925), Fiedler (1956) and Hill (1958), this species shows no sexual swellings and no sexual skin changes. Eckstein and Zuckerman (1960), however, mention a

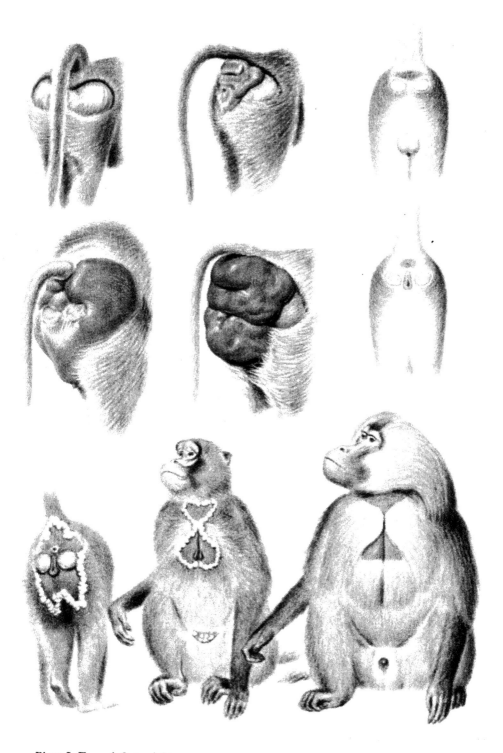

Plate I From left to right:

above: *Papio anubis* male; ditto, oestrous female; *C. aethiops* male
middle: *P. hamadryas* male; ditto, oestrous female; *C. aethiops* female in oestrous
below: Sexual skin of female gelada; female gelada from front; male gelada

slight sexual skin activity, a mere puffiness of the circumanal and vulvar regions, which are more dark purple than red. This is confirmed by Simonds (1965):

In bonnet macaque females, the sexual skin does not swell very much. Only one female out of 18 in the Somanathapur group had noticeable swelling. The oestrous and nonoestrous periods of the others could not be differentiated by this means. Variation in the colour of the sexual skin was also slight. Some did have bright red sexual skin, but it remained red throughout the month. Variation may have existed, but it was no greater than that caused by differences in lighting in the sun and shade.

Fig. 7 Posture of a high-ranking rhesus male (from Altmann)

Whether or not there is a colour change in the sexual skin correlated with other motivational events, as in the rhesus, is still unknown. Presentation and social mounting occur as in the rhesus monkey, but quite often clearly subordinate monkeys mount dominant ones in the bonnet macaque. Sometimes, the dominant monkey forces the subordinate to do so by backing into it (without presenting!) and reaching back to grab it (Simonds, 1965).

Presenting consists of raising the tail and turning the hind-quarters towards the monkey to which the presenting is directed, often with the tail held to one side. Presenting in dominance situations is similar to the position the receptive female takes as a male approaches to mount for copulation, but differs in that the presenting monkey in a dominance interaction is usually more tense. Both lip-smacking and presenting are clear indicators of subordination and the monkey to which they are directed is almost always the dominant monkey in that social situation (Simonds, 1965).

Macaca fuscata: This species is said to have a clear-cut breeding season. During this time, not only does the colour of the face and sexual skins of both sexes turn bright scarlet, but the sexual skins – and sometimes the faces – of young oestrous females swell markedy (Imanishi, 1963). Presentation represents subordinate behaviour and greeting behaviour among juveniles; sexual presentation is specific to the females of this species: 'A sexually receptive female orientates her buttocks to the approaching male while in a sitting posture' (Tokuda, 1961–2).

Macaca irus: This species seems to resemble closely the rhesus macaque in the characters here under discussion. The females exhibit a slight sexual swelling and a visible reddening of the sexual skin which, however, depends not only on the sexual cycle but on other factors as well. Adult males always show red hind-quarters and red faces (Spiegel, 1930).

Genus *Cynopithecus* + *Macaca* (=*Gymnopyga*) *maura*. The Celebes-macaques have been systematically arranged and re-arranged several times by Schlegel, Weber, Meyer, Matschie, Elliot and others, and their systematic position is still not clear. Büttikofer (1917) distinguishes eight species, which have been arranged in a series of geographic evolution by De Beaux (1929). According to the careful description and coloured illustrations of Büttikofer, these species differ distinctly in the form and colour of the ischial callosities and the surrounding naked or hair-covered areas. The remarkable differences in form can scarcely be interpreted as special adaptations to the inanimate environment. Here, more forcefully than in the case of the *Papio*-forms discussed above, one is given the impression that species-specific differences in signals are involved (which would run contrary to the inclusion of these forms within one species as suggested by Fiedler, 1956). Unfortunately, little is known about these animals.

The sexual skin in the females of all species shows conspicuous catamenial swellings, which are especially large in *M. maura* (though, in this species, the immediate surrounding of the vulva is not involved in the swelling). The callosities are flesh-coloured or red in the males; the male *M. maura* shows red bare skin on the hind-quarters.

Genus *Cercopithecus*. This genus seemed to be a crucial one with respect to the hypothesis explaining the males' anogenital coloration as being an imitation of the colour pattern of their females. It is known that in a number of *Cercopithecus* species the males exhibit very brilliant coloration of the penis, scrotum and sometimes the area immediately surrounding the anus. According to Pocock (1925), Eckstein and Zuckerman (1960) and others, there are no external cyclical changes in the sexual skin of the females. Hill (1958) mentions that in some cases there is a slight puffiness of the margins of the rima and the cutaneous field surrounding the vulva, but I was unable to find any discussion of possible colour changes in this area in the female *Cercopithecus*. I therefore readily took the chance to make a pilot-study on just one species, *Cercopithecus aethiops* (known as a rather aggressive species judging from observations and the torn coats of adult males: Booth, 1962). The species inhabits open bushland, by preference the small strips of trees and scrub growing along the banks of rivers in the savannah country, and spends at least part of the time on or near the ground. Sexual dimorphism is moderately developed. In accordance with the general concept mentioned earlier, this all seemed to characterize a phylogenetically more specialized species which might be expected to have developed a male imitation of the female's sexual skin coloration.

I observed two groups of *Cercopithecus* near Banagi in the Serengeti National Park (Tanganyika) for two months in 1964 and again for two weeks in 1965. The taxonomy of this species is still rather confused; according to Dandelot (1959) the animals observed belonged to *C. pygerythrus* Cuv.; according to Swynnerton's checklist ('Mammals recorded from the Serengeti National Park', 1958) it would have been *C. aethiops centralis* Neumann. The animals are rather shy, especially the females, and therefore it was impossible to obtain a useful colour photograph of their sexual skins. However, observations with binoculars were relatively easy. When one is familiar with the habits of these animals and remains still, they will climb up to the crowns of the acacias above the observer, particularly when the trees are still bearing young, green pods (readily eaten by the monkeys). In addition to this, the animals often seek out certain sites on trunks or strong branches of *Acacia xanthophloea*, where they have bitten away the bark and they repeatedly lap up the exuding sap. In such positions, the animals were easy to observe, since they were not partially obscured by leaves and branches. The (changing) sleeping-sites, upon which the animals stay for some time after sunrise before slowly departing, also provide good opportunities for observation. During the period of observation, two cases were seen where a female came into heat. Otherwise, adult females outside of the heat-period can be recognized by their long teats, which are located very close together with the distal ends convergent, and which hang prominently from the fur when the animal is on all fours. Out of heat, the perineal area of the female has a pale flesh-colour, the vulvar slit can scarcely be seen and the ischial callosities are grey-black. The fur on the inside of the thigh is yellowish. As the heat-period approaches, the perineal patch becomes redder and the vulvar slit margins become bluish; a female in oestrous has a deep red perineum and somewhat swollen, distinctly bright blue vulvar margins, from which the conspicuously red clitoris protudes a number of millimetres (Plate I). (Ischial callosities and fur-colour remain unchanged, of course.) A female in oestrous has reddish teats, which out of oestrous and during lactation are pale or even slightly blackish.

The anogenital region of the adult male shows noticeably exact correspondence with the colour-pattern of the same area in the oestrous female: grey callosities are likewise present, the perineum is scarlet, the scrotum is bright blue (sometimes with yellow hairs on the lower margin) and the penis is scarlet. When viewed from behind, the male colour-pattern is very similar to the heat-coloration of the female, since the tip of the bright red penis is just visible below the raphe scroti. This correspondence in pattern has, however, developed from non-homologous structures. The homologue of the scrotum is represented by the labia majora, which in Old World monkeys are vestigial or absent (Harms, 1959); the coloured structures surrounding the vulva being the labia minora. Further, the bright red patch around the male's anus is mostly due to a fringe of long, vivid red hairs; these are absent in the female, which shows instead a smaller red patch of naked skin. Also, the scrotum and penis of the male are larger than the labia and clitoris of the female. The convergencies show that the resemblance of pattern is not due to a mere lack of sexual dimorphism but strongly suggests a special evolutionary trend, which would then support the suggested signal function. (That true presenting nevertheless seems to be very rare in this species will be discussed later.)

What is now required is more accurate observations of other *Cercopithecus*-species (or subspecies), which differ in just this feature – the coloration of the male genitalia (Harms, 1956; Hill, 1958):

C. aethiops callitrichus scrotum: green		prepuce:	?
C. a. aethiops	„ slate blue	„	reddish-orange
C. albigularis kolbi	„ iron grey	„	slate blue
C. büttikoferi	„ ?	„	vivid lilac

It would be interesting to know whether comparable colour differences exist in the genitalia of the corresponding females.

Genus *Miopithecus*. The dwarf species *M. talapoin*, which Fiedler includes in the genus *Cercopithecus*, has a rather long, pendulous clitoris. The females exhibit a true sexual swelling (Tomilin, 1940), reddish in colour. The male scrotum is blue, the penis reddish. Seen from behind, the tip of the glans penis is to be seen below the scrotum, as in *C. aethiops*. I have found no resemblance in the genital coloration between the sexes.

Virtually nothing is known of the behaviour of this species, but I have been able to observe copulations in the Frankfurt Zoo, which take place in the usual manner: the male grips the lower hind-legs of the female with his feet. Intromission is followed by a series of pelvic thrusts, but whether or not ejaculation occurred could not be determined.

Genus *Erythrocebus*. The female patas monkey is said to exhibit no cyclic sexual skin activity. The males have a bright red perineum, a turquoise-blue scrotum and a pink flesh-coloured penis. Between the scrotum and the perineum is white fur. This species is said to lack the presentation gesture (Hall, Boelkins and Goswell, 1965), although this gesture is the only known type requiring the colour-pattern which is evident. However, 'presentation' is often ill-defined; standing or walking on all fours with the tail erect (Fig. 7), the animal automatically displays the colour-pattern.

Genus *Theropithecus*. The systematic position of this genus is still uncertain. Despite the popular name 'Bleeding-heart Baboon' for *T. gelada*, this is certainly no baboon but rather a relative of *Cercopithecus* or mangabeys. The highly-specialized sexual skin of the gelada female has been described and illustrated several times (Garrod, 1879; Pocock, 1925; Appelman, 1953, 1957; Matthews, 1956). The perineal sexual skin is a bare area between the pubes and the base of the tail, expanded into two conspicuous patches ventral to each calliosity and alongside the ventral vulvar commisure.

The cyclic changes involve a waxing and waning in the intensity of the red coloration and the tumescence of the bare areas. A very striking change is the development of numerous, more or less symmetrical, pearly cutaneous vesicles along the margins of the pigmented area. As menstruation approaches, the sexual skin loses its intense scarlet colour and fades to a delicate flesh pink; simultaneously, the vesicles become enlarged and fluid-filled. This is probably controlled by the luteal phase of the ovarian cycle; the subsequent intensified coloration of the skin, with diminishing fullness of the cutaneous vesicles, then being controlled by the follicular phase (Matthews, 1956). The ischial callosities are greyish in colour. Most extraordinary, however, is what might be called a 'secondary sexual skin'; a characteristic naked chest-space 'formed of two median triangular isosceles areas reversely directed, with their apices approximate' (Garrod, 1879). This cordate breast patch is of the same colour and shows the same colour changes as the sexual skin in the anogenital area. Moreover, it shows the same bordering with cutaneous, pearl-like blisters (Plate I). Comparing the two naked skin areas, one will find that the lateral incursion of fur on the chest (which divides the patch into two triangles) corresponds in colour and position to the ischial callosities. Further: the nipples, which in immaturity remain small and widely spaced, become enlarged, pear-shaped, and bright red from puberty on, and now lie in close proximity on each side of the midline, thereby corresponding in colour and relative position to the bright red vulva. Those who would still regard the improbable development of the sexual skin on the hind-quarters as without significance can scarcely describe this duplication of the pattern as pure chance. The correspondence has arisen (within the same individual) by convergence (fur = callosities; teats = vulva) and goes so far that one cannot but acknowledge the presence of imitation (Wickler, 1963). Simultaneously, this fact demonstrates that the pattern is significant as a signal, although the function of this signal is as yet unknown. The individual characters can also be found in other species: the macaques and *Cercopithecus* exhibit change in intensity of the red nipples during the sexual cycle, and in *C. aethiops* the nipples may lie so close together that the young can suck on both at once. But there is no other case in which these characters (whether 'by-products' or adaptations is still unknown) are combined in such a conspicuous pattern.

The hind-quarters of the gelada male show large ischial callosities with a large area of naked skin below, which contains two large, lead-coloured, soft, tough-skinned cushions (Pocock, 1925). There is no obvious similarity in coloration to the female sexual skin. The male has, however, the same double patch of pink naked skin on its throat and chest; the inconspicuous nipples are situated as in the female. The male has no carunculations of the skin along the border of the chest-patch. It seems that the male displays only one of the two female colour phases.

There is very little known of the social behaviour of the Gelada. What is known is that it has a threatening gesture during which the chest is displayed towards an opponent.

The Colobinae. The behaviour and the heat-symptoms of this subfamily are virtually unknown, particularly in the genera *Pygathrix*, *Rhinopithecus*, *Simias* and *Nasalis*. Several authors agree that no sexual swellings occur in the genus *Presbytis*; 'perineal swelling or sexual skin does not occur in langurs' (Jay, 1965). According to Hill (1958), in *Semnopithecus* (during oestrous) the bright pink clitoris protrudes some millimetres from the vulva and provides a sharp contrast with the deeply pigmented adjoining skin.

Genus *Colobus*. Fiedler includes three species in this genus, *C. polykomos*, *C. badius* and *C. (Procolobus) verus*, described as a subgenus. In the black-and-white colobus (*polykomos*), no catamenial swelling is known. The red (*badius*) and olive colobus (*verus*), however, exhibit remarkable perineal swellings, which may affect the circumanal skin in some subspecies of *badius* and regularly do so in *verus* (Hill, 1952). The swelling is not correlated with the sexual cycle. It takes place before puberty and is a more or less permanent character once it has developed, but it is reduced in gravid and lactating females (Booth, 1957). The swelling of the female *C. verus* overshadows the callosities to some extent. The structures involved are easiest to analyse at an intermediate stage (Fig. 8b). Following the detailed description of Hill (1952), there are one or two folds dorsal to the anus, a swelling of the perineal body and the tract between the vulva and the medial edges of the callosities, and a swelling of the intrinsic vulvar tissues; ventrally, the tumescence of the praeputium clitoridis builds up a 'pubic lobe'. With full swelling, the different parts around the vulva merge into one another.

In *Colobus verus*, the juvenile males are often mistaken for females, because they show much the same kind of swelling: the greater part of it affects the circumanal zone, deforming the anus into a transverse slit. The perineal body becomes a transverse depressed zone, a transverse sulcus; it separates the circumanal swelling from the dorso-ventrally elongated pseudo-vulvar mass which is divisible into three lobes: a pair of lateral swellings separated by a median vertical sulcus (blind at the bottom) and, ventral to this, a median pubic lobe greatly resembling a swollen praeputium but composed solely of oedematous subcutaneous tissues covered with delicate skin (Hill, 1952). The adult male shows only traces of the juvenile swelling. The same holds for *C. badius*, whose young males show even greater swellings.

The only previous explanation for this unusual male swelling was based upon abnormal hormone balance (probably an excess of oestrogens: Booth, 1957). This explanation concerns the causal aspect of the phenomenon, whilst we are here dealing with the functional significance. But even as a causal explanation, the suggestion is unsatisfactory, since the swelling of the male is even in minor details very similar to that of the female, although non-homologous structures are involved. There is no explanation for the fact that these structures and not those structures homologous to those of the female should respond to the hormone. The circumanal skin is involved in both sexes, but not the perineal body. The 'pubic lobe' arises in the female from the swollen praeputium, which is not involved in the male (the male penis and scrotum are completely hidden from view; the urine emerges from a rather inconspicuous opening well forward between the thighs; Fig. 8a). The same is true for the labial region. In the juvenile male, there is no scrotum and that part of skin which later grows out into it is not, according to Hill, involved in the swelling. Most remarkable is the deep but imperforate depression at the site corresponding to that of the vaginal opening in the female, forming a pseudo- or dummy-vulva from narrow sulci crossing the swollen area. Again, there is a close correspondence with the female pattern, which we can justifiably regard as a signal, since once again the correspondence is based upon non-homologous characters. This may represent an 'imitation' of the female characters which deceives the male in determining the true sex of a growing rival. Behaviour studies on this species are sorely needed.

SUMMARY OF THE COMPARATIVE SURVEY

Sexual presentation in its least intense form, that is with the female standing still and allowing the male to approach from behind to mount, is present in all Old World monkeys and apes. Some of the latter sometimes copulate face-to-face and this may

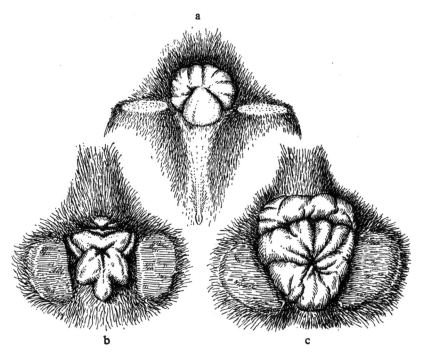

Fig. 8 *Colobus verus*. (a) juvenile male from below to show the site of penis; (b) and (c) early and full oestrous in the female (from Hill)

change the form of presentation. Often, presentation occurs in a more intense form which provides the typical presentation gesture usually described. This is still unknown for some species, but only in *Erythrocebus patas* is it explicitly said not to occur.

The same presentation at various intensities is known to occur in a number of species in both sexes as an appeasement or greeting gesture, usually directed towards higher-ranking group members (which need not be conspecifics; even human beings are presented to). The exact distribution of the 'social presentation' among Old World monkeys is not yet known.

Changes of the vulvar region in the female, correlated with the ovarian cycle and therefore called oestrous swellings, are known as early as in Insectivores, from which Lemurs and lower primates have been derived. Swellings and colour changes in the female ano-genital region are to be seen in Lemurs and *Tarsius* (Fig. 9). In *Tarsius*, the labia minora swell in the sexual phase and the brightly coloured clitoris becomes visible (Hill, Porter and Southwick,

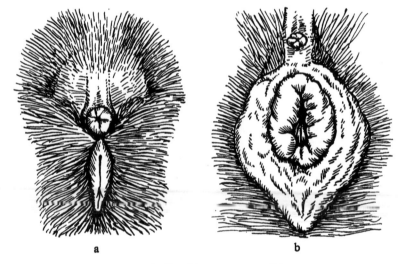

a   b

Fig. 9 Female *Tarsius* genitalia; (a) anoestrous, (b) oestrous

1952). In *Microcebus*, the vaginal opening is normally closed outside the reproductive period; during oestrous the vaginal skin swells and becomes pink and turgescent before the opening of the vulva, which lasts only for about ten days before closing again (Petter-Rousseaux, 1962, 1964). Among higher primates, sexual swellings are confined to the Catarrhines, but are widely distributed among this group. Sometimes there is simply a change of colour of the sexual skin, which may moreover be controlled by other 'emotional' states. Swelling and/or colour changes are often referred to as sexual skin activity. They usually, but not always, occur during oestrous periods. In *Macaca radiata*, *M. sinica*, *M. mulatta* and sometimes in baboons, they remain present during pregnancy as well. The sexual skin areas of female rhesus monkeys are bright red in late pregnancy or even during the whole gestation period (Hartman, 1928) and pale after parturition and during lactation (Rowell, 1963). In *Colobus*, the swelling shows no correlation with the sexual cycle and is most characteristic for young females. Similarly, it is most pronounced in the young females of *Macaca mulatta*; the reverse is true for the baboons. Among anthropoid apes, sexual swelling is well marked in *Pan* during oestrous, sometimes persisting through pregnancy (Hediger and Zweifel, 1962). In *Pongo*, a copious enlargement of the clitoris and the perivulvar secondary sexual tissue occurs only during late pregnancy (Graham-Jones and Hill, 1962). The gorilla merely shows a variation in the length of the vulvar opening, which is about 1 cm in menstruation and 3 to 4 cm in mid-cycle (Spiegel, 1954). Among the Hylobatidae, there are some minor changes in colour and eversion of the labia during oestrous.

Swelling and/or colour changes may be confined to the rims of the vulva or may involve in addition the circumanal region and/or the tail base and the skin surrounding the ischial callosities. In Gymnopyga, only the vulvar area is not affected. Other cutaneous structures may be involved in the colour-change: the teats, and in the gelada the naked skin on chest and lower vent.

The activity of the sexual skin areas is primarily correlated with the biological rôle of the animal as a female, in most cases with the special phase of it during which attraction of a male is necessary. It underlines the invitation posture, referred to as presentation. The presentation is more widely distributed than conspicuous sexual skin activity, and I would suggest that a tendency to develop sexual skin activity is common to the Catarrhines and has been several times independently exploited, rather than assume that sexual skin activity has been lost several times in phylogeny.

In some species, the male possesses sexual skin characters, which closely resemble a special phase of the female's sexual skin cycle (usually that of the receptive female). The conspicuous sexual skin of the male may be permanent or characteristic for a special life-phase but never exhibits periodical waxing and waning.

There is no sexual skin activity in females of a species which is known not to present. There is no species with a conspicuous male anogenital region known in which the female does not show a comparably conspicuous sexual skin activity, except in the patas monkey. In the talapoin, the coloration pattern of the male genitalia and of the female sexual swelling do not correspond. In both cases, the male colour pattern shows the blue-red contrast whose function will be dealt with later. It seems to be a rather ancient character and may have been retained independently of the female's sexual skin coloration.

The distribution of the similarities between female and male sexual skin coloration allows of two possible explanations: sexual skin coloration is either a primitive character that has been lost in several cases. (This is contradicted by the facts that (1) the sexual skins are very variable in form and can scarcely be referred to a common origin; (2) it is just those species which are now regarded as the most generalized which lack a conspicuous sexual skin), or a character which has been independently developed several times in parallel. A conspicuous sexual skin is more widely distributed amongst females as amongst males and is often formed from non-homologous structure in males and ·females. This leads to the conclusion that the correspondence arose by convergent evolution of the males to resemble the females.

Only the possible biological function of these signals have been discussed, whilst the underlying physiological causes have been considered only in passing. It may be that the signal-structures arose in females as a 'by-product' of physiological processes, but the development of analogous (and not homologous) structures in males would seem to be due to their specialized evolution as a signal apparatus.

The interpretation of a number of conspicuous primate characters suggested above has the following corollaries:

1. There is intraspecific morphological signal-imitation.

2. The primary signal is derived from the sexual behaviour and has been made available for social functions by alterations in function and motivation.

There are examples of both phenomena in other groups of animals, as will be briefly demonstrated. Such examples are still somewhat recent and they magnify the probability of the suggested hypothesis.

INTRASPECIFIC SIGNAL-IMITATION

Each case of specific adaptation of a specific signal to a biologically meaningful response from the signal-receiver to the signal raises the question of the origin of this adapted signal-complex. The signal and the response, pre-adapted as a functional duet, would hardly arise by chance. A possible evolutionary pathway is that along which an already evolved response (which is elicited by a sign stimulus) is transformed to another context and is possibly released by a new object, if the sign stimulus is incorporated into the new object. The response may be said to remain attached to the releasing stimulus – as in any artificial model-experiment – and can be transposed to a new object. A natural prerequisite is that the internal state of readiness of the signal-receiver to produce the appropriate response must be present and at a sufficient intensity to produce the response when the 'new' object is encountered. Such cases are best known where mimicry-effects are present, where constantly available flight or feeding responses are most commonly exploited. A mimicry system, however, usually involves more than one species; but if one does not regard this as essential for the application of the term, it is possible to describe the following examples as intraspecific mimicry. In addition, an intraspecific development of a deceptive signal automatically provides the proof (difficult to find in interspecific mimicry) that the deception carries a selective advantage.

Signal imitation is almost certainly present where the same unmistakable signal pattern arises by convergence, i.e. via non-homologous pathways. Such cases are really not so rare, even within a given species, but have been hardly investigated, although they are very important to the evolution of communication.

From pure morphological description, for example, emerges the fact that many Canids have a round black spot on the dorsal side of the tail, next to the root of the tail. This spot is located exactly above the black perineum covered by the tail and the perineum itself is a very important body area for social, naso-anal control. In sloths of the genus *Bradypus* the brilliant mid-dorsal spot of the males (formed from specialized hairs) exhibits the same contrasting colours in nearly the same arrangement as is found on the face of the same animals. The neck-spots of various (sub-) species vary in accordance with the faces of the same (cf. the excellent colour plates in P.Z.S. 1871). This replication alone allows of the assured conclusion that the facial patterns represent (probably intraspecific) signals. Nothing is known of the behaviour of these animals nor of the significance of these signals. Another example is the conspicuous naked anal and scrotal area in the male *Tragulus*, imitated by a psuedo-scrotal area in the female (Pocock, 1919).

Closer investigation has been made into the conspicuous pigmented spots present on the anal fin of many male Cichlids of the African *Haplochromis* group. These spots are replicas of the species' own eggs, laid by the female. Using a number of available Cichlid species, it proved possible to determine the origin and function of these 'egg-dummies', as has been exhaustively reported in a previous publication (Wickler, 1962): The females of these species take up the unfertilized eggs into the mouth immediately after spawning; the males spread the anal fin whilst ejecting sperm on to the vacant substrate and consequently display the egg-dummies; the female then tries to take the dummies, like real eggs, into the mouth and thereby sucks the sperm into the mouth and fertilizes the eggs. The response of the female to the egg-dummies is, like the response to the eggs, a brood-care response and is dependent upon brood-care motivation, but in effect this response functions as a fertilization pattern when directed to the egg-dummies. In addition, the males court every female which approaches, and this involves displaying the anal fin plus egg-dummies. Even at this early stage, the female responds with brood-care behaviour by following the counterfeit eggs and thus reaching the spawning pit of the male. In this case, the (brood-care) response functions as a pair-forming action. The perception-mechanism and the response-readiness of the signal receiver may remain unchanged; the functional transposition of the behaviour pattern can be brought about simply by the transmission of the signal to a new site. For all derived signals that have originally passed through a mimetic stage, during which they may be called 'deceiving signals', it is the signal-receiver that is the key-figure, not the signal-sender (Wickler, 1965c).

THE SOCIO-SEXUAL LINK

It has often been suggested that constant sexual motivation throughout the year is the basis of the primate's social life; this has been called Zuckerman's principle of sociology in primates and is still heavily discussed. Quite recently it has been argued that sexual attraction could not be the main determinant of social grouping in non-human primates, because at least some monkeys which live permanently in social groups have a clear-cut breeding season. Very often there is a more-or-less recognizable, preferred breeding season resulting in a birth peak during which most births occur. Some species, however, have a true seasonal limitation of reproductive activities, resulting in a discrete birth season to which all births are confined (Lancaster and Lee, 1965). Washburn and DeVore (see the discussion following the paper of Imanishi, 1960) claim: 'that sexual attraction is an entirely inadequate explanation for the troop is best shown by the fact that during most of her life a female is not involved in sexual activity. Most of the time, she is either juvenile, out of oestrous, pregnant, or lactating'. There are, however, in adult free-ranging rhesus females distinct oestrous cycles after conception and copulations may occur until six weeks before parturition (Conaway and Koford, 1964). Nevertheless, according to Petter (1962):

des recherches plus récentes ont cependant montré que la structure sociale restait constante au cours de l'année chez des espèces de Primates à reproduction nettement saisonnière comme le Macaque japonais. Les quelques faits que nous avons pu recueillir sur les sociétés de Lemur viennent s'ajouter à cette constatation. Bien que nos observations soient encore insuffisamment nombreuses et précises, il semble bien que, chez ces animaux, les groupes persistent aussi durant toute l'année malgré le caractère saisonnier de la reproduction. Il faut donc chercher la cause de la sociabilité de ces Mammifères.

These arguments – even including those of Zuckerman in his reply to Washburn and DeVore – are based upon the assumption that the main reason for constant sexual attractiveness is the recurring menstrual cycle. In fact, the menstrual cycle is frequently not actually investigated, especially since in some species lacking oestrous markings this is difficult to follow; instead, the distribution of births throughout the year is examined or, since this itself involves a rather long-lasting study, assumptions are actually based upon the age-constitution of the groups at any given time. The conclusion 'It is clear that constant sexual attraction cannot be the basis for the persistent social groupings of primates' (Lancaster and Lee, 1965) is thus reached without mention of the male monkeys and their possible cyclic spermatogenesis, thereby leaving out one of the main factors in the suggested mechanism of group cohesion. Zuckerman (1932) suggested a socio-sexual link, but not a menstruation-social one. His key-observation, like that of other students of monkey behaviour, was that sexual gestures are shown outside of the primary sexual context by both sexes towards higher-ranking group members, regardless of their sex. The responding animal often shows at least intention movements of the male sexual behaviour. As has been illustrated by the *Haplochromis* example cited above, it is most important if the sender's signal is shown in a new situation (and may therefore be entirely changed in motivation) that the receiver's action should remain motivationally unchanged, i.e. that he should be in some way deceived. In the primates, the sexual gesture occurring out of the sexual context is changed in motivation (see p. 106), but it is to be expected that the male's sexual interest in female-signals remains roughly constant. This assumption has not been tested, but to my knowledge spermatogenesis is continuous from puberty to senility and without seasonal interruptions or temporal cycles.

These arguments are, of course, no proof of the correctness of Zuckerman's suggestion; they only imply that the arguments brought forward against it are not entirely water-tight.

It is quite probable that behaviour patterns other than the sexual ones help to stabilize the troop. In this context, brood-care behaviour would be particularly concerned. Kühme (1965) has recently shown that this is true of the African hunting dog (*Lycaon pictus*). A careful examination of primate brood-care behaviour has not yet been undertaken, though a number of primary elements of this context are in fact known to occur out of their original context in social situations, e.g. lip-smacking, mouth-to-mouth greeting and grooming. We will see below that in human beings as well re-motivated brood-care behaviour elements are of specific social importance.

The socio-sexual link is also to be found in other vertebrates, which offer the opportunity to study this phenomenon in more detail. Quite recently we have become aware of a socio-sexual link comparable to that of the primates in the behaviour of a Cichlid fish from Lake Tanganyika, *Tropheus moorei*. This is a highly-specialized mouthbreeder and was therefore expected to show a division of labour between the sexes and a well-established morpho-logical sexual dimorphism. (Both peculiarities have been proven as typical of specialized mouthbreeders, which have been evolved independently at least three times among African Cichlids: Wickler, 1962.) But *Tropheus*, although possessing clear division

Plate II *above:* Resting male *Cercopithecus aethiops*

bottom: Resting male Miamin (Upper May River, New Guinea, photograph: D. v. Holst)

of labour (the female alone exhibits mouthbrooding and care of the young) entirely lacks a morphological sexual dimorphism: both sexes are of nearly the same size and of the same coloration, i.e. black with an orange-yellow band around the body, which flashes suddenly under certain conditions. Such colour-bands are common in Cichlids and inhibit biting by conspecifics as a result of specific motor patterns of display towards the potential attacker. *Tropheus* displays this band to an attacking conspecific by quivering the broadside in an S-configuration before the opponent, with all fins spread and the hind-quarters lowered. This motor pattern is the typical male courtship pattern, the 'jogging-dance' ('Rütteln') of the *Haplochromis* and related genera, and is in *Tropheus* also a courtship element and even a spawning pattern present in both sexes, i.e. true sexual behaviour. *Tropheus* is extremely aggressive towards conspecifics and lives in small bands, composed of individuals of both sexes which know each other personally. It is very difficult or impossible to add a foreign individual to an already established group. Foreign conspecifics in an aquarium usually fight to the finish. Males and females regularly exhibit the jogging-dance, independent from spawning, in cases where they are subject to aggressive threat from a group member, and this usually inhibits bites from the attacker except during intense rank-order fighting. As in the presentation of the baboon, 'Rütteln' (i.e. presentation of the coloured band) in *Tropheus* can occur with differing intensity, recognizable from the intensity of the yellow coloration, of the body contortion, of the fin-positions, of the frequency of quivering and from the deviation of the longitudinal body axis from the horizontal. It is shown at its lowest intensity by the fish in response to a neutrally approaching rank-superior from the group; the highest intensity is shown during spawning or in response to an intense attack from a group member. If the females were not in possession of this courtship pattern (in other genera typical of the male only), together with the requisite colour-pattern, they would not be able to survive in such a group. We do not know what selective advantage is possessed by small groups of personally acquainted animals in comparison with large anonymous swarms, but in comparison with the primates it is important to note that a sexual behaviour pattern has developed into a social greeting ceremony (Wickler, 1965b).

A further example is provided by the spotted hyaena (*Crocuta*

crocuta). It has been known from Aristotle that it is almost impossible to distinguish males and females on the basis of their external genitalia, since the females possess a pseudopenis, which resembles the penis of a male and may be likewise erected (for anatomical details see Matthews, 1941; Davis and Story, 1949). Penis and pseudopenis in this species represent one of the most important social signal mechanisms. Spotted hyaenas live in small colonies. Two hyaenas upon meeting extend the penis (this occurs even in quite young animals!), raise the hind-leg nearest the partner and offer the genital region for inspection, whilst simultaneously inspecting that of the partner. Animals of both sexes at any age perform this appeasing ceremony (Wickler, 1965b).

In *Tropheus* and *Crocuta* it is not obvious whether secondary convergence between the sexes, following former differentiation, has occurred, or whether differentiation of the two sexes was prevented from the outset. In principle, both pathways are equivalent and require equal selection pressures, since we know from comparable species that another selection factor operates towards sexual differentiation.

In the cormorant, a special greeting ceremony is clearly derived from the copulatory courtship, but it serves a different function, leading to social bond-formation in all age-classes (Kortlandt, 1959). In cats, lions, etc., the female pattern of sexual presentation of the vulva with tail raised has evolved into a social greeting ceremony for both sexes (Fig. 10) in much the same way as the presentation of primates (Leyhausen, 1960).

This demonstrates the frequency with which the use of the external genitalia and sexual behaviour patterns are incorporated into social life, with new significance. Mating is, of course, also social behaviour and a mated male-female-pair forms the smallest social unit. Mating in the widest sense of staying together at least a short time to achieve copulation is absolutely necessary for all mammals, including those which otherwise live solitarily. Therefore, sexual behaviour is necessarily bound to social behaviour; a socio-sexual link is present *a priori*. It is therefore difficult to distinguish which was originally more important; even in modern baboons, Bopp (1962) regards presentation as a social and interspecific behaviour pattern, primarily without sexual significance. The socio-sexual link becomes more conspicuous the more the partners tend to avoid each other, to enter into some sort of competition with each other, or to respond aggressively to each other. Such aggression towards the sexual partner must be suppressed, at least at times. If selection should favour large groups, then aggression among the group members must be likewise suppressed. Where rivalry is advantageous, but where the group represents the smallest viable unit under specific environmental conditions, then rivalry between groups is favoured. In this case, a necessary prerequisite is a group-specific signal (kinship recognition) or individual acquaintanceship, in order to restrict aggression

a   b

Fig. 10 (a) Social presentation of a young male lion towards a full-grown male; (b) sexual presentation of a female domestic cat (from Leyhausen)

towards non-group-members. Personal recognition, of course, sets a limit on group-size, but in itself does not prevent aggression. Apart from reduction of aggression, it is also necessary that flight of the group members from one another be reduced, so that the group does not split up into individuals, incapable of surviving alone. Aggressive and flight behaviour are classed together as agonistic behaviour, and therefore any factor operating against agonistic tendencies serves to stabilize such groups. It is particularly important that agonistic behaviour should be reduced if intra-specific competition is slightly in favour of older and more experienced animals and at the same time there is a certain degree of sociability. In this case, there is either a high probability of older individuals bearing favourable genes competing successfully with their own descendants also bearing these genes (the selective advantage of which is thus diminished) or some special mechanisms must arise that reduces competition between animals of the same family line, thus favouring the intra-family development of appeasement gestures (Hutchinson, 1963). The longer the young stay within the group, the greater this effect should be.

THE MOTIVATIONAL ASPECT

Judging from the present distribution of the individual characters, the presentation gesture and mounting were originally purely sexual behaviour patterns. But at the present time, if not at the point of origin, presentation is not exclusively sexually motivated, as has been best shown in the rhesus monkeys. Bernstein and Mason (1963) formed an artificial group from eleven foreign *Macaca mulatta* and observed the process of group-formation. These authors found, among other things, that the frequency of sexual behaviour during the initial period of group-formation was fifteen times greater than during any comparable subsequent period. Presentation in the rhesus monkey often takes place without the animal being mounted.

It is likely to appear in the behaviour of excited females in a variety of situations and has different functions, e.g. as a greeting gesture, as an expression of affinity or as part of what is called here 'sign equilibration'; when combined with outwardly directed threat, it serves to redirect the aggression of a more dominant animal. Finally, it appears to diminish threat especially when combined with facial gestures directed towards a more dominant animal, and may in this way provoke mounting by the dominant animal in place of aggression. The mounting of one animal by another, whether or not preceded by presentation on the part of the subordinate animal, is a part of the social relationships between all ages and sexes alike. In the behaviour of the adults, the act is usually followed by about half a dozen quick, vigorous pelvic thrusts. In the young, the thrusts are less vigorous, and the behaviour less definite. . . . It is almost certain that the majority of the mating gestures exhibited so frequently between these monkeys serve at least one function which is non-reproductive and non-sexual (Chance 1956).

Altmann in his excellent paper (1962) gives examples of mounting in the rhesus monkey occurring as a 'greeting' or as an indicator of non-aggression, as an indicator of relative dominance, and as a component of play. In the chacma baboon, presentation is

sometimes, but by no means always, a gesture that the female makes prior to copulation by the male. It is also, at least when performed by a female, effective in checking actual aggression by the dominant male. In by far the greater number of cases, however, the act clearly has a simple social function that has no direct reference either to mating or aggression intention (Hall, 1962).

In consequence, the following statements would appear to be justified:

1. Presentation occurs typically in the sexual context;
2. Presentation is to be seen much more often than in any mammal in which it serves sexual functions only;
3. Presentation can be employed by the animal not only when sexually aroused but also independently in a number of other social situations.

From this it follows that presentation is governed by at least one other source of motivation apart from sexual motivation. 'All sexual behavior may roughly be divided into two types, that motivated mainly by the sexual drive, and that motivated mainly by the dominance drive' (Maslow, 1936). Only a detailed motivational analysis can demonstrate exactly what this 'dominance drive' is. It is certain that presentation is not simply motivated by flight, since the animal does not flee but remains on the spot or even approaches the source of the threat. Zuckerman (1932) suggests that presentation in social situations is all that is left of the animal's original reaction of flight from the environment of fear and discomfort. But instead of being a vestige of a former flight behaviour pattern, it seems more probable that presentation is at least partially motivated by aggression and/or some other tendency that hinders fleeing. That aggression is involved is strongly suggested by the fact that (1) in some species the females in oestrous are more aggressive than others and (2) that presentation may be elicited when the human observer enrages the animal, as described for *Macaca mulatta* by Mason, Green and Posepanko (1960) and for the mandrill by Winkelsträter (1960). Whether or not a ritualization of this gesture has taken place need not concern us here. It seems unlikely to be a displacement reaction, as Reynolds suggested (1961). In addition, we may provisionally assume that the motivation of presentation in the social context is the same for both males and females of all species. This working hypothesis is probably an over-simplification, but the lack of data renders more complex proposals worthless at the present time.

The same conclusions with respect to motivation are true of presentation of the coloured band in *Tropheus* and of the genitalia in the spotted hyaena, and they may be true of sexual and dominance mounting in primates as well. Copulatory courtship and 'amorous courtship' or greeting in cormorants share largely the same overt motor patterns, but their internal motivating factors are different (Kortlandt, 1959). The form of the different presentation behaviour patterns in each case – within the frame of its scale of intensities – is always the same, whatever the source of motivation. This is exactly the reason for applying the term 'sexual' behaviour (or courtship) in sexual and social situations. The uniformity of form is important, since the signal-receiver (at least at the outset of the evolution of these now separately-motivated gestures) should not differentiate, i.e. should respond sexually even to non-sexually motivated gestures. It is, in fact, the sexual response itself which in this social situation has positive selective consequences, because it is incompatible with overt aggression. The simplest working hypothesis is the assumption that the *response* of the partner has remained sexually unaltered. On the part of the *signal-sender*, social and sexual presentation (of hind-quarters, coloured band or genitalia) are as different from one another as bite-intention in hunger and in threat.

MALE SEXUAL ACTIVITIES

So far, we have mainly considered female sexual behaviour, which is of course primarily directed towards a male. Presentation by both sexes is, however, observed in extra-sexual social contexts in interaction with either sex. The animal towards which presentation is directed responds to this in the same manner as to sexual presentation but usually within a lower range of intensity. This is true of both males and females. Since presentation is characteristic of the lower-ranking animal, the recipient animal usually occupies a higher rank. The fact that the typically male responses of the latter are truly social dominance gestures is proven by the occurrence of these responses in the absence of elicitation by presentation. In the social context, male sexual actions imply dominance and female sexual actions submissiveness – both sexes are capable of exhibiting both types of behaviour in interaction with an individual of either sex. If a male rhesus monkey mounts another male, it is the more dominant individual that plays the masculine rôle. 'Attacks may be foiled when the object of attack suddenly presents and thus diverts the aggressor from attacking and stimulates him to mount instead' (Chance, 1942).

According to Reynolds (1961), dominance mounting in the rhesus monkey differs from sexual mounting: in sexual mounting, the male grips the hind-legs of the female with his hind-feet, whilst the hind-feet remain on the ground in dominance mounting. On the other hand, other authors report dominance mountings with intromission and it would seem likely that Reynolds is regarding differences in *intensity* as differences in *quality*.

Dominance mounting of males and females on lower-ranking group-members has been observed in:

Papio hamadryas	by Kummer, 1957
Papio ursinus	Hall and DeVore, 1965
Papio anubis	Hall and DeVore, 1965
Papio sphinx[1]	Schloeth, 1956–7
Macaca mulatta	Carpenter, 1942; Chance, 1956; Hinde and Rowell, 1962; Altmann, 1962
Macaca radiata	Nolte, 1955; Simonds, 1965
Macaca silenus	Inhelder, 1955
Macaca irus[1]	Schloeth, 1956–7
Cercopithecus sp.	Hamilton (cited in Blin and Favreau, 1964)
Presbytis entellus	Jay, 1965
Pan troglodytes	Bingham, 1928; Crawford, 1942a; Galt, 1947

However, according to Simonds (1965), the mounting of one monkey by another in *Macaca radiata* is not a certain indication of dominance; in chimpanzees presentation and mounting in non-competitive situations do not always indicate submissiveness and dominance respectively (Crawford, 1942b). In this case, as in the case of *Papio hamadryas*, it is likely that gestures that are typically those of subordinate may be used as tranquillizing gestures by a

[1] ♀-mounting not clearly indicated.

monkey that is obviously dominant to the other monkeys ('self-handicapping', according to Altmann, 1962). Dominance mounting is well known to occur in other mammals too, e.g. in the yellow-bellied marmot (*Marmora flaviventris;* Armitage, 1962) and in male wolves (Schenkel, 1948); mounting is further known for females in oestrous in a number of domestic animals and is sometimes, but not always, a further example of dominance-demonstration (Alba and Asdell, 1946, have studied this in cows in connection with its dependence upon oestrogens).

These extra-sexual social implications of the male copulatory posture suggest some relationship between sex and aggression in male primates (cf. the observations of Antonius, 1929–30, on castrated hamadryas males). This hypothesis is supported by the frequent occurrence of 'Wut-Kopulationen' (rage-induced copulation) in the males of a number of Old World monkeys. It is particularly easy to elicit in captive animals by showing food which is then temporarily withheld, and can also occur in shock situations and the like, e.g. by *Papio ursinus* (see below) in response to a stuffed serval. Such 'Wut-Kopulationen' or aggressive mountings with females, whether in oestrous or not, give a strong impression of redirected activities and have been described for:

Papio papio	by Winkelsträter, 1960
Papio ursinus	Hall, 1963a
Macaca mulatta	Carpenter, 1942; Hinde and Rowell, 1962
Cynomolgus fascicularis	Winkelsträter, 1960
Nemestrinus nemestrinus	Müller-Using, 1952
Erythrocebus patas	Hall, Boelkins and Goswell, 1965
Pan	Kirchshofer, 1962

In such situations, *Papio hamadryas* exhibits erection of the penis, masturbation and friction-movements *in vacuo* (Winkelsträter, 1960); instead of copulating, male chimpanzees may perform wild capers with penis erect. According to Miller and Banks (1962), erections and rage-induced copulations may be elicited in the rhesus monkey by retention of previously exhibited food, but not by competition for a seat providing safety from electric shocks. In this latter situation, the animals are probably predominantly governed by fear. The situations in which male sexual behaviour components occur allow of the interpretation that they are linked with aggressive motivation and are therefore readily available as dominance gestures.

In the squirrel monkey, *Saimiri sciureus* (a Platyrrhine), the males display the erect phallus under varying conditions of courtship, aggression and social greeting (Fig. 11). One variety at least shows this display towards its own mirror-image: it spreads one thigh and makes a number of forward thrusts with the tumescent phallus

Fig. 11 Genital display of a young *Saimiri* male (from Ploog)

(MacLean, 1964). If the display is made towards another monkey, the two animals may be either in contact or a few inches apart from each other. Within a group of squirrel monkeys, the alpha-animal signifies dominance by display of penile erection with thigh-spreading directed towards the face of another, lower-ranking, monkey. 'Genital display is a ritual, derived from sexual behaviour, but serving social and not reproductive purposes'; it is the most effective signal with respect to group hierarchy, but definitely separated from reproductive activity (Ploog, Blitz and Ploog, 1963). Lower-ranking group members show much less genital display.

The correlation between penile display and rank-position and its use as a form of greeting towards strange monkeys suggests that it may serve a function comparable to that of presentation in Old World monkeys. In the latter, the animals concerned emphasize their submissiveness by employing a gesture derived from female courtship behaviour. In *Saimiri*, the animals emphasize dominance by employing a gesture derived from the male copulatory pattern. Both types of group contain representatives of both sexes. In the Old World monkeys, the males also employ the female pattern in a social context. Likewise, the female *Saimiri* uses the 'Male' genital display in social contexts as a dominance gesture as well. During this display, the clitoris may be visibly enlarged (Ploog, 1963; Ploog, Blitz and Ploog, 1963). The *Saimiri* female has a relatively very large clitoris and, dorsal to its root, a pseudoscrotal enlargement; the whole vulva thus recalls the genitalia of the male (Hill, 1958). Many Platyrrhine monkeys show paradoxical similarities between the sexes, due chiefly to the great development of the clitoris and sometimes of pseudoscrota in the females; sometimes this makes it difficult to tell males and females apart in the field. It may well be that the conspicuous pseudoscrota and the extremely long clitoris of some species (*Ateles, Alouatta*) serve some related social function, though the long, pendulous clitoris in these species has no erectile tissue. It has been impossible until now to explain these strange morphological characters of the females, but it is justifiable to assume that an obvious character such as this must have some social function. Lack of sexual differentiation is certainly no explanation, since the pseudoscrotum is derived from the labia minora (Hill, 1958). Little is known of the social and sexual behaviour of these species, since they rarely breed in captivity. Carpenter (1935) observed males of *Ateles geoffroyi* manipulating the greatly developed clitoris of the female and describes this as secondary sexual behaviour. Sanderson (1949–50) regards the pseudoscrota of *Hapale* as an adhesive structure, a kind of suction cup, which functions as an aid to copulation whilst squatting on narrow branches. He postulates the same function for the independently evolved structures of similar form in the African lemur *Perodicticus*.

Cebus males show the same kind of penile display towards their mirror-image as *Saimiri* and some regularly show erections in 'greeting' a human being (Hediger and Zweifel, 1962). Erections in response to the mirror-image has also been described for *Erythrocebus patas* by Hall, Boelkins and Goswell (1965).

GENITAL DISPLAY, THREAT AND TERRITORY MARKING

The external genitalia of mammals also function as discharge organs for urine and this double function is further reflected in social behaviour. Many mammals employ urine for territory marking (Hediger, 1944), particularly these species with predominantly olfactory orientation. Although both sexes are capable of marking, it occurs more frequently and/or more pronouncedly in the male. The best known case is that of urine-marking in the dog, in which the typical male micturition pattern with elevation of one leg develops under the influence of the male sex hormone (Berg, 1944). Stags, reindeer and other ungulates at rutting-time often urinate in the slough or against their own abdomens (Hediger, 1944; Espmark, 1964); horses and male pigs spray urine in the neighbourhood of sexual objects. In all cases, urine-marking is distinct from normal micturition: in normal micturition the urine is expelled continuously and uniformly; in marking, urine is sprayed briefly and sporadically, in fact more-or-less ejaculated. The correspondence in pattern suggests that the ejaculation mechanism is partially involved. Urine can likewise be expelled in fright, but as a rule in the form of droplets expelled in a trickle, this probably being the origin of trail-marking in an unknown environment.

It is highly probable that scent-marking originally served for orientation of the individual, as in the rodents; the animals leave trails and mark specific points of their daily pathways. Such marked tracks can even aid foreign conspecifics in the navigation of previously unknown areas (Eibl-Eibesfeldt, 1958). Among the primates, urine-marking is known for *Galago, Loris, Nycticebus, Cebus, Saimiri, Alouatta* (Hill, 1938; Eibl-Eibesfeldt, 1953; Ilse, 1955; Kirchshofer, 1963). These species can either expel urine drop-wise in rhythmic fashion on to the substrate whilst running (*Loris, Nycticebus, Alouatta*) or take up urine on the plantar surfaces of the hands and feet and then deposit it on the substrate, or occasionally on the coat (*Galago, Loris, Saimiri, Cebus, Alouatta*). Such marking of tracks, is, however, confined to nocturnal

Prosimians. *Cebus, Alouatta* and *Saimiri* mark specific sites, but it is questionable whether this serves only for individual orientation. Apart from urine, marking can also be carried out with the faeces, and in both cases specialized scent-glands in the anal and genital regions may be involved. Among the Prosimians, *Lemur catta* exhibits particularly obvious marking (Fiedler, 1957; Bolwig, 1960), and this case is exceptional in that the females mark more than the males. The females press the anogenital region and the long clitoris against the substrate. In *Lemur mongoz*, the males often expel a small quantity of urine during scent-marking (Bolwig, 1960).

Scent-marking (probably secondarily) is more frequently directed towards the demarcation of territories between con-specifics than for individual orientation and in this context occurs mainly at rutting-time. In this case, the urine-marking is quite often applied to objects connected with the sexual interest of the male. Some ungulates spray urine on the faeces or over the urine of females in heat. According to Schmidt (1956), male dogs in court-ship direct the urine spray towards the female. Quite often the female herself is marked with urine. In *Lemur catta* and *L. mongoz* the males mark the females in addition to other objects (Bolwig, 1960). Schloeth (1956–7) postulates a general social 'demonstration value' for the marking of inanimate objects by Prosimians. In a number of lagomorphs and rodents epuresis or enurination (that is emitting a jet of urine at a conspecific) has been described (Southern, 1948; Kirchshofer, 1960; Dathe, 1963; Kunkel and Kunkel, 1964). This was mainly observed in males, but in the rabbit, the guinea-pig and the mara (*Dolichotis pata-gonum*), this occurs in both sexes. There is a difference in the orientation of the motor patterns concerned: the rabbit and the guinea-pig aim backwards, whilst the acouchi (*Myoprocta pratti*) and the Canadian porcupine (*Erethizon dorsatus*) aim forwards. In the mara, males and juvenile females aim forwards and adult females backwards (Kirchshofer, 1960). In some species, the retracted penis is directed backwards and must be at least partially erected for aiming forwards. The enurinating acouchi is typically situated behind the female, so that the powerful jet of urine shoots over her head (Morris, 1962).

The squirting of urine at the partner usually occurs in a sexual context, but in the agouti, guinea-pig, mara, and probably also in the porcupine, this behaviour may appear in other contexts as well. According to Kunkel, this phenomenon is underlain by weak or inhibited aggression. The same would seem to be true of the mara (see Kirchshofer) and of *Erethizon* (Shadle, Smelzer and Metz, 1946). In the latter species, a distinct connection with the sexual patterns is indicated, since the penis of the enurinating male is as fully erected as that of a male with the penis inserted in the vagina of the female (Dathe, 1963). In all species, there is a distinct relationship between such behaviour patterns and sexual and agonistic behaviour. Fully developed male dogs in fear situations exhibit revived puppy-urinating (Berg, 1944), which in some species (e.g. *Lycaon pictus*) seems to act as a submissive gesture in adults (Kühme, 1965); the typical tree-marking of dogs is an expression of 'self-confidence'. The 'Urinwaschen' (urine-washing) of *Saimiri*, *Cebus* and *Alouatta* occurs in association with anger, strong sexual arousal and also in fright (Kirchshofer, 1963). *Saimiri* may expel a few spurts of urine when confronted with its own mirror-image and responding with penis-demonstration as described above (MacLean, 1964).

It is possible to see from this survey that excitation of the urogenital region also occurs in specific situations separate from sexual behaviour, where rank-order or territory play a part. Micturition has an obvious function for nocturnal animals and/or animals with predominant olfactory orientation. But even in the rodents, there are cases where the male employs the external genitalia as optical signals. The testes of the threat-displaying guinea-pig descend from the abdominal cavity exactly as in courtship and come to occupy the very conspicuous scrotum. The testes are then displayed to the rival, especially in threat, by means of a contorted body-posture. In copulation, on the other hand, the testes remain in the abdominal cavity (Kunkel and Kunkel, 1960). Expert opinion is still divided over the significance of the descent of the testes, which takes place at quite different times in different primates (Schultz, 1951–2). However, Ottow (1955) has demonstrated the improbability of the temperature-dependence theory, according to which the testes avoid an unfavourably high temperature in the abdominal cavity. In some animals, active *ascent* of the testes into the inguinal canal and sometimes even into the abdominal cavity occurs. In addition, there is a relationship with social behaviour, since ascent often occurs in frightened animals and, conversely, descent takes place in situations evoking 'self-confidence' (Back-house, 1959). Sade (1964) describes a relationship between the colour of the sexual skin and the size of the testes in the male *Macaca mulatta*; when closely approached by a dominant monkey, the testicles of an adult male may even retract temporarily into the inguinal canal and the scrotum then appears as an empty fold of skin (Altmann, 1962). Mortally frightened human males also show remarkably large reductions in size and weight of the testes within short spaces of time (Stieve, 1952).

In spite of this obvious relationship between social dominance and behaviour of the male genitalia, it is still not apparent whether the scrotum and the descent of the testes together have any significance in communication, although it is a rather unlikely assumption that there is no significant function where the scrotum is very conspicuously coloured. It is a necessary assumption that animals with such optical acuity as the higher primates must learn to appreciate the significance of these situation-dependant changes in the genitalia in the same way as the human observer, even if these changes occur simply as physiological by-products.

When the close connection between ownership of territory, territorial demarcation, social dominance and the male genetalia is considered, it is logical to expect that the predominantly optically-oriented higher primates would emphasize the optical features of the male genitalia. The fact is that the most obvious examples of cutaneous coloration in the mammals are indeed provided by the genitalia of the Old World monkeys. No previous explanation of this phenomena is to be found, but an evaluation of the well-known field studies to date readily offers a possible explanation.

In the African bush-zone, a number of males from a group of *Papio* (*anubis* or *doguera*) can regularly be seen sitting on mounds of earth, tree-stumps and in the forks of branches. These males are conspicuously situated and gaze around whilst the remaining members of the group are eating, often turning their backs to the group. They sit with thighs spread and the penis hangs extended but usually not actually erected (Fig. 12c). In *Papio anubis* and *P. doguera*, the penis shows up bright pink against the dark fur. (This sitting posture is known to the author from observations of captive *Papio hamadryas* and was even depicted by the Ancient Egyptians as early as the 18th Dynasty, Fig. 13.) The same sitting posture can be seen in males of *Cercopithecus aethiops*. Semi-tame

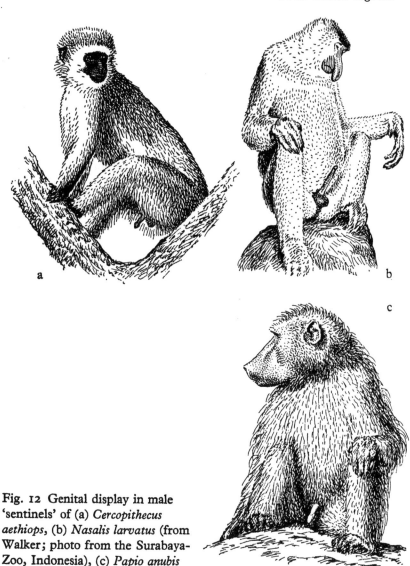

Fig. 12 Genital display in male 'sentinels' of (a) *Cercopithecus aethiops*, (b) *Nasalis larvatus* (from Walker; photo from the Surabaya-Zoo, Indonesia), (c) *Papio anubis*

or tame males simply sitting quietly in familiar territory usually sit with spread thighs, but with the penis retracted (Plate II). Wild males keeping watch beside the group sit in exactly the same posture, facing away from the group members, but the penis is usually erected (Fig. 12a) and may be repeatedly but irregularly

struck against the stomach with a jerky action. This occurs sometimes in situations of high arousal (approach of a human observer;
approach of other long-tailed monkeys). These distinctive movements serve to emphasize the already conspicuous coloration of
the male genitalia.

These males, which sit in a somewhat remote position, have been
referred to as sentinels, guards, leaders, etc. However, Hall (1960)
summarized the previously-assembled data and postulations on
this subject and rejected all of these terms, since they suggest
interpretations which up to the present time are not warranted by
the facts. Hall goes on to provide his own, more accurate observations on the vigilance behaviour of *Papio ursinus*. In fact, these

Fig. 13 *Papio hamadryas*, XVIII Dynasty
bas relief (from Huard, *Art mobilier de
l'Egypte ancienne*)

supposed sentinels do not give noticeably frequent warning against
enemies or approaching human beings, but *immediate* warning is
given on the approach of another troop of *Papio ursinus*. Apart
from this, Hall, Boelkins and Goswell (1965) also report that adult
male patas monkeys keep watch to prevent the approach of foreign
conspecifics to their social group. My own observations showed
that baboons and vervets never warned of my approach, but
withdrew in conspicuous silence when I approached too closely.

The sum of the evidence allows of the more probable explanation that these 'look-out posts' function as optical markers of the
presence of the group or of its territorial boundaries, largely as a
warning to conspecifics. These higher primates have retained the
basis of the response, which served for urine-marking of territory
in the lower primates, but actual urine-discharge is lost and

replaced by optical conspicuousness of the genitals in accordance with predominant optical orientation. (Even in *Saimiri*, urine-discharge is of low intensity and therefore presumably of no significance.) In addition, the morphological features of the genital signals vary widely from species to species, as is common for other intraspecific signals. The behaviour pattern 'Zur-Schau-Stellen' (i.e. the display action itself), on the other hand, is markedly uniform throughout and the relationship with social dominance and aggressive behaviour also shows general persistence.

Red is a very conspicuous colour commonly found in the pigmentation of the penis in Old World monkeys, and this is usually contrasted with blue cutaneous coloration of the scrotum. Hill (1955) has shown with injection experiments that the red and blue cutaneous coloration in several Catarrhine monkeys is under the influence of haemoglobin circulating in the cutaneous capillaries, and this renders the dependence of colour-intensity upon the vegetative arousal-state of the individual animal observed in some cases comprehensible. There is only one more striking case of cutaneous coloration among the primates, which is represented by the face of the male mandrill (*Papio sphinx*), but the increase in effect is entirely due to the larger surface areas concerned; the colours themselves and their arrangement are very similar to the patterns described for the genital region of the long-tailed monkeys: dorsum nasi scarlet, commencing abruptly below nasion and extending rostrally to apex nasi, thence downwards to within 1 cm of labial margin; paranasal eminences with parallel longitudinal sulci bright cobalt blue (Hill, 1955). The colour and pattern of the mandrill's face therefore show a striking degree of similarity with those of the genital region of *Cercopithecus* and of the mandrill itself, although in the latter case the genital colours are paler than the facial colours. The resemblance is borne out by the following comparison:

Length of nose: scarlet	Penis: scarlet
Paranasal areas: bright blue	Scrotal area: pink with bluish suffusion
Chin area: orange-yellow	Tuft of hairs below scrotum: orange-yellow

I have never observed the mandrill to display the genitalia in the manner described above for *Cercopithecus*. The conspicuous face-pattern is, however, automatically displayed in the direction in which the animal is looking. Since the animal looks very attentively in a particular direction when its suspicion is aroused, the conspicuous colour pattern is automatically displayed towards a possible opponent or rival. It should prove easier to test the demoralizing effect of this display using experiments with models than would be the case if attempting to test Darwin's assumption (1876) that these colours arose primarily through sexual selection. The view expressed in this paper also provides a possible explanation for a number of minor details, such as the peculiar bare spot on the lower vent of the female *Theropithecus gelada*. In the sitting animal, this spot corresponds in position to the male genitalia, which are displayed as in baboons. Together with the large red cutaneous area on the ventral thorax, the colour pattern of the sitting female is thereby rendered very similar to that of the sitting male.

DISCUSSION

It is well known that the brightest skin colours among mammals occur in some male Old World monkeys. These coloured areas are situated in the ano-genital region and in the face of the mandrill. If we seek the biological significance of these brilliant colours, this of course presupposes the assumption that they do indeed have a biological significance. If so, this will very probably be at least in part a signal function. It might well be difficult to prove this deduction, but it would surely be far more difficult to disprove it. The only advantage in assuming instead from the very start that these colours have no biologically relevant signal function would be to avoid further research. If facing up to the difficulties of proving or disproving the proposed deduction, its value may be estimated by testing how many details it will be able to explain to us.

A valuable method for finding out the 'meaning' of a suggested signal-character has been developed during our behavioural studies in different animals; it may be called the 'search for the model'. It is based on the fact that far more 'new' signals than hitherto realized began their evolution as imitations of signal-characters already responded to in a definite way by the animal. This response often changes its function merely if the releasing sign-stimulus (to which it remains coupled) is transmitted in a new situation or context.

With respect to the conspicuous hind-quarters of some male monkeys, the possible model is easy to find: it is represented by the female's hind-quarters, which are known to act as an oestrous signal by undergoing a conspicuous colour-change or cyclic swelling. The significance of this signal is approximately known. It is displayed by the female with a certain gesture, that of 'presentation'. This gesture is known to exist in males as well. Since a special explanation would be required if one wished to postulate that this colour pattern acts as a signal in females and not in males, it is simpler to assume that the male colours act in the same way.

In baboons, the sexual skin swelling of the female is merely an enlargement of the anogenital area, very variable in form. It differs from species to species but is even understood interspecifically (e.g. in macaques, Tinklepaugh, 1928). It therefore seems reasonable that the imitation of the male hamadryas need not be very accurate in details (especially in the third dimension).

In other cases, however, the correspondence between male and female hind-quarter markings is obvious even down to minor details (Vervet, Procolobus). The fact that the conspicuous structure and coloration of the male hind-quarters is arrived at in a different way from that of the female very strongly suggests that both the model and the mimic-pattern act as signals and that the similarity between male and female is not simply a character common to both sexes, i.e. not a mere absence of a sex-distinguishing feature. The proposed explanation that regular hormonal disturbances cause the close similarity of males to females (in monkeys and hyaenas) does not touch upon the problem of their possible signal value.

It was then further suggested, as a working hypothesis, that conspicuous ornamentations of the male monkey's hind-quarters may generally be interpreted as imitations of (or at least as showing close similarities to) the respective female's oestrous markings. This led to the discovery of the female vervet's oestrous markings; but in other species, such as the patas monkey, there is no female equivalent to the male's conspicuous genital coloration, according to the literature. It is, of course, possible to assume that in this case an historical carry-over of the male colour-signals has outlived the retention of the female signals in phylogeny. It may also be that the pure imitation-hypothesis is too simple.

These sexual signals also have social significance, presentation acting as a greeting ceremony. In the female the sexual skin activity primarily underlines the sexual presentation. In males, however, the corresponding ornamentation underlines the social presentation. Though, in agreement with the hypothesis, the female vervet monkey indeed has oestrous markings resembling the conspicuous male anogenital colours, presenting in these animals seems nevertheless to be a rather rare event. It must, however, be pointed out that these conspicuous colour markings are displayed without special gestures when the animal simply moves on all fours, carrying the tail above the horizontal, as is usually the case.

Moreover, the colour pattern of the male genitalia is often displayed from the sitting position, this probably being derived from an ancient urine-marking pattern, functioning as a component of aggressive and territorial behaviour. In this case, the significance of the colour-pattern – possibly derived from that of the female – may be superimposed on the already available sitting posture. Much more detailed studies on this are needed.

The general concept is that the male's conspicuous hindquarters have, by a process of adaptation, become closely similar to the sexual skin of the female in order to elicit in a conspecific a reaction originally directed only towards a female. The signal-receiver is, therefore, in a certain sense deceived. One might argue that such a keen-sighted animal would appear to be too intelligent or too observant to be misled about the true sex of the signal-sender by such a signal. However, neither keen-sightedness nor intelligence is an argument against the fact (see *The Hidden Persuaders* by V. Packard) that an article bearing only the photographic dummy of an attractive girl on the cover is paid for in preference even by human beings. Reynolds (1961) argues: 'It seems unlikely that the sight of a small male monkey presenting would be a sufficient stimulus to change the motivation of an adult pursuer from aggression to sex'; but no total change of motivation is needed, the signal will have achieved its effect as soon as it reduces the probability of an attack.

There are some possible pathways by which this presumably deceptive signal may operate, and these pathways are not mutually exclusive; in fact, if combined, they would heighten the selective advantage of such a signal. This signal might be of intra- as well as of inter-troop significance, e.g. by furthering attractiveness to troop-members and repulsion to foreign conspecifics. According to the most widely accepted interpretation, the immediate assumption would be that the signal acts up the rank-order and reduces aggression. But there seems also to be an effect acting down the rank-order, which reduces the tendency to flee and which might be brought about by balancing attractive or 'encouraging' female characters against fear-arousing male characters. This would fit in with the fact that the largest and highest-ranking baboon or rhesus males possess particularly obvious development of the female signal (in contrast to *Colobus* males). This, then, would fall into Altman's category of 'self-handicapping'. Again, we can only enumerate the possibilities; much more information is needed and a comparative study of societies of the different species (e.g. baboon versus *Colobus*) should bring interesting facts to light.

As has been mentioned earlier, the signal-sending has been re-motivated although the signal itself has retained its original form. The signal still elicits a male sexual response, however. Here the male sexual behaviour (present in excess) is exploited. Male sexual behaviour in excess is also known for a number of other animals and especially well known in some tropical teleosts such as viviparous tooth-carps (Swordtail *Xiphophorus*, Guppy *Lebistes*) and some Glandulocaudine characids with internal fertilization. Nelson (1964) has discussed the significance of the fact that these males spend much time throughout the year engaged in elaborate courtship activities, very rarely leading to true copulations. He suggests this to be a result, at least in part, of the temporal dissociation between mating and spawning in the female, which favours a continuous presence of both mature sperm and readiness to mate in the males. There is at least the males' continuous sexual interest in females which, as a by-product, keeps the two sexes together. This may, however, achieve a selective advantage of its own; the excess male sexual behaviour or sexual interest can be exploited in evolution for social cohesion. The males' continuous readiness to mate seems to have originated in the tropics and the association of both sexes seems to have been necessary to assure fertilization. In fact, most of the monkeys mentioned, as well as *Tropheus* and *Crocuta*, breed throughout the year. But in some monkeys the males' readiness to mate has been retained throughout the year, although the females show a clear-cut breeding season. In this case, the maintenance of the sexual bond seems to be the most important function and the continuous presence of mature sperm (which may be wasted in unreproductive ejaculations) seems to be the price which must be paid for continuance on the evolutionary pathway towards sociality.

SOME REMARKS BEARING ON HUMAN SOCIAL BEHAVIOUR

The importance of primate studies in anthropology has very often been emphasized (Hooton, 1954; Grimm, 1963), and I propose to mention a few possible parallels between human social behaviour and that of the monkeys discussed above, not in order to make definite statements but rather to direct attention into a direction which up to now has escaped investigation.

Man is a social being. He lives in groups whose members are individually acquainted with each other. He behaves differently towards group members as compared to foreigners. Man's aggressiveness towards conspecifics is well known and is today not well compensated (Lorenz, 1963). Furthermore, it is generally accepted that the family. is the basic institution of human societies (see Spiro, 1954). It has been shown that some non-human prim-ates, a cichlid fish species and the spotted hyaena have developed the particular use of sexual gestures for social communication as an adaptation to the same basic social traits; these gestures are now shown with much greater frequency than in less social species and the motivation is at least partially changed. If we confine ourselves to the problems mentioned above, and if we assume that the human societies are comparable to those of the sub-human vertebrates, one may ask whether in humans, too, there are sexual signals comparable to those of the non-human primates, and whether sexual behaviour patterns might serve another, social function; whether they occur more often than would be expected according to their primary function and whether they might have changed in motivation. Of course, a motivational analysis in the ethological sense has not been carried out in humans and will prove to be extremely difficult. We must therefore make the best of the other indications we have.

Oestrous signals, consisting of a change in colour and/or form of the immediate periphery of the vulva, are to be seen in the females of some insectivores and from there onwards in the Lemuriformes and Tarsiiformes through the Cercopithecoidea up to Hominoidea. Usually, it is mainly the labia minora which are involved, the labia majora being vestigial or absent in Old World monkeys. This oestrous signal may be subdivided into two components, which may occur singly or combined: the colour change and the swelling, both of which affect separate, overlapping or identical areas. The enlargement of the labia minora in sexually receptive females is still to be seen among humans, in the Koisanids, where it is known as Prolixitas labiorum minorum (Fig. 14) ('Makronymphie' or 'Hottentot's skirt'). This is a perfectly natural phenomenon, originating with the onset of puberty (Pöch, 1911; Schultze, 1928; Gusinde, 1954). In the non-human primates, the homologous enlargements usually wax and wane with the sexual

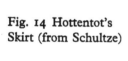
Fig. 14 Hottentot's Skirt (from Schultze)

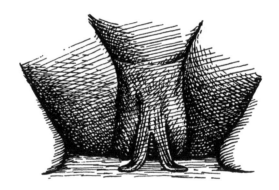

cycle, although there are exceptions where the conspicuous stage is dependent on additional factors or may even be independent of the sexual cycle. The Koisanid's tablier, which is not merely a result of swelling, never disappears entirely. It does, however, show a marked change of colour from the normal pale rose to purple-red during sexual excitement (Fischer, 1955). Nothing is known about the dependence of this character on hormones in human beings. It may be added, however, that in normal human females large doses of androgen induce swelling and vascular congestion of the clitoris and labia majora (Bishop, 1953).

That 'Hottentot's Skirt' indeed acts as a signal to conspecifics is proven by the fact that in other peoples, in which an enlargement of the labia minora does not naturally occur, the same effect is convergently brought about by artificial manipulations, e.g. in the Batetela (Torday and Joyce, 1922), the Basutos and the people of

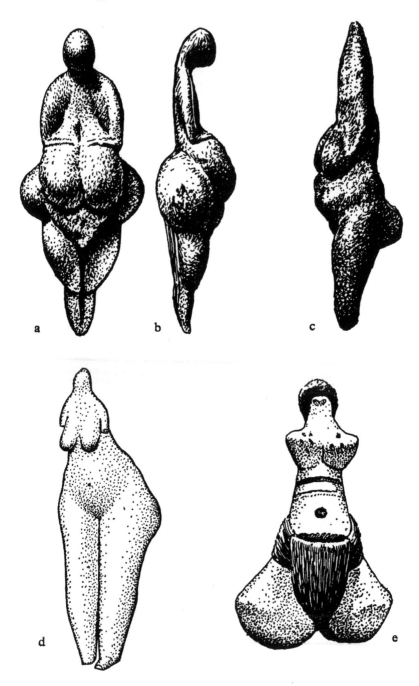

a   b   c

d   e

Fig. 15 (a, b) Stone-age figurine from Laspugue, Haute-Garonne
(c) Stone-age figurine from Savignano, Italy
(d) modern Kaffa figurine (Africa)
(e) modern Caraja-Indian figurine (Brazil)
(f) Hottenton female from Bersaba (from Schultze)
(g) Bustle dress (1882)

Ponapé (Reitzenstein, 1931), and in some Bantu tribes (Hokororo, 1960).

The sexual swelling in non-human Primates may affect different parts of the female's hind-quarters. This suggests the interpretation that the conspicuous steatopygy, known particularly well in the Koisanids, represents the same type of socio-sexual signal, though it is here brought about by subcutaneous fat deposition. Local depositions of subcutaneous fat in human beings act as sign-stimuli in other cases, too, e.g. in the female breasts and – as an important part of the 'Kindchenschema' – in the Corpus adiposum buccae of the child. That the prominent buttocks actually act as a social signal and are preferred by human males is, again, proven by its artificial imitation in fashions through the ages and by the fact that this feature is always emphasized in female figurines dating back to early man (see Fig. 15). Arguments derived from historical art studies and from psychiatry (Schlosser, 1952; Leonhard, 1964) also show that the most important sexual sign stimuli in the human female are the breasts (which are omitted from the present discussion) and the hind-quarters. An important additional stimulus is provided by movement (as can be seen from Baumann, 1930). High-heeled shoes cause women to walk in such a way that just this movement is emphasized. Apart from this, high-heels (because of the alteration of the centre of gravity of the body relative to the supporting surface) give rise to a posture in which the pelvis between the femoral caps is curved backwards and the lumbar vertebrae form a backwards-directed hollow, simultaneously giving rise to a compensatory forwards thrust of the thoracic vertebrae (quoted from Veit-Stoeckel in Saller, 1964). High-heeled shoes as a means of displaying three important socio-sexual sign-stimuli (movement, hind-quarters and breasts) in the human female might therefore be expected to be favoured by cultural selection.

Presentation of the hind-quarters in the social context is still present as a behavioural feature even in human beings, although usually overlain by cultural taboos. However, such presentation may be compelled by high-ranking group members for the purpose of punishment. If the victim is forced to unclothe in such a situation, this has a far more demoralizing effect than the actual beating which may be administered (see Hävernick, 1964). In former times, social presentation was of more common occurrence, even as a part of European cultural tradition, and has been linguistically symbolized as the well-known 'kiss my back-side'. This latter expression, because of the similarities between the eliciting situations, was used in the earliest descriptions of monkey presentation as 're-directed' interpretation (e.g. by Konrad Gessner).

The human penis is under the control of the same physiological mechanisms as those discussed above. Erections outside of the sexual context may occur in aggressive or threat situations, even 'the nursing babe will angrily fight the breast if no milk is forthcoming and at the same time develop penile erection' (MacLean, 1962). In some 'primitive' people, the penis is conspicuously ornamented in Africa as well as in South America and in Melanesia (Plate II). The human male genitalia are displayed in sitting postures as already described for monkeys (Fig. 16), the former being different from the definitive feminine postures (Hewes, 1957). As is well known, the erect penis may symbolize dominance (power,

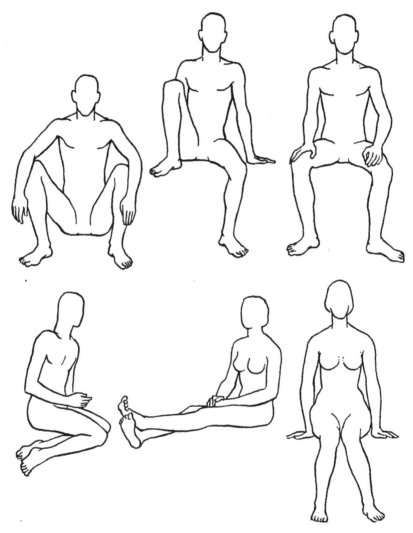

Fig. 16 Human male and female sitting postures (from Hewes)

'Herrschaft'), and there exists a quite extensive phallus symbolism which, according to some authorities, even includes sceptre, mace, etc. The Egyptian King of the Gods, Amûn-Rê, is depicted with

ig. 17 Pharaoh
lessed by a deity
uxor-Temple, Egypt;
om Eibl-Eibesfeldt)

a very large erect phallus (e.g. in the huge temple at Karnak) (Fig. 17). Lastly, a special type of sculpture, the so-called ithyphallic herme ('Herme') has been known since ancient times. These sculptures were discussed by chroniclers as early as Herodotus, who was of the opinion that this form of sculpture had been

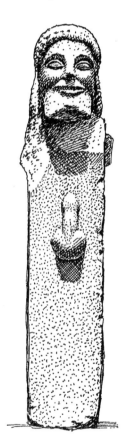

Fig. 18 Hermes of Siphnos, 490–80 BC
Athens, National Museum

carried over from far earlier times. These sculptures (Fig. 18) consist of a square stone pillar surmounted by a man's bearded head and bearing, in front, male genitalia with erect phallus. These sculptures have been acknowledged as a specific class by archaeologists (Lullies, 1931), but their significance has remained a mystery. Their usage is, however, documented: they were used as border-stones or, more commonly, as 'house-guards', which as late as Greek and Roman times stood just outside the doors of houses and shrines and (an important factor, I think) always facing away from the guarded premises – just as in the sentinels or 'lookouts' in Old World monkeys in the above descriptions (p. 118). The same type of sculpture is to be found in present cultures, e.g. on Borneo or Nias (Wickler, 1966) as well as on Bali and the Nikobar Islands as fetishes against the spirits of the Dead (Eibl-Eibesfeldt, p.c.).

Since in man the sexual propensities of the female are not dependent, as in subhuman Primates, on fluctuating anatomical or behavioural features, her attractiveness for the male originates from the socio-sexual status of her behaviour. This represents the culmination of the process of the modification of sexual gestures and postures into overt forms of social communication, establishing socio-sexual status, which has originated in the primates (Chance and Mead, 1953).

In subhuman primates this 'leads to the modification of sexual gestures, and even copulation itself, into a system of socially specialized forms of overt behaviour' (Chance and Mead); in human beings the pelvic thrusts of the copulation pattern (here present in both sexes) have become emancipated and are displayed in a more or less ritualized form during certain social dances without overt sexual function (e.g. Baumann, 1930, modern 'twist', etc.).

A number of other gestures exists in monkeys and apes, involving either sexual functions or direction towards erogenous zones of the body occurring in social (often greeting) situations as well. Chimpanzees, in greeting, may touch the genital area or the scrotum of another with the finger or the hand; to touch the lips or the scrotum serves as appeasement behaviour (Goodall, 1965), often confirming the social status of the greeted animal. This gesture seems to exist in human beings, too, as e.g. in Genesis 24.2, where Abraham says to his eldest servant: 'Put, I pray thee, thy hand under my thigh. And I will make thee swear . . . that thou shall not take a wife unto my son of the daughters of the Canaanites.' (Thigh here is used as a euphemism for the genitals; see MacLean, 1962.)

It is well known that sexual behaviour, whether in the narrower or in the broader context, occurs much more frequently in human beings than would be necessary purely for the provision of further generations. The so-called hypersexualization effect usually becomes apparent where a number of human 'troops' are forced to live in close contact. There is, therefore, good indirect evidence that elements of sexual behaviour in humans (as in non-human vertebrates) not only play an important rôle in personality-formation (Spitz, 1962) but also serve extra-sexual social functions. This would then indicate that a number of activities, gestures and signal-displays, usually referred to as 'sexual', might have become emancipated and attached to non-sexual motivation, in order to serve the function of establishing and maintaining good relations between the individual members of the community, possibly by compensating for their aggressive (or agonistic) tendencies. These gestures and activities may be expected to be very similar to their respective, specifically sexually motivated, actions (for reasons discussed earlier). If this could be shown to be true, then placing these gestures under the same moral taboo as the true sexual basic activities would at least add some new obstacles to the maintenance of the society and social relations. That basic reproductive behaviour elements in human beings in fact serve social functions and have become emancipated is easier to show in the brood-care behaviour which is free from taboo. Caring for the offspring is a relatively extended process in Primates. Rhesus females exhibit parental behaviour for three years and the sons of highest-ranking females may thereby attain high rank, even to the point of becoming male No. 2 in the group without once engaging in a serious fight (Koford, 1963). Parental behaviour is shown by both sexes and is sometimes very pronounced in adult males (Itani, 1959; DeVore, 1963). Such behaviour is not confined to the actual parent-offspring relation, but may include other troop-members. Protection of the young by aunts is something which can always occur in the rhesus monkey (Rowell, Hinde and Spencer-Booth, 1964), and in the langurs the newborn infant is held and carried by several other females, whose interest decreases with increasing age of the infant for the first days of its life; male langurs are indifferent to the infants (Jay, 1963). The invitation of a female rhesus or baboon to the young to jump on her back, in order to be carried away, very much resembles the invitation for mounting towards a male (mentioned by Carpenter, 1942; Reynolds, 1961; Kummer, 1957). In captive hamadryas baboons, infantile behaviour is prolonged into the subadult stage (Kummer, 1965); this may even occur in captive foxes (Tembrock, 1964). There exists a wealth of data showing that behaviour patterns of the parent-offspring relations may be incorporated into other social relations; this is well known in birds, where the begging of the nestling may function as a gesture with which the male attracts the female to the nest (e.g. in doves: Nicolai, 1959) or, more commonly, as a female gesture soliciting copulation and may then be indistinguishable for the responding partner from true food-begging (see Sibley, 1952). In monkeys and apes, grooming and lip-smacking are primarily brood-care behaviour patterns, but both are also very important gestures of greeting and pacification among adults (Jay, 1963; DeVore, 1963; Goodall, 1965). The motivational aspect has not in these examples been subjected to further research. Basic patterns of human parental behaviour are represented e.g. by caressing, embracing, tender conversation (from high- to low-pitched), leading hand-in-hand, lip-smacking and kissing (originally derived from mouth-to-mouth feeding: Ploog, 1954). All of these behaviour patterns are also present in non-human primates. It is probably a commonplace to point out that these interactions are to be seen both between adult human beings (e.g. in courtship) and even inversely between children and grandmother, in which case it is clearly not brood-care which is motivated. Nevertheless, these 'displaced' brood-care responses serve an important function in stabilizing the society and the pair-bond between adults. These responses, in their derived state, are still under the influence of their primary eliciting sign-stimuli for human parental responses (Lorenz, 1943); the human female remaining from puberty onwards more representative of the juvenile form than the male (Jürgens, 1960). In this way, a more babyish-looking girl may be preferred by courting males or even by bosses looking for a new secretary. Because mongoloids are even more paedomorphic, this sometimes influences the choice of partner so that European men sometimes prefer Japanese women, etc.

Human parental reactions to particular sign-stimuli of the baby appear to be innate (Lorenz, 1943; Spindler, 1961); the same also seems to be true for the solicitation of the main sign-stimuli by the human female (Lorenz, 1943; Leonhard, 1964). However, many additional factors are learnt, and these acquired factors can vary greatly between different human populations (Mead, 1949; Gajdusek, 1963). What can be said is that a constant readiness to react to both primary and derived signals in social situations is a necessary precondition to the effectiveness of these signals. But if these signals have an at least partially non-biological basis and must be learned, then artificial signals will automatically achieve socio-sexual significance, as is the case with the many and varied fashions of primitive and civilized human populations. Our clothing, whether as a by-product or as direct result of the reason for its existence, does conceal a number of primary, socially valid biological signals. This also hinders any adaptive process involving these signals and renders even the slightest display of them more conspicuous.

Living in the modern world, clothed and muffled, forced to convey our sense of our bodies in terms of remote symbols like walking-sticks and umbrellas and hand-bags, it is easy to lose sight of the immediacy of the human body plan. But when one lives among primitive peoples, where women wear only a pair of little grass aprons . . . and men wear only a very lightly fastened G-string of beaten bark . . . and small babies wear nothing at all, the basic communications between infant, child, and adult that are conducted between bodies become very real (Mead, 1949).

The comparison with animal societies may help us to understand how and why the derived functions of such behavioural elements and signals have been achieved and what may be the social consequences of encouraging or discouraging their display or of ignoring the motivational state (in the ethological sense) of signal sender and receiver.

SUMMARY

In this paper, it is assumed that:

1. In restricted groups of individually acquainted members, a definite need exists for some kind of appeasement behaviour to counterbalance intra-group agonistic tendencies.
The intention is to show that:
2. Sexual signals *do* serve additional social functions.
3. In order to serve this function, they will keep their original form but change (at least in part) in motivation on the part of the sender.
4. The main source for counterbalancing intra-group agonistic tendencies are behaviour patterns and/or signals from the sexual and brood-care context.
5. The socio-sexual link is made available to all species in

Evolution in combating courtship and parental behaviour with overt aggression, because of the necessity for each species to reproduce itself.

6. This explains the social significance of both female and male sexual organs and gestures in widely different species.

It is suggested that:

7. The social signal-function of male or female sexual characters favours the evolution of corresponding dummy-characters in the opposite sex;

8. Some of the basic requirements of sociality and some elements of social behaviour in man are the same as in non-human primates.

9. There is a continuous evolutionary line from lower mammals to man, leading in the females from oestrous swellings to the Koisanids' Prolixitas labiorum minorum and to true as well as imitated steatopygy, and in males from the laying of urine-marked tracks, through general urine-marking, territory marking, threatening and dominance demonstration to various forms of phallus-symbolism.

The possibility is discussed that:

10. Derived sexual and parental behaviour patterns may be incorporated into the normal family and social life of humans. A possible mechanism is suggested.

REFERENCES

ALBA, J. DE and S. A. ASDELL (1948). 'Estrous behavior and hormones in the cow', *J. Comp. Psychol.* **39**, pp. 119–23.

ALTMANN, S. A. (1962). 'Social behavior of anthropoid primates: analysis of recent concepts', *Roots of behavior*, ed. E. L. Bliss, New York.

—— (1962). 'A field study of the sociobiology of rhesus monkeys, (*Macaca mulatta*)', *Ann. N.Y. Acad. Sci.* **102**, pp. 338–435.

ANTONIUS, O. (1929–30). 'Einige Beobachtungen an Affenkastraten in Schönbrunn', *D. Zool. Garten* (NF) **2**, pp. 134–5.

APPELMAN, F. J. (1953). 'Über Theropithecus gelada Rüpp.', *D. Zool. Garten* (NF) **20**, pp. 95–8.

—— (1957). 'Noch einige Worte über Theropithecus gelada Rüpp.', *D. Zool. Garten* (NF) **23**, pp. 246–7.

ARMITAGE, K. B. (1962). 'Social behaviour of a colony of the yellow-bellied marmot (*Marmota flaviventris*)', *J. Anim. Behav.* **10**, pp. 319–31.

BACKHOUSE, K. M. (1959). 'Testicular descent and ascent in mammals', *XVth Intern. Congr. Zool. London*, pp. 413–15.

BARTELMEZ, G. W. (1953). 'Factors in the variability of the menstrual cycle', *The Anat. Rec.* **115**, pp. 101–20.

BAUMANN, H. (1930). 'Tänze der Tschokwe in N.O.-Angola', *Film C* 110, Inst. f.d. wissenschaftl. Film, Göttingen.

BEAUX DE, O. (1929). 'Rettifica, descrizione e deduzioni sul genere Cynopithecus Is. Goffr', *Boll. Musei. Zool. Anat. Comp. R. Univ. Genova*, IX, pp. 1–35.

BERG, I. A. (1944). 'Development of behavior: The micturition pattern in the dog', *J. Exper. Psychol.* **34**, pp. 343–68.

BERNSTEIN, I. S. and W. A. MASON (1963). 'Group formation by rhesus monkeys', *J. Anim. Behav.* **11**, pp. 28–31.

BIRCH, H. G. and G. CLARK (1950). 'Hormonal modification of social behavior. IV. The mechanism of estrogen-induced dominance in chimpanzees', *J. comp. physiol. Psychol.* **43**, pp. 181–93.

BISHOP, P. M. F. (1953). 'Sex hormones and human behaviour', *J. Anim. Behav.* **1**, pp. 20–2.

BLIN, P. C., J. A. FAVREAU et J. M. FAVREAU (1964). 'Considérations su les Anomalies de Comportment sexuel chez l'Animal', *Psychiatrie Animale*, ed. A. Brion et H. Ey, Paris.

BOLWIG, N. (1959a). 'A study of the behaviour of the chacma baboon (*Papio ursinus*)', *Behaviour* **14**, pp. 136–63.

—— (1959b). 'Observations and thoughts on the evolution of facial mimic', *Koedoe*, pp. 60–9.

—— (1960). 'A comparative study of the behaviour of various lemurs', *Mém. Inst. Sci. Madagascar*, sér. A, **14**, pp. 205–17.

BOPP, P. (1953). 'Zur Abhängigkeit der Inferioritätsreaktionen vom Sexualzyklus bei weiblichen Cynocephalen', *Rev. suisse de Zool.* **60**, pp. 441–6.

—— (1962). 'Einige Beobachtungen über inferiores und sexuelles Verhalten bei Pavianen in Zoos', *D. Zool. Garten* (NF) **26**, pp. 279–83.

BOOTH, A. H. (1957). 'Observations on the natural history of the olive colobus monkey, *Procolobus verus* (van Beneden)' *Proc. Zool. Soc. Lond.* **129**, pp. 421–30.

BOOTH, C. (1962). 'Some observations on behavior of Cercopithecus monkeys', *Ann. N.Y. Acad. Sci.* **102**, pp. 477–87.

BÜTTIKOFER, J. (1917). 'Die Kurzschwanzaffen von Celebes', *Zool. Mededeelingen* (Leiden) **3**, pp. 1–86.

CARPENTER, C. R. (1942). 'Sexual behavior of free ranging rhesus monkeys (*Macaca mulatta*). I. Specimens, procedure and behavioral characteristics of estrus. II. Periodicity of estrus, homosexual, autoerotic and non-conformist behavior', *J. Comp. Psychol.* **33**, pp. 113–62.

CHANCE, M. R. A. (1956). 'Social structure of a colony of Macaca mulatta', *J. Anim. Behav.* **4**, pp. 1–13.

CHANCE, M. R. A. and A. P. MEAD (1953). 'Social behaviour and primate evolution', *Symp. Soc. Exper. Biol.* VII, 'Evolution', pp. 395–439.

COLLINGS, M. R. (1926). 'A study of cutaneous reddening and swelling about the genitalia of the monkey, M. rhesus', *Anat. Rec.* **33**, pp. 271–8.

CONAWAY, C. H. and C. B. KOFORD (1964). 'Estrous cycles and mating behavior in a free-ranging band of rhesus monkeys', *J. Mammal.* **45**, pp. 577–88.

CRAWFORD, M. P. (1940). 'The relation between social dominance and the menstrual cycle in female chimpanzees', *J. Comp. Psychol.* **30**, pp. 483–513.

—— (1942). 'Dominance and social behavior, for chimpanzees, in a non-competitive situation', *J. Comp. Psychol.* **33**, pp. 267–77.

—— (1942a). 'Dominance and the behavior of pairs of female chimpanzees when they meet after varying intervals of separation', *J. Comp. Psychol.* **33**, pp. 259–65.

DANDELOT, P. (1959). 'Note sur la classification des Cercopithèques du groupe Aethiops', *Mammalia* **23**, pp. 357–68.

DARWIN, CH. (1876). 'Sexual selection in relation to monkeys', *Nature* **15**, pp. 18–19.

DATHE, H. (1963). 'Vom Harnspritzen des Ursons (*Erethizon dorsatus*)', *Z. f. Säugetierkde.* **28**, pp. 369–75.

DAVIS, D. D. and H. E. STORY (1949). 'The female external genitalia of the spotted Hyena', *Fieldiana Zoology* **31**, pp. 277–83.

DEVORE, I. (1963). 'Mother-infant relations in free-ranging baboons', *Maternal behavior in mammals*, ed. H. L. Rheingold, New York, pp. 305–35.

DEVORE, I. and S. L. WASHBURN (1963). 'Baboon ecology and human evolution', *African ecology and human evolution*, ed. F. C. Howell, in press.

DEVORE, I. and K. R. L. HALL (1965). 'Baboon ecology', *Primate Behavior*, ed. I. DeVore, New York, San Francisco, Toronto, London.

HOOTON, E. (1954). 'The importance of primate studies in anthropology', *Human Biology* **26**, pp. 179–88.

HUTCHINSON, G. E. (1963). 'Natural selection, social organization, hairlessness, and the Australopithecine canine', *Evolution* **17**, pp. 588–9.

ILSE, D. R. (1955). 'Olfaktory marking of territory in two young male Loris (*Loris tardigradus lydekkerianus*) kept in captivity in Poona', *J. Anim. Behav.* **3**, pp. 118–20.

IMANISHI, K. (1960). 'Social organisation of subhuman primates in their natural habitat', *Current Anthropology* **1**, pp. 393–407.

—— (1963). 'Social behavior in Japanese monkeys (*Macaca fuscata*)', *Primate social behavior*, ed. C. H. Southwick, Princeton, Toronto, London, New York, pp. 68–81.

INHELDER, E. (1955). 'Zur Psychologie einiger Verhaltenweisen-besonders des Spiels—von Zootieren', *Z. Tierpsychol.* **12**, pp. 88–144.

ITANI, J. (1959). 'Paternal care in the wild Japanese monkeys', *Primates* **2**, pp. 61–93.

JAY, P. (1963). 'Mother-infant relations in langurs', *Maternal behavior in Mammals*, ed. H. L. Rheingold, New York, pp. 282–304.

—— (1965). 'The common langur of North India', *Primate Behavior*, ed. I. DeVore, New York, San Francisco, Toronto, London.

JÜRGENS, H. W. (1960). 'Über sexualdifferenzierte Proportions-veränderungen beim Wachstum des Menschen', *Z. Morph. Anthr.* **50**, pp. 210–19.

KAUFMANN, J. H. and A. KAUFMANN (1963). 'Some comments on the relationship between field and laboratory studies of behaviour, with special reference to Coatis', *J. Anim. Behav.* **11**, pp. 464–9.

KIRCHSHOFER, R. (1960). 'Über das "Harnspritzen" der Großen Mara (*Dolichotis patagonum*)', *Z. Säugetierkd.* **25**, pp. 112–27.

—— (1962). 'Beobachtungen bei der Geburt eines Zwergschimpansen (*Pan paniscus Schwarz* 1929) und einige Bemerkungen zum Paarungsverhalten', *Z. Tierpsychol.* **19**, pp. 597–606.

—— (1963). 'Einige bemerkenswerte Verhaltensweisen bei Saimiris im Vergleich zu verwandten Arten', *Z. Morph. Anthr.* **53**, pp. 77–91.

KNOTTNERUS-MEYER, T. (1901). 'Allerlei Beobachtungen aus dem Affenhause des Hannoverschen Zoologischen Gartens', *D. Zool. Garten* **42**, pp. 353–71.

KOFORD, C. B. (1963). 'Rank of mothers and sons in bands of rhesus monkeys', *Science* **141**, pp. 356–7.

—— (1965). 'Population dynamics of Rhesus monkeys on Cayo Santiago', *Primate behavior*, ed. I. DeVore, New York, San Francisco, Toronto, London.

KORTLANDT, A. (1959). 'Analysis of pair-forming behaviour in the cormorant (*Phalacrocorax carbo sinensis*) (Shaw and Nodd)', *XVth Intern. Congr. of Zoology*, London.

KÜHME, W. (1965). 'Freilandbeobachtungen zur Soziologie des Hyänenhundes (*Lycaon pictus lupinus Thomas 1902*)', *Z. Tierpsychol.* **22.**

KUMMER, H. (1956). 'Rang-Kriterien bei Mantelpavianen. Der Rang adulter Weibchen im Sozialverhalten, den Individualdistanzen und im Schlaf', *Rev. Suisse Zool.* **63**, pp. 288–97.

—— (1957). 'Soziales Verhalten einer Mantelpavian-Gruppe', *Z. f. Psychol. u. Anwendung*, Suppl., **33**, Bern and Stuttgart.

—— (1965). 'A comparison of social behavior in captive and free-living hamadryas baboons', *The baboon in medical research*, ed. H. Vagtborg, University of Texas Press. (1. 'Symp. on the Baboon and its use as an exp. animal', San Antonio, Texas, 1963.)

KUMMER, H. und F. KURT (1963). 'Social units of a free-living population of Hamadryas Baboons', *Folia primat.* **1**, pp. 4–19.

KUNKEL, P. und L. (1964). 'Beiträge zur Ethologie des Hausmeerschweinchens Cavia aperea f. porcellus (L.)', *Z. Tierpsychol.* **21**, pp. 602–41.

LANCASTER, J. B. and R. B. LEE (1965). 'The annual reproductive cycle in monkeys and apes', *Primate behavior*, ed. I. De Vore, New York, San Francisco, Toronto, London.

LEONHARD, K. (1964). *Instinkte und Urinstinkte in der menschlichen Sexualität*, Ferdinand Enke Verlag Stuttgart.

LEYHAUSEN, P. (1956). *Verhaltensstudien an Katzen*, Berlin, P. Parey (*Z. Tierpsychol.* Suppl. 2).

LORENZ, K. (1943). 'Die angeborenen Formen möglicher Erfahrung', *Z. Tierpsychol.* **5**, pp. 235–409.

—— (1963). *Das sogenannte Böse*, Borotha-Schoeler-Verlag, Wien.

LULLIES, R. (1931). 'Die Typen der griechischen Herme', Königsberg (Pr.).

MACLEAN, P. D. (1962). 'New findings relevant to the evolution of psycho-sexual functions of the brain', *J. nervous and mental disease* **135**, pp. 289–301.

—— (1964). 'Mirror display in the squirrel monkey (*Saimiri sciureus*)', *Science* **146**, pp. 950–2.

MASLOW, A. H. (1936–7). 'The rôle of dominance in the social and sexual behavior of infra-human primates', *J. Gen. Psychol.* **48**, pp. 261–77, 310–38; **49**, pp. 161–98.

MASON, W. A. (1960). 'The effects of social restriction on the behavior of rhesus monkeys. I. Free social behavior', *J. Comp. Physiol. Psychol.* **53**, pp. 582–9.

MASON, W. A., P. C. GREEN and C. J. POSEPANKO (1960). 'Sex differences in affective-social responses of rhesus monkeys', *Behaviour* **16**, pp. 74–83.

MATTHEWS, L. H. (1941). 'Reproduction in the spotted hyaena (*Crocuta crocuta*) (Erxleben)', *Philos. Trans. Roy. Soc. London,* ser. B **230**, pp. 1–78.

—— (1956). 'The sexual skin of the Gelada Baboon (*Theropithecus gelada*)', *Transact. Zool. Soc. London* **28**, pp. 543–52.

MAYR, E. (1960). 'The emergence of evolutionary novelties', *Evolution after Darwin,* Vol. I, ed. S. Tax, Chicago, pp. 349–80.

MEAD, M. (1949). *Male and female. A study of the sexes in a changing world,* New York.

MICHAEL, R. P. and J. HERBERT (1963). 'Menstrual cycle influences grooming behavior and sexual activity in the Rhesus Monkey', *Science* **140**, pp. 500–1.

MILLER, G. S. (1928). 'Some elements of sexual behavior in primates and their possible influence on the beginnings of human social development', *Journ. Mammal.* **9**, pp. 273–93.

MILLER, R. A. (1945). 'The ischial callosities of Primates', *Amer. J. Anat.* **76**, pp. 67–91.

MILLER, R. E. and J. H. BANKS (1962). 'The determination of social dominance in monkeys by a competitive avoidance method', *J. Comp. Physiol. Psychol.* **55**, pp. 137–41.

MORRIS, D. (1962). 'The behaviour of the green acouchi (*Myoprocta pratti*) with special reference to scatter hoarding', *Proc. Zool. Soc. Lond.* **139**, pp. 701–32.

MÜLLER-USING, D. (1952). 'Über einige bisher unbeachtete Übersprungsbewegungen bei höheren Säugern', *Z. Tierpsychol.* **9**, pp. 479–81.

MURIE, J. (1872). 'Observations on the Macaques. III. The Formosan or round-faced Monkey', *Proc. Zool. Soc. Lond.,* pp. 771–80.

NELSON, K. (1964). 'Behavior and morphology in the Glandulocaudine fishes (*Ostariophysi, Characidae*)', *Univ. Calif. Publ. Zool.* **75**, pp. 59–152.

NESTURKH, M. F. (1946). 'Observations on menstrual cycles in lower catarrhine monkeys of the subfamily of marmosets (*Cercopithecinae*)', *Trudy Moskovskogo zooparka* **3**, pp. 33–83, Moscow.

NICOLAI, J. (1959). 'Aus dem Leben des Diamanttäubchens', *Orion* **14**, pp. 812–17.

NOLTE, A. (1955). 'Freilandbeobachtungen über das Verhalten von Macaca radiata in Südindien', *Z. Tierpsychol.* **12**, pp. 77–87.

NOWLIS, V. (1942). 'Sexual status and degree of hunger in chimpanzee competitive interaction', *J. Comp. Psychol.* **34**, pp. 185–94.

OTTOW, B. (1955). *Biologische Anatomie der Genitalorgane und der Fortpflanzung der Säugetiere*, Jena.

PARKES, A. S. (1960). Ed., *Marshall's Physiology of reproduction*, 1, Part I, London.

PETTER, J. J. (1962). 'Recherches sur l'écologie et l'éthologie des Lémuriens Malgaches', *Mém. Mus. Nat. Hist. Nat. us. sér. A Zoologie* **27**, 1, pp. 1–146.

PETTER-ROUSSEAUX, A. (1962). 'Recherches sur la sociologie de la reproduction des primates inférieurs' (Théses Fac. Sci. Univ. Paris), *Mammalia Suppl.* **3**, Paris.

—— (1964). 'Reproductive physiology and behavior of the Lemuroides', *Evolutionary and genetic biology of Primates*, ed. J. Buettner-Janusch, New York, London. Vol. II, pp. 91–132.

PLOOG, D. (1963). 'Vergleichend quantitative Verhaltensstudien an zwei Totenkopfaffen-Kolonien', *Z. Morph. Anthr.* **53**, pp. 92–108.

—— (1964). 'Verhaltensforschung und Psychiatrie', *Psychiatrie der Gegenwart; Forschung und Praxis*, Vol. I/1, Berlin, Göttingen, Heidelberg.

PLOOG, D. W., J. BLITZ UND F. PLOOG (1963). 'Studies on social and sexual behavior of the squirrel monkey (*Saimiri sciureus*)', *Folia primat.* **1**, pp. 29–66.

POCOCK, R. J. (1906). 'Notes upon menstruation, gestation, and parturition of some monkeys that have lived in the Society's Gardens', *Proc. Zool. Soc. Lond.*, pp. 558–67.

—— (1919). 'On the external characters of existing Chevrotains', *Proc. Zool. Soc. Lond.*, pp. 1–11.

—— (1925). 'The external characters of the catarrhine monkeys and apes', *Proc. Zool. Soc. Lond.*, pp. 1479–579.

PÖCH, R. (1911). 'Die Stellung der Buschmannrasse unter den übrigen Menschenrassen', *Korrespondenz-Blatt der dtsch. Ges. f. Anthropol., Ethnol. u. Urgeschichte* XLII, pp. 75–80.

RAND, R. P., A. C. BURTON and T. ING (1965). 'The tail of the rat, in temperature regulation and acclimatization', *Canad. J. Physiol. Pharmacol.* **43**, pp. 257–67.

REITZENSTEIN, FRHR. F.V. (1931). *Das Weib bei den Naturvölkern*, Neufeld und Henius, Berlin.

REYNOLDS, V. (1961). 'The social life of a colony of Rhesus monkeys (*Macaca mulatta*)', Ph.D. Thesis, Univ. of London.

REYNOLDS, V. and F. REYNOLDS (1965). 'Chimpanzees of the Budongo forest', *Primate behavior*, ed. I. De Vore, New York, San Francisco, Toronto, London.

ROWELL, T. E. (1963). 'Behaviour and female reproductive cycles of Rhesus macaques', *J. Reprod. Fertil.* **6**, pp. 193–203.

ROWELL, T. E., R. A. HINDE and Y. SPENCER-BOOTH (1964). ' "Aunt"– Infant interaction in captive Rhesus monkeys', *J. Anim. Behav.* **12**, pp. 219–26.

SADE, D. S. (1964). 'Seasonal cycle in size of testes of free-ranging *Macaca mulatta*', *Folia primat.* **2**, pp. 171–80.

SANDERSON, I. T. (1949–50). 'A brief review of the mammals of Suriname (Dutch Guiana), based upon a collection made in 1938', *Proc. Zool. Soc. Lond.* **119**, pp. 755–89.

SALLER, K. (1964). *Leitfaden der Anthropologie*, Stuttgart.

SCHENKEL, R. (1948). 'Ausdrucksstudien an Wölfen', *Behaviour* **1**, pp. 81–129.

—— (1959). 'Lebensforschung im sozialen Feld und menschliche Sprache', *Homo* **10**, pp. 129–53.

SCHLOETH, R. (1956–7). 'Zur Psychologie der Begegnung zwischen Tieren', *Behaviour* **10**, pp. 1–80.

SCHLOSSER, K. (1952). *Der Signalismus in der Kunst der Naturvölker*, Kiel (Arbeiten a.d. Museum f. Völkerkunde 1).

SCHMIDT, H. D. (1956). 'Das Verhalten von Haushunden in Konflikt-situationen', *Z. f. Psychol.* **159**, pp. 161–245.

SCHULTZ, A. H. (1951–2). 'Vergleichende Untersuchungen an einigen menschlichen Spezialisationen', *Bull. Schweiz. Ges. f. Anthropol.* **28**, pp. 25–37.

SCHULTZE, L. (1928). 'Zur Kenntnis des Körpers der Hottentotten und Buschmänner', *Denkschr. med.-naturwiss. Ges. Jena* **17**, pp. 145–227.

SCHWARZ, E. (1926). 'Die Meerkatzen der Cerpithecus aethiops-Gruppe', *Z. Säugetierkd.* **1**, pp. 28–47.

SHADLE, A., R. SMELZER and M. METZ (1946). 'The sex reactions of porcupines (*Erethizon d. dorsatum*) before and after copulation' ; *J. Mammal.* **27**, pp. 116–21.

SIBLEY, C. S. (1952). 'Wren-tit attempts copulation with begging fledgling', *The Condor* **54**, p. 117.

SIMONDS, P. E. (1965). 'The bonnet macaque in south india', *Primate behavior*, ed. I. De Vore, New York, San Francisco, Toronto, London.

SOUTHERN, H. N. (1948). 'Sexual and aggressive behaviour in the wild rabbit', *Behaviour* **1**, pp. 173–94.

SPIEGEL, A. (1930). 'Beobachtungen über den Sexualzyklus, die Gravidität und die Geburt bei Javamakaken (*Macaca irus mordax Thomas and Wroughton (cynomolgus L.)*)', *Arch. Gynäkol.* **142**, pp. 561–91.

—— (1954). 'Beobachtungen und Untersuchungen am Javamakaken', *D. Zool. Garten* (NF) **20**, pp. 227–70.

SPINDLER, P. (1961). 'Studien zur Vererbung von Verhaltensweisen. 3. Verhalten gegenüber jungen Katzen', *Anthrop. Anz.* **25**, pp. 60–80.

SPIRO, M. E. (1954). 'Is the family universal?', *Amer. Anthropol.* **56**, pp. 839–46.

SPITZ, R. A. (1962). 'Autoerotism re-examined. The rôle of early sexual behavior patterns in personality formation', *The Psychoanalytic study of the child*, **17**, pp. 283–315.

STIEVE, H. (1952). *Der Einfluß des Nervensystems auf Bau und Tätigkeit der Geschlechtsorgane des Menschen.*, Stuttgart, G. Thieme.

TEMBROCK, G. (1964). *Verhaltensforschung*, 2. Aufl., Jena, Gustav Fischer.

TINKELPAUGH, O. L. (1928). 'The self-mutilation of a male Macacus rhesus monkey', *J. Mammol.* **9**, pp. 293–300.

TOKUDA, K. (1961–2). 'A study on the sexual behavior in the Japanese Monkey troop', *Primates* **3**, pp. 1–40.

TOMILIN, M. I. (1940). 'Menstrual bleeding and genital swelling in Miopithecus (*Cercopithecus*) talapoin', *Proc. Zool. Soc. Lond.* **110**, pp. 43–5.

TORDAY, E. and T. A. JOYCE (1922). 'Notes ethnographiques sur des populations habitant les bassins de Kasai et du Kwango Oriental', *Ann. Mus. Congo Belge, Ethnogr. Anthropol.*, sér. 3, Bruxelles.

ULLRICH, W. (1961). 'Zur Biologie und Soziologie der Colobusaffen (*Colobus guereza caudatus Thomas* 1885)', *D. Zool. Garten* (NF) **25**, pp. 305–68.

WASHBURN, S. L. (1950). 'Analysis of primate evolution', *Cold Spring Harbor Symps. quant. Biol.* 15 (Origin and evolution of man), pp. 66–77.

—— (1957). 'Ischial callosities as sleeping adaptations', *Am. J. phys. Anthropol.* (n.s.) **15**, pp. 269–76.

WASHBURN, S. L. and I. DE VORE (1962). 'Social behavior of baboons and early man' (MS).

WICKLER, W. (1962). 'Zur Stammesgeschichte funktionell korrilierter Organ- und Verhaltensmerkmale: Ei-Attrappen und Maulbrüten bei afrikanischen Cichliden', *Z. Tierpsychol.* **19**, pp. 129–64 u. 256.

—— (1963). 'Die biologische Bedeutung auffallend farbiger, nackter Hautstellen und innerartliche Mimikry der Primaten', *Naturw.* **50**, pp. 481–2.

—— (1964). 'Vom Gruppenleben einiger Säugetiere Afrikas', *Mitt. d. Max-Planck-Ges. H.* **5-6**, pp. 296–309.

—— (1965a). 'Die äußeren Genitalienals soziale Signale bei einigen Primaten', *Naturw.* **52**, p. 269.

—— (1965b). 'Die Evolution von Mustern der Zeichnung und des Verhaltens', *Naturw.* **52**.

—— (1965c). 'Mimicry and the evolution of animal communication', *Nature* **208**, pp. 519–21.

—— (1966). 'Ursprung und biologische Deutung des Genital-präsentierens männlicher Primaten', *Z. Tierpsychol.* **23**, pp. 422–37.

WINKELSTRÄTER, K. H. (1960). 'Das Betteln der Zoo-Tiere', *Suppl. Schweiz. Z. f. Psychol. u. Anwendg.* **39**, Bern und Stuttgart.

WISLOCKI, G. B. (1936). 'The external genitalia of the Simian Primates', *Human Biology* **8**, pp. 309–47.

YERKES, R. M. (1940). 'Social behavior of chimpanzees: dominance between mates in relation to sexual status', *J. Comp. Psychol.* **30**, pp. 147–86.

YERKES, R. M. and J. H. ELDER (1936a). 'Oestrus, receptivity, and mating in Chimpanzee', *Comp. Psychol. Monographs* **13**, No. 65.

—— (1936b). 'The sexual and reproductive cycles of Chimpanzee', *Nat. Acad. Sci.* **22**, pp. 276–83.

YOUNG, W. C. and W. D. ORBISON (1944). 'Changes in selected features of behavior in pairs of oppositely sexed chimpanzees during the sexual cycle and after ovariectomy', *J. Comp. Psychol.* **37**, pp. 107–43.

ZEEB, K. (1961). 'Der freie Herdensprung bei Pferden', *Wiener Tierarztl. Monatsschr.* **48**, pp. 90–102.

ZUCKERMAN, S. (1932). *The social life of monkeys and apes*, London.

Chapter Four

Allogrooming in Primates: a Review

JOHN SPARKS

INTRODUCTION

IF the skin and all of its various allied structures like scales, hair, and feathers, are to carry out their various functions efficiently, they must be kept in good condition. Animals have evolved an extensive repertoire of behaviour patterns of varying complexity which help to keep the surface of the body free of parasites, and to prevent the plumage or pelage from becoming entangled with dirt. As a general rule, an individual will clean itself, but the individuals of certain species spend some of their time in grooming or preening another member, usually of the same species. This kind of allelomimetic behaviour has been referred to under a variety of names, such as social, mutual or reciprocal grooming; 'flea catching', 'sham louse picking', or simply the 'picking reaction', have been used with reference to primate behaviour. A number of objections may be raised in connection with these terms, and a number of them are misleading; for example, mutual grooming should only apply to the behaviour when the participants are simultaneously grooming each other. In the interests of standardization, it has been found convenient to use a terminology based upon that suggested by Cullen (1961), which has been applied to most of the avian studies published since then. The behaviour observed when one individual grooms or preens another, may be called allogrooming or allopreening, as opposed to self or autogrooming behaviour. Mutual grooming, referring to the simultaneous grooming between two individuals, and reciprocal grooming, when first one and then another of a pair groom each other but not simultaneously, add a further refinement to this terminology.

THE DISTRIBUTION OF ALLOGROOMING BEHAVIOUR

Many invertebrates, some fish, and perhaps birds enter into often very specialized cleaning relationships, which are essentially inter-specific and symbiotic. The cleaner species removes detritus of food value from the host species, which in turn benefits by the removal of ecto-parasites and debris from the surface of the body; these kinds of relationships are beyond the scope of this paper.

Allogrooming occurs in the social Hymenoptera; ants lick each other, and this behaviour has been studied in *Formica fusca* by Wallis (1962); Allen (pers. comm.) has reported allogrooming in the bee (*Apis mellifera*); individuals will remove pollen grains from the wing bases of other bees, and it is an interesting occurrence because this area of the body is beyond the reach of an auto-grooming bee. Lizards (*Lacerta viridis*) have been seen to groom each other in a perfunctory manner but it is in the birds and mammals where allopreening or allogrooming reaches its highest development, presumably because of the more elaborate behaviour involved in skin care in these orders. Sparks (1964a) and Harrison (1965) have reviewed the distribution of this behaviour in birds, and it was found that allopreening tended to be correlated with clustering or clumping behaviour; that is, in those species which do not maintain an 'individual distance' but tend to sit flank-to-flank in groups (see Hediger, 1942). Thus, families of predominantly 'contact species' such as the Estrildidae, Psittacidae, Timaliidae, and Zosteropidae, are noted for their allopreening behaviour. This correlation between allopreening and clustering behaviour is not absolute, since some Corvids which are otherwise 'distance species' may preen each other (Lorenz, 1931).

It should be pointed out at this stage that allogrooming (licking and combing) are characteristic patterns of parental behaviour in most mammals, and some birds which do not allopreen normally, will preen their young. However, in mammals, parental allo-grooming tends to decrease as the young become independent of the parents. Allogrooming behaviour between adult mammals in non-sexual and non-parental contexts appears, on the whole, to be less common than in birds. It has been observed in a wide variety of ungulate and artiodactyl species in zoos, but the individuals may have been paired in most cases. Many members of the Felidae lick each other, often mutually and reciprocally, and in the domestic cat this behaviour is seen when friendly individuals meet. It seems as though allogrooming may be widespread in rodents, and in rats (*Rattus rattus*) and domestic mice (*Mus musculus*) it occurs especially in conflict situations, although it is by no means restricted to them (Scott and Fredericson, 1951; Barnett, 1963; Grant and Mackintosh, 1963). In the highly social marmots, the grooming of one by another is a common occurrence; Armitage (1962) has described this behaviour for the yellow-bellied species (*Marmota flaviventris*). Among the Prosimian primates, the members of the Tupaiidae and *Tarsius* are reputed not to allogroom (Andrew and Buettner-Janusch, 1962; Zuckerman, 1933). Of *Daubentonia madagascariensis* nothing is known of any allogrooming behaviour. However, in the members of the families Lemuridae and Lorisidae, allogrooming is a common social activity and has been studied by Andrew and Buettner-Janusch (*loc. cit.*), Bishop (1962), and Manley (in press). In all of the higher primates comprising the sub-order Anthropoidea, the absence of allogrooming in any species is the exception, although the frequency with which this behaviour may be seen varies from species to species. Carpenter has stated that social grooming was less common in the New World monkeys than in the Old World representatives. Apart from scratching, either with the front or hind-legs, most of the skin care requirements of a member of a monkey troop are provided by a companion or grooming partner. Indeed, the picking through the fur of another individual is perhaps the most characteristic of all primate behaviour and no doubt led Watson (1908) to state of rhesus macaques that it was 'the most fundamental and basal form of social intercourse' between monkeys. Fitzgerald (1935) has recorded some interesting observations on the marmoset (*Callithrix jacchus*), and Nolte (1958) has described allogrooming behaviour in *Cebus apella*. Field observations by Carpenter (1935) showed that the Spider monkeys (*Ateles geoffroyi*) do groom each other to some extent, but that this behaviour was not recorded for the howler monkey (*Alouatta palliata*), (Carpenter, 1934). So far as the families Cercopithecidae and Pongidae are concerned, the practice of grooming another member of the troop is widespread, and has been recorded for all species which have been studied in any detail so far.

DESCRIPTION OF ALLOGROOMING BEHAVIOUR

Invitation Behaviour

The term 'invitation' behaviour is used here to refer to the characteristic behaviour of the groomee which may precede grooming by its partner, without implying any deliberate intention on the part of the presenting individual. Species which have evolved allogrooming or allopreening behaviour have ritualized invitation behaviour which often immediately precedes the partner starting to allogroom or allopreen. In birds, this may consist of a ruffling of the body plumage, as in the white-eyes, *Zosterops* sp., or, more usually, the pilo-erection may be restricted to the head plumage. Morris (1956) has pointed out that the stimulus of parted feathers may be an important factor in eliciting cleaning behaviour in the preener.

Pilo-erection appears to be unimportant in the grooming invitation behaviour of mammals, but, instead, is replaced by a characteristic body posturing on the part of the potential groomee. Thus, a marmot which approaches another, and rolls over on to its back, stands a good chance of being allogroomed. Higher primates seem to have no ritualized invitation behaviour, in so far as the behaviour is rigidly stereotyped. It may involve an individual approaching a potential groomer, and sitting down squarely in front of it but facing sideways, or presenting its back. On other occasions, an individual may be groomed after it has approached its grooming partner; it may lay lengthwise along the perch with its body half rolled over, often in the opposite direction to that of the groomer, and an arm may be extended loosely in any of a number of directions. The sequence will usually be relaxed but nevertheless giving the impression of carefulness in avoiding quick or erratic movements. A female Moor monkey (*Macaca maurus*) at the London Zoo used to stand on all four legs and place her crown squarely on the ground, so that she would be looking backwards through them; this would precede grooming by another individual. Reynolds (1961) has described 'blocking' by rhesus monkeys (*M. mulatta*) as grooming invitation behaviour; this behaviour is very similar to sexual presentation except that the orientation of the 'blocking' animal is different. Schaller (1963) has described grooming invitation behaviour of the mountain gorilla (*Gorilla gorilla*); one would move up to its potential groomer and swing itself around so that it presented its buttocks to the other; again, this behaviour is rather similar to sexual presentation. Once the gorillas started allo-grooming, the submitting individual displayed different parts of the body for inspection; Schaller did not observe juveniles invite grooming.

If an individual does behave as though it is soliciting grooming, the behaviour may have a number of characteristics. It usually involves presenting the back to the groomer; Reynolds (1961) concluded that the sight of another's back was an important stimulus initiating allogrooming in rhesus monkeys. However, presenting the back to the groomer may be part of a wider feature of facing away from the grooming partner; a monkey inviting allogrooming will not fixate or look at the groomer, but will usually turn its head away, either looking downwards, upwards, or sideways. This kind of behaviour is characteristic of appeasement gestures in species where fixating an opponent is a sign of dominance or threat (Moynihan, 1955; Tinbergen, 1954; *et al.*). Facing away has also been described in the preening invitation postures of birds (Morris, 1956; Sparks, 1962, 1964a, b, c, 1965; Harrison, 1965).

Presenting the chin seems to be an important feature of allo-grooming behaviour of Patas monkeys (*Erythrocebus patas*). Also, the way in which one of these monkeys solicits the attentions of another member of the group depends upon its status within the group; thus a subordinate individual will never invite allo-grooming from a supine position (Hall, Boelkins, and Goswall 1965).

Once the groomer has started to manipulate the fur, the groomee will make constant adjustments and present different areas of the body to receive attention. This behaviour may involve body rolling, and arm or leg stretching, but throughout the course of this behaviour, the groomee will avoid looking at the face of the groomer, although it may look surreptitiously out of the corner of its eyes at the hands of the groomer. Should their eyes meet, then the groomee will usually lip-smack vigorously and look away once more (pers. obs.). While being groomed, monkeys will often close their eyes, and a two-year-old rhesus monkey at the London Zoo would predictably fall asleep when she was groomed (by humans) and would wake up with a start as she slipped off her perch. Fitzgerald (1935) noted that marmosets closed their eyes in response to being groomed, and she interpreted this as being due to an 'ecstatic' disposition!

Grooming Behaviour

Primate grooming behaviour has been the subject of many descriptions (see Yerkes, 1933; Zuckerman, 1932 and 1933; Ewing, 1935). In most mammals, and in the lower primates, cleaning of the body surface is effected by the mouth, either through licking or combing with the teeth. In lemurs, the specialized procumbent incisors of the lower jaw act as a scraper or comb. A mutual grooming bout in *Lemur fulvus* may be divided into three phases, namely (1) licking, (2) licking and scraping, and finally (3) scraping and vocalizing (Buettner-Janusch and Andrews, 1962). With the evolution of increasing dexterity, the higher primates have largely given up using the mouth, although mother langurs (*Presbytus entellus*) and baboons (*Papio ursinus*) will lick their young (Jay, 1963; DeVore, 1963a). Allogrooming is accompanied by fixation on the area receiving attention, and this gives the impression that the grooming monkey is exercising extreme concentration. However, Reynolds (*loc. cit.*) has made the interesting observation on *M. mulatta* that, whereas a subordinate monkey will groom one of higher rank with busy movements and constant adjustment to the groomee's changes of position, a dominant individual will allogroom one of lower rank with less vigour and may often break off in order to scratch itself. Allogrooming is generally accompanied by lip-smacking (see van Hooff, 1962). In this context, its form seems to vary but little from one species to another, although Hall *et al.* (1965) report that lip-smacking is less frequent and slower in the patas monkey. The jaws make small amplitude movements in a rhythmic manner, allowing the tongue to protrude between the lips to a greater or lesser extent. In the Celebes ape (*Cynopithecus niger*), the rate of lip-smacking is not necessarily constant, but it occurs in bouts at a frequency of up to seven to eight lip-smacks per second. If any pieces of scale or hair are conveyed to the mouth, sometimes the lip-smacking is not interrupted; on the other hand, the movements may be frozen with the tongue extended to receive the piece of foreign material brought from the skin or hair. The frequency of lip-smacking may increase momentarily after something has been conveyed to the mouth. The tongue may also be extended now and again and brought into contact with an area of skin or fur, but this is not necessarily the area which is being allogroomed at the time. Licking (defined as the drawing of the tongue repeatedly over a particular spot) seems to be uncommon in the Cercopithecidae, although a female *Cynopithecus niger* did lick an open cut on the author's hand on one occasion; however, the taste of blood may have stimulated the female to lick. Occasionally a group of hairs may be taken into the mouth and nibbled between the incisors, and blemishes on the skin may be similarly treated.

It seems clear that the lip-smack is derived from a licking or lip-moistening movement, and is functionally related to grooming behaviour, but it has undergone some evolutionary change and appears even in the grooming context to have evolved a 'typical intensity' (Morris, 1957). There can be little doubt that it serves to reassure other members of the troop, or that it may be used in the 'assuagement of submissive behaviour' in other contexts (see van Hooff, *loc. cit.*; Altmann, 1962). Because of its importance in social interactions, the rhythmic slapping sound coming from the groomer during an allogrooming bout may be important in preventing the groomee from withdrawing or reacting in a hostile manner to the meandering hands of the allogroomer.

In the higher primates the hands are used to groom, but for any given species there is a bewildering variety of ways in which the hands may be employed to manipulate the fur, and makes it difficult to describe their rôle. The behaviour may be roughly categorized according to effect they have on the fur of the groomee.

Drawing hair(s) through forefinger and thumb. This is a very basic pattern and is self-descriptive. In the Colobinae, the first finger may be opposed against the nubbin, which is all that remains of the thumb. The various species of *Macaca* and *Cynopithecus* which have been studied do not seem to be able to exercise any independent movement between the forefinger and the other three fingers so that as the first finger closes with the thumb on to the hair(s), the others move in also; if the hand is close to the skin, the nails tend to scratch and rake the surface, thus loosening and removing any foreign material. Often, hairs may be pulled out as a result of this nipping and pulling action between the thumb and forefinger.

Raking/combing/smoothing. The projecting thumb of the grooming hand may be moved against the general direction of the hair tracts with a rotary motion of the wrist. The fingers may also be used but tend to be curled under the palm and raked downwards or sideways. There is a great deal of variation in these movements and it is dubious as to whether these slight variations have any profound significance; the important point is that there is a raking or combing motion, the result of which is to loosen scale, ecto-parasites, and to dislodge loose hairs; the hairs will also be raised so that the thumb and forefinger may be brought to bear upon them.

Collecting/sheafing. In species like the Gelada baboon (*Thero-pithecus gelada*), which have relatively long hair, the hair may be bundled together by an allogrooming individual with a downward sheafing action of the forearm, so that the region of the pellage is not an encumbrance to the other hand which may be performing picking movements. This kind of behaviour was also observed in a female moor monkey, a relatively short-coated species, which was allogrooming the 24 cm long hair of a female laboratory assistant at the London Zoo. Not only was the hair pulled downwards with wide sweeps of the arms (the plantar surfaces of the hands were at right angles to the head), but it also looked as though the macaque was flicking the hair upwards and away with quick movements of the wrist. It seems as though the pattern is very adaptable to the particular circumstances and problems posed by the groomee.

Miscellaneous patterns. Kummer (1957) noted one-fingered allo-grooming in the Hamadryas baboon (*Papio hamadryas*) and he called it formal grooming. However, this was also recorded by Schaller (1963) for the gorilla; in *Cynopithecus niger*, it takes the form of one-fingered combing or scraping with the nail, or occa-sionally drawing the side of the index finger over the surface of the skin. The mobility of the lips of the gorilla and chimpanzee (*Pan troglodytes*) is prodigious, and Schaller and Carpenter (1937) saw the former species manipulate hair with the lips during allo-grooming bouts. This behaviour has also been recorded in captive chimps. Indeed, Carpenter (1937) thought that chimps 'lipped' hair more frequently than gorillas.

The Sequence of Motor Patterns during Allogrooming
If one hand is used, then it usually alternates between 'picking' and combing or scraping. Usually both hands are used in a well-defined sequence. One hand 'picks' and conveys any particles to the mouth, while the other is used to scrape or comb. While one

hand is in action, the other is usually redundant. It is not altogether unusual for the hands to change their rôles during the course of a grooming sequence. A typical sequence would begin

<pre>
        RH   LH   RH LH
. . . pick pick pick/comb comb/pick/comb/pick
        RH   LH
      pick . . . convey to mouth/comb comb . . .
</pre>

Both hands may be engaged in combing for several seconds, one alternating with the other, with the groomer looking intently at the fur as it is raked upwards or downwards.

When they are allogrooming, lemurs tend to grasp the fur with both hands, then lick or scrape the area between them. Andrew (in press) states that the Cercopithecoid/Ceboid gesture of parting the fur with both hands could have evolved directly from lemuroid behaviour. Indeed, the need for efficient grooming of the hair could have led to the independent control of the thumb in the Cercopithecoidea. Also, since grooming forms one of the chief social bonds in both lemur and monkey troops, Bishop (1962) postulated that strong selection for more efficient and pleasurable allogrooming would lead to increasing fine control and tactile sensitivity.

A female Celebes ape at the London Zoo used to grasp her mate's fur with both hands, rotate the wrists outwards, and lick or bite the parted hair in between, thus almost simulating the lemuroid method of allogrooming.

Parts Allogroomed

Lemurs can autogroom the tail, flanks, belly and limbs; and the flanks, armpits and side of the head may be scratched with the claws of the hind feet. A lemur may, however, be allogroomed in any region of the body, although it tends to be concentrated in areas inaccessible to the groomee, particularly in *Galago* species (Buettner-Janusch and Andrew, 1962). An allogrooming gorilla tends to confine its attentions to the parts of the pelage which cannot be reached by the submitting member of the troop (Schaller, *loc. cit.*). On the whole, this was also true of various species of baboons and macaques which were watched in the London Zoo. There was a tendency for the back to be allogroomed, together with the upper arms, armpits and the fur around the neck. Of *M. mulatta*, Reynolds (*loc. cit.*) stated that only when two individuals formed a 'strong partnership' were the head and ears allogroomed as well. Schaller (*loc. cit.*) found that female gorillas paid particular attention to the rectal regions of their offspring when allogrooming, but that this was not true of adult-adult allogrooming encounters. Zuckerman (1932) found that in consort pairs of hamadryas baboons, a considerable amount of allogrooming time was spent in examining the genital area of the partner; this activity was often accompanied by sexual arousal. In Celebes apes, the sitting male would sometimes grab the female around her waist, pull her towards him, and proceed to pick at her vulval regions; penile erection often followed.

In birds, allopreening tends to be restricted to the head region, although the extent to which this statement is true varies from one species to another (see Kunkel, 1962; Sparks, 1964a). Thus, this kind of cleaning behaviour does tend to be restricted to those regions which are inaccessible to the bill or mouth. Since higher primate cleaning patterns involve the close co-ordination between both hands and mouth, then an individual monkey will be unable to clean itself as effectively as it could with the help of another member of the troop.

GROOMING RELATIONSHIPS WITHIN TROOPS OF HIGHER PRIMATES

Not all of the higher primates appear to show non-parental allogrooming behaviour, and in those species which have been studied there would appear to be considerable variation in the frequency with which this behaviour may be recorded. In the Ceboidea, Carpenter (1934) did not observe any social grooming in troops of howler monkeys, and only a few instances of one individual picking through the fur of another were noted in the red spider monkey (Carpenter, 1935). So far as the Cercopithecoidea are concerned, Haddow (1952) states of the red-tailed monkey (*Cercopithecus ascianus schmidti*) that they did not participate in much social grooming; although this behaviour was recorded in green monkeys (*Cercopithecus aethiops*) by Hall and Gartlan (1965); this species showed a peak in the amount of allogrooming both in the morning and late afternoon, very little occurring during the midday 'siesta'. However, the more terrestrial patas monkey (*Erythrocebus patas*) spends more time in allogrooming than the other guenons (Hall, pers. comm.). The Hamadryas baboon spends much of its leisure time in allogrooming (Zuckerman, 1932; Kummer, 1957). The same is true of *P. ursinus* (Washburn and DeVore, 1961), and Bolwig (1959) states that it is a common activity while the troop is stationary. Allogrooming is a conspicuous activity in macaques; troops of *M. radiata* start the daily routine with 'mutual grooming', and this behaviour occupies a considerable part of the day (Nolte, 1954). It is clear from the studies of Carpenter (1942), Altmann (1962), Chance (1955) and Reynolds (1961) that the members of a troop of rhesus macaques indulge in much allogrooming activity, and this has also been noted and studied by Furuya (1957) in troops of the Japanese macaque (*M. fuscata*).

Despite the prevalence of this behaviour in the baboons and macaques, the frequency with which it occurs in the apes appears to be less marked under natural conditions. Field studies on the Lar gibbon (*Hylobates lar*) by Carpenter (1940) reveal that they do allogroom, but within the Pongidae, this behaviour seems to be less important. In the Budongo forest, Reynolds and Reynolds (1956) only witnessed 57 clear-cut examples of allogrooming behaviour in chimpanzees over the course of 300 hours of observation; they did state that in this habitat, chimpanzees have little time for relaxed social activities after seeking and ingesting food. Schaller (*loc. cit.*) also found that allogrooming 'was not a prominent activity among gorillas'. Each bout usually lasted less than one minute, and over the course of the study period, he saw 134 instances of this behaviour, which turns out to be 0·28 bouts/ hour of observation; this compares with approx. 0·2 hour for Reynolds and Reynolds' data. Of the gorilla figures, approximately one half were cases of mothers grooming their infants, and thus many could be considered as parental allogrooming which is common to most mammals.

One observation may be made at this stage, namely that the frequency of allogrooming in the higher primates of the Old World appears to be roughly correlated with the kind of social structure prevailing within the species. As a generalization, it may be said that those species which have a rigidly maintained dominance hierarchy, like most of the macaques and baboons, show more allogrooming behaviour than those in which there is a less well-defined hierarchical organization, like the red-tailed guenon, gorilla, and chimpanzee. Marler (1965) has also come to the conclusion that allogrooming is particularly prominent in species in which dominance relations play a significant rôle. A similar correlation can also be drawn between dominance mounting and the extent to which a hierarchy is maintained. These different kinds of organizations are adaptions to predation and available food supply (Chance, 1959; Hall, 1965). Aggression plays an important rôle in maintaining these hierarchies, and the more aggressive species like the Hamadryas baboon, for example, enforce very rigid social orders. Thus, in the Old World species which have been studied under natural conditions, there appears to be a correlation between the amount of allogrooming behaviour and the steepness of the 'dominance slope' within the troop.

The allogrooming relationships in a monkey troop are not random. This is also true of allopreening behaviour within a flock of red avadavats (*Amandava amandava*), an Estrildid species which has been extensively studied by Sparks (1962, 1964a, b, 1965); in this species, each bird tends to narrow its choice of clumping partners within the flock, and only individuals which form a clumping group will tend to allopreen each other. Within monkey troops, mothers will groom their offspring, but outside these parental relationships, a number of factors such as age, sex, and dominance, have a large bearing upon the extent to which a monkey will allogroom another member of the troop.

Infant monkeys do not allogroom, and this may be due to the fact that, in *M. mulatta* for example, the thumb and forefinger cannot be opposed until the age of nine months or thereabouts (Hines, 1942). Gorillas do not groom each other to any great extent until they are over three years of age, according to Schaller (Table I). Yerkes (1933) states of chimpanzees, that an individual will allogroom more as it becomes older. Reynolds (1961) has quantified the increase in allogrooming behaviour in captive rhesus monkeys with age, and the information is reproduced in Table II.

In species where sufficient details are available, it is clear that allogrooming is predominantly characteristic of mature females (see Table III for *P. ursinus*); males tend to do less allogrooming than the females within the troop. (Zuckerman, 1932; Carpenter, 1942; Schaller, 1963; Washburn and DeVore, 1961; Bolwig, 1959; Reynolds, 1961; Reynolds and Reynolds, 1965.) With the possible exception of the langur, *Presbytis entellus*, studied by Jay (1963;

TABLE I

NUMBER OF OBSERVATIONS OF ALLOGROOMING IN THE VARIOUS AGE AND SEX CLASSES IN THE GORILLA

(After Schaller, 1963)

ANIMALS GROOMED/ ANIMALS GROOMING						TOTAL
Silver-backed male (10+ yrs) .	0	0	0	0	3	3
Black-backed male (6–10 yrs) .	0	0	0	0	0	0
Female	0	1	5	12	76	95
Juvenile (3–6 yrs) . . .	1	1	9	10	12	33
Infant (0–3 yrs) . . .	0	0	2	0	1	3
TOTAL .	1	2	16	23	92	134

TABLE II

AVERAGE AMOUNT OF ALLOGROOMING BY AGE GROUPS IN CAPTIVE RHESUS MONKEYS

(After Reynolds, 1961)

	Adult	*3 yrs*	*2 yrs*	*1 yr*
FEMALE . .	198·4	62·25	77·5	31·0
n . .	5	4	2	1
MALE . .	118·3	54·0	—	18·0
n . .	3	2	0	1

TABLE III

NUMBER OF OBSERVATIONS OF ALLOGROOMING IN VARIOUS AGE AND SEX CLASSES IN PAPIO URSINUS

(After Bolwig, 1959)

GROOMED/ GROOMING	♂	♀	Y	TOTAL
♂ . .	0	7	0	7
♀ . .	33	27	5	65
Y . .	0	4	3	7
TOTAL .	33	38	8	79

1965), most primate troops are patriarchal, with certain males acting as leaders, and these will have precedence over most other members of the troop in feeding, sexual, and spatial situations (Zuckerman, 1933; Washburn and DeVore, 1961; etc.). Males, then, tend to be dominant over females, and in *Papio ursinus*, for example, the dominant male of the troops studied by Washburn and DeVore (*loc. cit.*) was the most frequently groomed individual.

Superficially, it thus appears as though animals of lower status tend to allogroom those higher in the social hierarchy. Bolwig (1959) noticed that in grooming groups of female baboons, the younger ones tended to groom the older individuals, and that these bouts were longer than those between other female partnerships; Schaller's work on the gorilla also confirms the inverse relationship between dominance and allogrooming. Crawford (1942) measured the relative dominance between two chimpanzees by the amount of food they took in a competitive situation, and she found that the more dominant individual invited grooming more often than the subordinate one, which rarely refused to comply. The field studies of Reynolds and Reynolds (1965) supports the view that animals of lower status within the troop do relatively more allogrooming than higher ranking individuals.

The situation in rhesus monkeys is rather more complex. Maslow (1936) obtained conflicting results about the relationship between allogrooming and status; in one set of experiments, the dominant individuals tended to groom, and in another set, allogrooming was characteristic behaviour of the subordinate ones. From this it was concluded that the connection between dominance status and allogrooming was 'so tenuous . . . that nothing as yet can be said about it'. Since then, more naturalistic studies by Carpenter (1942), Altmann (1962), Reynolds (1961), Southwick, Beg and Siddiqi (1965) have more or less confirmed that allogrooming relationships tend to go against the dominance slope with a few major exceptions. For example, Reynolds (*loc. cit.*) analysed allogrooming behaviour in some detail in captive rhesus macaques, and found that allogrooming behaviour in the males was positively correlated with rank, with the dominant male allogrooming most of all among the males. It may be relevant here to note that even so, he was groomed far more by his consort of adult females. Among the adult females, a different situation prevailed, and those which were not in the hierarchy allogroomed most of all.

An interesting case was noted in this colony, of a low-ranking female which had a grooming relationship with another of similar status; the former one then moved up the hierarchy, gaining favour with the dominant male, but she still continued this partnership, showing that despite the change in other relationships, allogrooming ones can remain stable. Southwick, Beg and Siddiqi (*loc. cit.*) found that adult female rhesus monkeys commonly allogroomed the males; indeed, females together with a few juveniles usually formed resting and grooming clusters around an adult male. Although juveniles often allogroomed males, the reverse was seen only on a few occasions. Allogrooming activity of adult males was typically confined to the sexual consort period. The inverse relationship between allogrooming and status was not found in Rowell, Hinde and Spencer-Booth's (1964) studies of 'aunt' rhesus monkeys; on the contrary, they found that dominant 'aunts' allogroomed more, and were often more aggressive than rhesus monkeys of lesser status.

Furuya (1957) found that in *M. fuscata*, the 'male's object of grooming changes from infant to female according to how his social status rises from young male to boss'.

The relationship between sexual arousal and allogrooming behaviour has been noted in many species. In Estrildid birds, and in particularly, *Uroloncha striata*, mounting and copulation may follow a bout of allopreening behaviour; in this case, the allopreening male mounts. Marler (1965) points out that a combination of tactile and sexual behaviour occurs in macaques and baboons to produce an extraordinarily rich repertoire of signals used for greeting. In *Papio hamadryas*, the performance of the 'picking reaction' seems to have special significance because it is sometimes accompanied by direct sexual activity (Zuckerman, 1933). In this and other species, perineal examination which may not be strictly allogrooming behaviour, may lead to sexual arousal, and in the formerly mentioned species, a considerable amount of allogrooming between pairs is concentrated in the genital regions (Zuckerman, *loc. cit.*). Dominant males in macaque and baboon troops are accompanied by a number of females in oestrus, and these spend some of the time in grooming the overlord; it is not, then, surprising that copulation is temporarily associated with allogrooming. From the work of Michael and Herbert (1963), it is apparent that the amount of grooming behaviour displayed by female rhesus monkeys fluctuates throughout the course of the menstrual cycle; the time these allogroom males in test situations fluctuated rhythmically and reached a minimum in mid-cycle. At the same time, the males' behaviour reached a maximum. Allogrooming cycles were abolished by bilateral ovariectomy. It was concluded that the amount of heterosexual allogrooming in primate societies may depend upon the endocrine status of the females. The results of these experiments conflict with the observations by Carpenter (1942) on free-ranging rhesus monkeys. He noted an increase of allogrooming during oestrus, but he also noticed that autogrooming activity likewise increased.

There seems little doubt that allogrooming is very important in pair behaviour in primates although it is unlikely to be sexual in its motivation. Parents groom their young, and it has been pointed out that sexual behaviour may be stimulated in this way. Parental allogrooming may be an important factor in the development of sexual arousal in rhesus monkeys (Harlow and Harlow, 1962).

ALLOGROOMING AS CLEANING BEHAVIOUR

No doubt the stimulus situation eliciting allogrooming behaviour in primates is complex, but Knotters-Meyer (1928) thought that 'crystallized secretions of the skin whose salty flavour tickles their palates' may be an important factor. Ewing (1935) however, could find no such particles in the fur, even when examined under a hand lens. Evaporation of 1 ml of sweat collected from the forearm, produced crystals which were only visible under a microscope. He also found that '*M. syrichta*', showed a strong response to salt and confined its allogrooming to a spot which had been salted. Pieces of epidermis were rarely taken, but loose hairs were conveyed to the mouth (see also Reynolds, 1961). A three-year-old moor monkey and a pair of Celebes apes of similar age showed no predilection for crystals of common salt presented upon a bared forearm, but if sawdust was rubbed into the arm, these individuals would show an increased interest and intensively allogroomed the arm; pieces of dust would be picked out and some would be taken to the mouth (pers. observs.). Of chimpanzees, Falk (1958) thought that abrasions, any unevenness of texture, 'especially a slightly raised mole' were particularly effective in eliciting allogrooming. Sparks (unpublished observations), found that in monkey-man allogrooming relationships, both moor monkeys and Celebes apes respond to 6 cm long hair on the head rather than that on the fore-arm which was approximately 1 cm long. Higher primates may react to fur innately by grooming it (Zuckerman, 1932), but the observations reported above do indicate that particles enmeshed in the fur may be important in restricting the response to certain areas.

Under natural conditions, allogrooming may be important in the removal of ecto-parasites from the pelage (Carpenter, 1940; Washburn and DeVore, 1961; Zuckerman, 1933; Yerkes, 1933). Indeed Schaller (1963) thought that for gorillas, no further explanation is necessary. Simonds (1965) noted that injured bonnet monkeys 'presented' for grooming and that dirt would be extricated from the wounds, which would then be licked clean. However there may be reasons to suppose that the purely functional demands of the coat are exceeded. It is difficult to believe, for instance, that the pelage of the more dominant males in troops of *Papio* species needs more attention than that of low-ranking individuals who are allogroomed less. Indeed, for baboons a further explanation is needed (DeVore, 1963a). Workers in the field of primatology since Watson (1908) have generally accepted the fact that allogrooming has some kind of social significance apart from its purely cleansing activity; this is not in any way disputed.

However, because allogrooming is part of an important social activity, Rowell and Hinde (1963) have suggested that it should be regarded as a different type of activity from skin care patterns common to all mammals. Primates are unusual however, in so far as the majority of the cleansing behaviour is of the allogrooming variety. Although most captive higher primates, will pick through their own fur, Nolte (1954) noted that bonnet monkeys did not autogroom under natural conditions. In most other species of mammals and birds which allogroom or allopreen, this activity does not take the place of autopreening or autogrooming behaviour.

In primates, two groups of skin-care patterns can be recognized. There are patterns related to searching through the fur with the fingers and most of this is in the form of allogrooming. Secondly, there is body shaking and scratching which may be carried out with the fore or hind feet. The latter response looks as though it may be elicited by local irritation and occurs frequently in lone monkeys which are infected with lice. Scratching is also given in conflict situations, indeed, Rowell and Hinde (1963) found that allogrooming tended to decrease in stressful situations whereas 'brief skin care patterns' like scratching increased. These results are not unexpected. Similarly in chaffinches (*Fringilla coelebs*) shorter cleaning patterns occurred when the motivational balance is changing fast in stressful situations (Rowell, 1961). Allogrooming between friendly individuals tends to occur in stable primate societies, and usually in stress-free circumstances. The Harlows (Harlow and Harlow, 1962) showed that rhesus monkeys brought up in isolation, when introduced to one another, were unable to form a dominance hierarchy very quickly, fought more, and allogroomed less than wild ones. Allogrooming was registered at normal frequency after nineteen socially deprived individuals had been together for a few days in a large compound.

DISCUSSION

Allogrooming in primates is obviously important as cleaning behaviour, but it does seem to go beyond the functional demands of cleansing in some species. The amount of allogrooming also varies from one species to another, and between different individuals of any given species, depending upon their position in the dominance hierarchy; these variations must be explained.

It is necessary here to examine allogrooming behaviour in birds and non-primate mammals, and to see how it compares with that described for primates. An Estrildid bird allopreens its clumping partner under fairly well-defined circumstances. (a) It may occur as an ambivalent preening response. While sitting in bodily contact, a bird may first groom itself, then its clumping partner, and then once more preen itself. Thus, when its own preening drive is aroused, it may simply preen its partner because of its closeness, and in a sense treat the neighbouring individual as an extension of its own body. (b) A bird may preen its clumping partner for no apparent reason, and the precedent and subsequent behaviour of the actor will give no clue as to its motivation. (c) Some allopreening bouts are preceded by aggressive behaviour; thus, a number of hostile pecks may change into allopreening. It seems as though the raising of the head feathers by the clumping partner may serve to elicit a conciliatory response (allopreening) in the potential attacker. In this way, hostile behaviour is re-motivated by the allopreening invitation posture into a preening response. From the evolutionary point of view, the last context may have been more important. Allopreening tends to occur only in 'contact' species, and the allopreening invitation posture may have been of selective advantage because it tended to re-motivate a potentially hostile clumping partner by preventing the arousal of aggression. This behavioural mechanism would thereby facilitate clumping behaviour (Sparks, 1964a, b, c, 1965). This appears to work even in some 'distance species', and Selander and La Rue (1961) give an instance of a house sparrow's (*Passer domesticus*) attack on a cowbird (*Molothrus bonaviensis*) being changed into an allopreening response by the latter adopting a ruffled head posture.

It is clear that there is a close relationship between aggression and allogrooming in birds, and that these tendencies may interact to a greater or lesser extent in the motivation of this behaviour. Furthermore, grooming invitation behaviour tends to serve an appeasement or conciliatory function in both birds and rodents (pers. obs.; R. Sadlier, pers. comm.).

A general statement may now be made about allogrooming and invitation behaviour of birds and non-primate mammals. In an allogrooming partnership, there is a tendency for the more dominant or aggressive individual to perform the grooming while the one of lower status assumes the grooming invitation posture and is allogroomed (Sparks, *loc. cit.*; Eisenberg, 1962). The allogrooming or allopreening invitation behaviour includes many elements of appeasement, such as facing away from the groomer, inverted body posture, and closure of the eyes (see Chance, 1962; Sparks, 1964a). The grooming individual also adopts behaviour which could be interpreted as potentially hostile to the inviting one; for example, it fixates the groomee, and places its teeth or bill in contact with the body of the grooming partner.

Old World primates allogroom in a number of contexts. It occurs for no apparent reason sometimes, and it must be remembered that Zuckerman (1932) thought that primates had an innate tendency to allogroom fur, and Reynolds (1961) indicated that the sight of another's back served an almost irresistible stimulus to allogroom. There may be no signs of dominance during these kinds of allogrooming encounters. Parental grooming is presumably influenced by the hormonal condition of the mother and will not be discussed further. There are instances in the literature of hostile behaviour seemingly changed into an allogrooming response. Reynolds and

Reynolds (1965) refer to a case between two male chimpanzees and a female; male 1 had just copulated with the female who then went off and solicited in front of male 2, whereupon the first male approached and allogroomed him; this may have been caused by 'displaced aggressive tensions'. Reynolds (1961) has described 'blocking' where a rhesus monkey boldly walks up to another which responds by picking through its fur. There may be element of provocation in this kind of allogrooming soliciting behaviour, and similar cases have been recorded in birds ('Butting' – Sparks, 1964a, b). Although sexual presenting is an important appeasement display in monkeys, an individual which is attacked may occasionally allogroom the attacker; a female Celebes ape would attempt to allogroom in 'play' situations if her play-mate became too rough (pers. obs.). Of rhesus monkeys, Carpenter (1942) states that allogrooming may also function as a means of foiling an attack. Allogrooming as appeasement may be seen more clearly in the examples given by Rowell, Hinde and Spencer-Booth (1964); female rhesus monkeys would often allogroom mothers as a means of touching their babies. (See also Simonds, 1965, for Bonnet monkeys.) Allogrooming appeased the parent and prevented her from checking the advance of the aunt. Oestrus females and males will enter into reciprocal allogrooming relations and this may act as mutual conciliatory behaviour; the allogrooming by the female may keep the male at her side (Furuya, 1958). A female rhesus monkey may allogroom the male as a means of overcoming his aggression (Carpenter, 1942).

Many allogrooming bouts are initiated by invitation of the dominant animal. Now, on comparative grounds, it would seem reasonable that higher-status individuals should allogroom those of lower status. But it has already been shown that the majority of the allogrooming bouts in Old World primates are against the dominance slope of the hierarchy prevailing in these communities. Another interesting feature is that, again, on comparative grounds the higher-status individual which solicits grooming from a subordinate does so in a manner which appears to be indicative of submission. At least, of the patas monkey, Hall, Boelkins and Goswell (1965) stated that an animal soliciting grooming places itself in danger should the other animal attack. By facing away from the subordinate individual, and perhaps by rolling on its side, it is taking up a position which may appear indicative of lower status. There is also a correlation between the rigidity of the hierarchy and the amount of allogrooming behaviour. Thus, apes, with their relatively non-hostile in-groupings, exhibit less allogrooming than the more aggressive and hierarchical baboons and macaques.

Washburn and DeVore (1961) noticed bickering and fighting in only 15 per cent of *Papio ursinus* troops, but they point out that the hierarchy depends upon aggression and this ultimately leads to peace and order. However, a monkey troop is a cohesive unit, and Altmann, for rhesus macaques, points out that there may be a purely social attraction between individuals. It seems that there may be conflicting tendencies within macaque and baboon troops; that of social cohesion, but at the same time, an individual must realize its place in the hierarchical scheme, and act according to a convention, the transgression of which will elicit aggression. To all the members of the troop excepting the most dominant ones, life will be a matter of constant readjustment to, and perhaps avoidance of, those of higher status within the troop. Under this pressure higher primates have evolved various ways of appeasing such as lip-smacking and presenting (see Zuckerman, 1932). Individuals highest in the hierarchy of a primate society may have a different problem: that of maintaining social contact with those of lower status. Because of its aggressiveness (potential or actual) a higher-status monkey may tend to be avoided in most situations and it is in this connection that allogrooming behaviour may be important. It has been shown that dominant individuals may approach subordinate ones in order to be allogroomed, in a manner which seems to be the opposite to that of threat. Furthermore, allogrooming in most other mammals and birds tends to be characteristic of the more 'aggressive' individuals. It is in relation to this apparent reversal of rôles in primates that the following observations are interesting: Altmann (1962) thought that allogrooming or the presentation for allogrooming by a dominant animal could be considered as part of the behavioural complex which functions to 'assuage submissive behaviour'; in the hierarchical Japanese macaque studied by Furuya (1957), it was thought that allogrooming by a subordinate could serve to 'ease the tension caused by the difference of ranking' between the two individuals. In the light of the comparative data presented above, these two statements are very significant and in effect say the same thing.

If a troop leader approaches an individual of lower rank, the latter would normally give way and avoid contact; should the former animal present for grooming, and the subordinate does not retreat, then it seems likely that the lower-ranking animal's avoiding response was either reduced or thwarted by an increased tendency to approach and allogroom. In this context, the effects of dominance had been temporarily reduced by the behaviour of the higher-status individual. It is interesting that the males of many hierarchical primates (e.g. hamadryas and gelada) are furrier than the females, and therefore present better stimuli for grooming. On the other hand, it could be argued that the dominant animal has precedence and gets its own way on most occasions, and that this applies to allogrooming also. In other words, if it presents for grooming in front of a subordinate, the latter will respond appropriately for fear of retaliation.

It is suggested here that presenting for grooming in the higher primates, can, in certain contexts, act as a pattern which works against the enforcement of the hierarchy, serving to reduce avoidance or fear responses in subordinate individuals. This works by arousing responses such as allogrooming which are incompatible with aggression and fear (see Marler, 1965). This contrasts with the function of the allopreening invitation posture in birds, where it may help to reduce overt aggression rather than avoidance. There are a number of interesting implications for primates, because if it tends to promote cohesion in troops which maintain a steep dominance hierarchy, then, as a behavioural mechanism, it should be less predominant in those species where dominance is less important; thus, species which do not have a steep dominance gradient between individuals have little need of reassuring behaviour. The hypothesis is partly corroborated, because in the great apes, and the howler monkey, and many guenons, allogrooming is not as common as in troops of baboons and macaques.

At the same time, the observations on bonnet monkeys by Simmonds (1965) and in langurs by Jay (1965) do not seem to fit in with this scheme. Langurs spend up to five hours each day in allogrooming and yet their social organization is very relaxed with little fighting or exerting of dominance.

Although bonnet monkeys have a social system very similar to that reported for macaques and baboons, Simmonds (1965) states that the allogrooming activity of males is equal to that of females, both in intensity and frequency. However, in this species, males are not greatly dominant to females, as in many other *Macaca* species.

Man is a primate and it is not unreasonable to suppose that he should show some kind of behavioural traits of allogrooming, which is, after all, very characteristic of primate social interaction. It is apparent that 'de-lousing' behaviour is common among Polynesians; a companion's hair is searched with the aid of the fingers, and when a parasite is found, it is popped with the finger nails and thereupon carried to the mouth (Kroeber, quoted in Yerkes, 1933). In the Tobrianders similar behaviour is reported by Malinowski (quoted in Yerkes, 1937); in this case, the participants are not necessarily of different sexes. Yerkes (1933) has postulated that hospital care, skin adornment and beauty culture may have evolved from pre-hominoid allogrooming behaviour; however, most of this behaviour could have evolved through rational or cultural processes, but man's tendency to stroke (as reassurance to pet animals, for example) or his propensity for enjoying skin massage may derive from his innate tendency to allogroom or to be allogroomed, as practised far back in his evolutionary history. There is certainly evidence that massage is one of the oldest forms of medical treatment.

Furthermore, it may not be accidental that hair-dressers hold a prominent position in human societies. But perhaps the most interesting observation on human allogrooming is reported in Russell (1959). It concerned three Belgian youths, one of whom was obviously the leader of the group. The individual of the least status suddenly brandished a comb and started to rake through the leader's hair, who looked completely detached about the whole business. Not to be out-favoured, the second youth in the hierarchy snatched the comb away from the third, and took over the hairdressing himself. It has already been shown that allogrooming can be used as an appeasement gesture in lower primates, and this incident seemed to reflect a currying of favour with the leader by a subordinate.

It is also interesting to reflect upon the incidence of 'allogrooming' as between master and slave, and the relative status between males and females in different human societies.

ACKNOWLEDGMENTS

I have discussed the subject of this paper with a number of people; their criticism proved to be most valuable. I should particularly like to thank Professor K. R. L. Hall, Dr Fae Hall, Christine King, Dr G. Manley, Dr D. Morris, Drs V. and F. Reynolds.

Some of the work reported here was carried out in the Ethology Laboratory of the London Zoo, and I am grateful to the Scientific Director, Dr L. Harrison-Matthews, F.R.S., for allocating me generous space. Miss Laura Cannon and Miss Anita Hudson looked after the laboratory animals with considerable expertise and the D.S.I.R. financed the research.

I should also like to record my gratitude to Miss Valerie Tombs for taking such care in the preparation of the manuscript.

REFERENCES

ALTMANN, S. A. (1962). 'A field study of the Sociobiology of rhesus monkeys (*Macaca mulatta*)', *Ann. N.Y. Acad. Sci.* **102** (2), pp. 338–435.

ANDREW, R. J. and J. BUETTNER-JANUSCH (1962). 'The use of the incisors by Primates in grooming', *Ann. J. Phys. Anthrop.* **20** (2), p. 129.

ARMITAGE, K. B. (1962). 'Social behaviour of a colony of the yellow-bellied marmot', *Anim. Behav.* **10**, pp. 319–31.

BARNETT, S. A. (1963). *A Study in Behaviour*, London, Methuen.

BISHOP, A. (1962). 'Control of the hand in Lower Primates', *Ann. N.Y. Acad. Sci.* **102** (2), pp. 316–37.

BOLWIG, N. (1959). 'A study of the behaviour of the Chacma Baboon (*Papio ursinus*)', *Behaviour* **14**, pp. 136–63.

CARPENTER, C. R. (1934). 'A field study of the behaviour and social relations of howling monkeys (*Allouatta palliata*)', *Comp. Psychol. Monogr.* **10** (48), pp. 1–168.

—— (1935). 'Behaviour of red spider monkeys in Panama', *J. Mammal.* **16**, pp. 171–80.

—— (1937). 'An observational study of two captive mountain gorillas (*Gorilla berengi*)', *Human Biol.* **9** (2), pp. 175–96.

—— (1940). 'A field study in Siam of the behaviour and social relations of the gibbon', *Comp. Psychol. Monogr.* **16** (5), pp. 1–212.

—— (1942). 'Sexual behaviour of free-ranging rhesus monkeys (*Macaca mulatta*)', *J. Comp. Psychol.* **33**, pp. 113–42.

CHANCE, M. R. A. (1956). 'Social structure of a colony of *Macaca mulatta*', *Anim. Behav.* **4**, pp. 1–13.

—— (1959). 'What makes monkeys sociable?', *New Scientist* **5**, pp. 520–3.

—— (1962). 'An interpretation of some agonistic postures; the rôle of "cut-off" acts and postures', *Symp. Zool. Soc.* **8**, pp. 71–89.

CRAWFORD, M. P. (1942). 'Dominance and the behaviour of pairs of female chimpanzees when they meet after varying intervals of separation', *J. Comp. Psychol.* **33**, pp. 259–65.

CULLEN, J. M. (1963). 'Allo, auto, and heteropreening', *Ibis* **105**, p. 121.

DeVORE, I. (1963a). 'Mother-infant relations in free-ranging baboons', *Maternal Behavior in Mammals*, ed. H. L. Rheingold, New York, pp. 305–35.

—— (1963b). 'Comparative ecology and behavior of monkeys and apes', *Classification and human evolution*, ed. S. L. Washburn, Viking Fund Publications in Anthropology, No. 37, New York, pp. 301–19.

EISENBERG, J. (1962). 'Studies on the behaviour of *Peromyscus maniculatus gambeli* and *Peromyscus californicus parasiticus*', *Behaviour* **19**, pp. 177–207.

EWING, H. E. (1935). 'Sham louse-picking, or grooming among monkeys', *J. Mammal.* **16**, pp. 303–6.

FALK, J. L. (1958). 'The grooming behaviour of the chimpanzee as a reinforcer', *J. Exp. Anal. Behaviour* **1**, pp. 83–5.

FITZGERALD, A. (1935). 'Rearing marmosets in captivity', *J. Mammal.* **16**, pp. 181–8.

FURUYA, Y. (1957). 'Grooming behaviour in the Wild Japanese monkeys', *Primates* **1** (1), pp. 47–68 (in Japanese).

GRANT, E. C. and J. H. MACKINTOSH (1963). 'A comparison of the social postures of some common laboratory rodents', *Behaviour* **21**, pp. 246–59.

HADDOW, A. J. (1952). 'Field and laboratory studies on an African monkey (*Cercopithecus ascianus schmidti*) Matschie', *Proc. Zool. Soc. Lond.* **122**, pp. 297–394.

HALL, K. R. L. (1965). 'Social organisation of the old world monkeys and apes', *Symp. Zool. Soc. Lond.* **14**, pp. 265–89.

HALL, K. R. L. and J. S. GARTLAN (1965). 'Ecology and behaviour of the vervet monkey (*Cercopithecus aethiops*), Lolui Island, Lake Victoria', *Proc. Zool. Soc. Lond.* **122**, pp. 37–56.

HALL, K. R. L., R. C. BOELKINS, and M. J. GOSWELL (1965). 'Behaviour of patas monkeys (*Erythrocebus patas*) in captivity, with notes on the natural habitat', *Folia primat.* **3**, pp. 22–49.

HARLOW, H. F. and M. K. HARLOW (1962). 'Social deprivation in monkeys', *Sci. Amer.* **207**, pp. 137–46.

HARRISON, C. J. O. (1965). 'Allopreening as agonistic behaviour', *Behaviour* **24**, pp. 161–209.

HEDIGER, H. (1942). *Wildtiere in Gefangenschaft.*

HINES, M. (1942). 'The development and regression of reflexes, postures, and progression in the young macaque', *Contrib. Embryol.* **30** (196), pp. 153–209.

JAY, P. (1963). 'Mother-infant relations in langurs', *Maternal behavior in mammals,* ed. H. L. Rheingold, New York.

—— (1965). 'The common langur of North India', *Primate behavior,* ed. I. DeVore, New York, Holt, Rinehart and Winston.

KNOTTERS-MEYER, T. (1928). *Birds and beasts of the Roman Zoo,* New York, Century Co.

KUMMER, H. (1957). 'Soziales Verhalten einer Mantelpavian-Gruppe', *Schweiz-Zeitsch. Psychol.* No. **33**.

KUNKEL, P. (1962). 'Bewegungsformen, Sozialverhalten, Balz. und Nestbau des Gangesbrillen vogels (*Zosterops palpebrosa* Tenam)' *Z. f. Tierpsychol.* **19**, pp. 559–76.

LORENZ, K. (1931). 'Beiträge zur Ethologie Sozialor Corviden', *J. Orn.* **79**, pp. 67–120.

MARLER, P. (1965). 'Communication in monkeys and apes', *Primate behavior,* ed. I. DeVore, New York, Holt, Rinehart and Winston.

MASLOW, A. H. (1936). 'The rôle of dominance in social and sexual behaviour of infra-human primates. IV. The determination of a hierarchy in pairs in a group', *J. Genet. Psychol.* **48**, pp. 278–309; **49**, pp. 161–98.

MORRIS, D. (1956). 'The feather postures of birds and the problem of the origin of social signals', *Behaviour* **9**, pp. 75–113.

—— (1957). 'Typical Intensity and its relation to the problem of ritualisation', *Behaviour* **11**, pp. 1–12.

MOYNIHAN, M. (1955). 'Types of hostile display', *Auk* **72**, pp. 247–59.

MICHAEL, R. P. and J. HERBERT (1963). 'Menstrual cycle influences grooming behaviour and sexual activity in the rhesus monkey', *Science* **140**, pp. 500–1.

NOLTE, A. (1955). 'Field observations on the daily routine and social behaviour of common Indian Monkeys, with special reference to the Bonnet Monkey (*Macaca radiata* Geoffrey)', *J. Bomb. Nat. Hist. Soc.* **53**, pp. 177–84.

—— (1958). 'Beobachtungen über das Instinktverhalten von. Kapu-zineraffen (*Cebus apella*) in der Gefangenschaft', *Behaviour* **12**, pp. 183–207.

REYNOLDS, V. (1961). 'The social life of a colony of rhesus monkeys (*Macaca mulatta*)', Ph.D. Thesis, Univ. of London Library.

REYNOLDS, V. and F. REYNOLDS (1965). 'Chimpanzees of the Budongo Forest', *Primate behavior*, ed. I. DeVore, New York, Holt, Rinehart and Winston.

ROWELL, C. H. FRASER (1961). 'Displacement grooming in the chaffinch' *Anim. Behav.* **9**, pp. 38–63.

ROWELL, T. E. and R. A. HINDE (1963). 'Responses of rhesus monkeys to mildly stressful situations', *Anim. Behav.* **11**, pp. 235–43.

ROWELL, T. E. and Y. SPENCER-BOOTH (1964). ' "Aunt"; infant inter-action in captive rhesus monkeys', *Anim. Behav.* **12**, pp. 219–226.

RUSSELL, W. M. S. (1959). 'On comfort and comfort activities in animals', *Univ. Fed. Anim. Welfare Courier* **16**, pp. 14–26.

SCHALLER, G. (1963). *The mountain gorilla: ecology and behavior*, Univ. of Chicago Press.

SCOTT, J. P. and E. FREDERICSON (1951). 'The causes of fighting in mice and rats', *Physiol. Zool.* **24**, pp. 273–309.

SELANDER, R. K. and C. J. LA RUE (1961). 'Interspecific preening invitation display of parasitic cowbirds', *Auk* **78**, pp. 473–504.

SIMMONDS, P. E. (1965). 'The bonnet macaque in South India', *Primate behavior*, ed. I. DeVore, New York, Holt, Rinehart and Winston.

SOUTHWICK, C. H., M. A. BEG, and M. R. SIDDIQI (1965). 'Rhesus monkeys in North India', *Primate behavior*, ed. I. DeVore, New York, Holt, Rinehart and Winston.

SPARKS, J. H. (1962). 'Clumping and social preening in red avadavats', *Birds Illustrated* **8**, pp. 48–9.

—— (1964a). 'Ethology of the red avadavat with particular reference to social and sexual behaviour', Ph.D. Thesis, Univ. of London Library.

—— (1964b). 'Flock structure of the red avadavat with particular reference to clumping and allopreening', *Anim. Behav.* **12**, pp. 125–36.

—— (1964c). 'Contact behaviour in the Cuban finch *Tiaris canora* (Gmellin')', *Bull. B.O.C.* **84**, pp. 164–9.

—— (1965). 'Allopreening invitation as appeasement behaviour in red avadavats: with comments on its evolution in the Spermestidae', *Proc. Zool. Soc. Lond.* **145** (3), pp. 387–403.

TINBERGEN, N. (1954). 'Courtship and threat display', *Ibis* **96**, pp. 233–250.

VAN HOOFF, J. A. R. (1962). 'Facial expressions in higher primates', *Symp. Zool. Soc. Lond.* **8**, pp. 97–125.

WALLIS, D. I. (1962). 'Behaviour patterns of the ant *Formica fusca*', *Anim. Behav.* **10**, pp. 105–12.

WASHBURN, S. L. and I. DeVore (1961). 'Social life of baboons', *Sci. Am.* **204** (6), pp. 62–71.

WATSON, J. B. (1908). 'Imitation in monkeys', *Psychol. Bull.* **5**, pp. 169–178.

YERKES, R. M. (1933). 'Genetic aspects of grooming, a socially important primate behaviour pattern', *J. Soc. Psychol.* **4**, pp. 3–25.

ZUCKERMAN, S. (1932). *The social life of monkeys and apes*, London.

—— (1933). *Functional affinities of man, monkeys and apes*, London.

Chapter Five

Play Behaviour in Higher Primates: a Review

CAROLINE LOIZOS

INTRODUCTION

THE subject of play, probably more than any other area of animal behaviour, is open to confusion, misinterpretation and armchair theorizing. Opinions differ even as to its existence. Anyone who has ever owned a pet has remarkably determined views as to when it is and is not playing; on the other hand hard-headed mechanists regard references to animal play as examples of imperfectly understood behaviour. This may often be accurate, as an example quoted by Beach (1945) will show. Various fishes have been observed to leap over free-floating objects in the water, such as sticks and reeds. This was considered to be playful activity with no practical value until Breder (1932) suggested that what the fish were doing probably served to dislodge encrustations of ectoparasites from their undersides.

The main problem in studying animal play has resulted from a basic misconception in the approach, arising purely from our use of the word *play*. Play is a human concept, used of activity that is other than, or even opposed to work. By analogy the word has come to be applied to behaviour in animals which cannot be seen to have any immediate biological end, any obvious survival value. Bierens de Haan's (1952) statement that 'animal play is useless' is fairly typical of this attitude. The implication is simply that animal play cannot be serious, since if it were, if it had a function, it would not be play.

But for animals, of course, it is different. Since they do not work, at any rate in our sense of the word, they cannot really be said to play – in our sense of the word. The problem is not solved by thinking of an alternative word for play. The fact that even an untrained observer may be quite accurate in determining when an animal is playing suggests that somewhere in the complex of behaviour called play there is a fundamental similarity with the same kind of activity in human beings. What this similarity might consist of will be suggested later. The word play, then, is a useful shorthand description of a characteristic motor quality common to behaviour occurring in certain situations.

It is probably more useful to change the direction of approach. Instead of saying that human play appears to be without survival value and that therefore the same must be true of animal play, let us assume that animal play *has* survival value and that it has possibly become divorced from some of its original functions by the time it occurs in human beings – at any rate adult human beings. It would probably be rash to assume that it does not have survival value in animals, and is not in this sense serious, since amongst other things the amount of time and energy spent in play by, for instance, the carnivores or the primates, would surely put these animals at a disadvantage if their play were totally without function.

Most of the earlier approaches to the subject of play, although fascinating and often fruitful, were made on the basis of anecdote and observation alone, with no attempt to provide experimental backing for the hypotheses arising from this mass of data. Now psychologists in particular have begun the attempt to analyse play behaviour in a quantitative manner. Their emphasis has been primarily on its causation rather than its function. This chapter will provide a review of this more recent work with particular reference to primates; for a clear summary of the earlier work on play, and discussion of the problems arising from it, the reader is referred to Beach (1945). Welker (1961) has also provided a review and discussion of some aspects of play, in particular those forms related to investigation and exploration. A brief review is provided at the end of this chapter of field observations on play in higher primates.

THE NATURE OF PLAY

Before discussing to any great effect current theories of the function and causation of play, it is necessary to indicate the kind of behaviour that is going to be referred to in this paper as play. One of its immediately noticeable characteristics is that it is behaviour that borrows or adopts patterns that appear in other contexts where they achieve immediate and obvious ends. When these patterns appear in play they seem to be divorced from their original motivation and are qualitatively distinct from the same patterns appearing in their originally motivated contexts. Lorenz (1956) points out that although in play-fighting movements occur which are only seen at the peak of intensity in a serious fight, the next moment the animal demonstrates that such specific motivation is lacking by switching to behaviour seen in defensive or grooming, or other unrelated situations. Thus a *re-ordering* of the original sequence is one of the ways in which play behaviour differs from the source of its motor patterns. All the other ways in which it differs are of economy, or degree; and it is here that we come back to a statement made earlier, about the fundamental similarity to the observer between human and animal play. This similarity lies in the exaggerated and uneconomical quality of the motor patterns involved. Regardless of its motivation or its end-product, this is what all playful activity has in common; and it is possible that it is all that it has in common, since causation and function could vary from species to species. Beach puts this more strongly: '. . . no single hypothesis can be formulated to explain all forms of play in every animal species'.

These are some of the ways in which motor patterns may be altered and elaborated upon when transferred to a playful context:

1. The sequence may be *re-ordered*.

2. The individual movements making up the sequence may become *exaggerated*.

3. Certain movements within the sequence may be *repeated* more than they would normally be.

4. The sequence may be broken off altogether by the introduction of irrelevant activities, and resumed later. This could be called *fragmentation*.

5. Movements may be both *exaggerated and repeated*.

6. Individual movements within the sequence may never be completed, and this incomplete element may be repeated many times. This applies equally to both the beginning of a movement (*the intention element*) and to its ending (*the completion element*).

In every case, during play, the performance of the original

movements, those from which the play is derived, is uneconomical and therefore would be inefficient in terms of the original motivating context. It might, of course, be possible to consider that it is the exaggerated movements of play that are refined and economized, and used in chasing, wrestling, biting, jumping, chewing and so on, instead of the reverse process. However, since the patterns of, for example, aggression and defence occur in the phylogenetic scale long before unequivocal play behaviour makes its appearance, it seems fair to assume that the 'original' context – in this case at any rate – is that of aggression. Aggression is also more basic to survival. The same is true of most other motor patterns that are employed in play: investigation of objects occurs both phylogenetically and ontogenetically before play with objects.

Thus playful patterns owe their origin to behaviour that appeared earlier phylogenetically and for purposes other than play. It follows that just as patterns of fight, flight, sexual and eating behaviour are species-specific, so will the play behaviour making use of these same patterns be species-specific.

However it is clear that the motor differences between a pattern used in play and the same pattern occurring in its original context are not exclusive to play. Ritualized behaviour shares many of the characteristics described as typical of play; and social play clearly contains a strong element of ritualization. Morris (1956) quotes the following examples of some of the ways in which basic patterns can be modified – or ritualized – to form signals: (a) threshold lowering, (b) development of rhythmic repetition, (c) differential exaggeration of components, (d) omission of components, (e) change in sequence of components, (f) changes in component co-ordination, (g) increase or decrease in speed of performance, (h) change in vigour of movements.

Although Morris (1954) has shown that the sequence of events in a highly ritualized piece of behaviour such as the stickleback's courtship dance is not nearly as rigid as it was once considered to be, it is now suggested that the most likely area in which the precise differences between play and other forms of ritualized behaviour will be isolated is that of relative rigidity in the ordering of the sequence. It is suggested that play has no formalized sequence of events, such that action A will always be followed by actions B, C or D. In play, depending upon the feedback from the object or the social partner, A may be followed with equal likelihood by B or by Z; anyway by a far greater range of responses than are seen in other forms of ritualized behaviour. In short, it may be that in play the number of combinations or permutations of the available motor patterns is greater than in almost any other form of behaviour.

For the moment it might be useful to approach play in a roundabout way, in terms of some of the conditions necessary for its occurrence, and some of the features which invariably accompany it. There are motivating conditions which are necessary for the occurrence of play, though as they consist almost entirely of the absence of other conflicting sources of motivation, they cannot be said in any way to be sufficient. As far as can be told, play only occurs when the animal is free of environmental pressures, such as heat, cold, wet and the presence of predators; and free of physiological pressures such as the need for food, drink, sleep or a sexual partner (Lorenz, 1956; Bally, 1945). Thus play is often most characteristic of young animals whose needs are taken care of by their parents, and of animals in captivity for whom the same functions are served by their guardians. Again, however, this is only part of the story, since there is some evidence from observation of zoo animals (Morris, 1964) that there exists a positive need to engage in certain types of play. Play probably does not occur solely as the result of the absence of conflicting drives.

It is not behaviour therefore, that appears to be concerned with immediate survival – probably the single factor that has contributed most to its abuse as a category of behaviour. However, it may be that the motor patterns employed in play were at one time related to survival. This point may be illustrated by referring to the example quoted by Beach mentioned earlier, in which behaviour in fish thought to be playful was shown to have an immediate and valuable cleaning function. It would be interesting to speculate on what might happen to this particular behaviour were the species to rid itself permanently of these parasites. It is quite possible that the habit of leaping over floating objects in the water would remain, and ethologists in the next century would certainly be tempted to call it play once more. Tembrock (1960) has pointed out that in the arctic fox (*Alopex lagopus*) certain behaviour which now appears only in playful contexts probably originally had specific adaptive value in a particular earlier environment; when freed of the necessity to perform this particular function, perhaps through a change in the environment, such behaviour could be incorporated freely into the repertoire of play patterns. Thus it is likely that some play consists of the vestiges of phylogenetically very old behaviour; in some cases it may be no more than a trace, but in others, complete sequences may have become 'fossilized' and preserved whole. One cannot do more than make informed guesses about the origins of some play patterns. On the other hand there are probably many more instances of what we now think of as play that will turn out to have quite specific and other functions, since the concept has, certainly until very recently, always been used as the wastepaper-basket of imperfectly understood animal behaviour.

Attempts to control animal play by means of specific kinds of deprivation or reinforcement have on the whole been unsuccessful. Schiller (1957) reported that '... with no incentive the chimpanzee displayed a higher variety of handling objects than under the pressure of a lure which they attempted to obtain'. In his case, the attempt to direct play by reinforcing the animals for this behaviour resulted in its inhibition.

However, these animals which play may be encouraged to play by the presentation of suitable stimuli in a suitable manner. Moreover, they may be *invited* to play by a conspecific, or even a member of another species, as, for example, will often happen with humans and domestic animals. A mother cat lying on her side and twitching the tip of her tail on which her kittens will pounce might well be an example of such a play invitation. This kind of invitation – what Altmann (1962) called metacommunication, or a signal about the quality of the communication which is to follow – may be seen in many mammals. Cats and dogs both have preliminary play movements, play intention movements in fact, which consist of a half-crouch with fore-legs extended stiffly, combined with wide-open eyes and ears pulled forward. Brownlee (1954) has described this phenomenon in domestic cattle. It occurs most noticeably in chimpanzees and other primates, but at this level on the phylogenetic scale the signal area has been reduced to the face, and a special facial expression indicating a playful mood is sometimes used as a kind of shorthand for the full motor play invitation. Similarly with humans: if one is punched by someone with a broad grin on their face one will at least hesitate before interpreting it as an aggressive act. The interesting thing in this situation is that given the choice of two conflicting signals to attend to, the one that is usually given priority is the one announcing that this is play, even though the punch may have been hard enough to hurt. Play signals seem to be very powerful and unambiguous. One can observe young chimpanzees putting up with treatment from each other that is quite rough enough to cause pain, provided it has been made clear at the outset that they are playing. Another interesting feature of the play signal or invitation is that it appears to be interspecific, or easily understood between species. In this it differs from, for example, courtship behaviour which is very often individual to the species in question. The fact that the play signals are so clearly understood between species, combined with the fact of their priority over other conflicting signals, makes it more difficult for the observer to remain objective than when watching, for example, the reproductive behaviour of the ten-spined stickleback, which bears little or no relation to our own species' courtship behaviour. The observer of play is in fact under constant pressure to join in.

Play is described, therefore, as a positive approach towards, and non-rigidified interaction with, any feature of the animal's environment, including conspecifics, involving stimulation through most available sensory modalities. The behaviour patterns occurring in such activity are motorically similar to patterns occurring in contexts in which they serve an immediate and specific biological function, but they differ qualitatively from these patterns in that the movements are exaggerated and uneconomical; they would therefore be *inefficient* in terms of the originally motivated context. In the context of play these patterns serve different ends, none of which appear to be related to immediate survival. Play patterns may also be remnants of phylogenetically ancient behaviour that have become freed through environmental changes from their original functions. Play is not associated with states of physiological deprivation or of environmental stress.

Both the motoric and the motivational differences between serious and playful chase behaviour may be illustrated by a condensed description of social interactions observed and recorded in the chimpanzee colony at the London Zoo (June–July, 1965). During a serious chase, the pursuer rarely takes his eyes off the pursued animal. His fur is erect and bristling, his lips are pursed and protruded in the aggressive manner. His movements are rapid and economical; if he catches the animal he is pursuing, the sequence of events will be snatch, grab, pull-towards-mouth and bite, at whatever portion of the offender is most readily available. Such interactions are often ended by the pursued animal's renewed escape, accompanied by loud screams and possibly by the fear grimace, in which both rows of teeth are bared, often including the gums.

A play-chase, on the other hand, takes place at an altogether different tempo. The initiator may approach another animal by walking or trotting towards him with a highly characteristic bounce to the gait; the head bobs up and down, his gaze may not be directed at the animal he approaches, he is often wearing the playface, and soft guttural exhalations may be audible. (The playface is a special expression indicating playful intent in which only the lower teeth show. It appears to occur most often at the beginning of a playful interaction, the point at which it is most necessary to avoid being misunderstood.) As he reaches the animal he is approaching and gains his attention he turns around and makes off at a slow lolloping pace in the opposite direction, looking back over his shoulder to see if the other animal is following. His head is still bobbing, he is still wearing the playface. If the other animal responds and takes up the chase the tempo may quicken slightly and the playface become less evident (since it has served its function of letting the partner know the intent of this particular interaction). A play-chase may occur in one direction only, or pursuer and pursued may switch rôles several times in the course of a few minutes. Quiet guttural exhalations are often heard throughout. If a play-chase ends in direct contact (not by any means inevitable), play-wrestling may occur, differing from the grab-pull-bite sequence of the serious fight both in intensity and in the order of this particular sequence. Pulling alone may occur, or a great deal of biting of each other's fingers or toes, or of the area between the neck and the shoulder. Visitors to the Zoo have been overheard to mistake a serious chase for play behaviour, but never a play-chase for one more serious. This fact again points up the interspecific quality of much play behaviour. However, sometimes the behavioural differences between playful behaviour and the same patterns occurring in their originally motivated contexts may be very minor, and take an expert – possibly an expert conspecific – to distinguish them. Bolwig (1963a) notes that if it were not for the fact that his young patas monkey was exhibiting the playface, it would be difficult to tell his playful attacks from those more seriously intended. (Further discussion of the signal value of the playface, and the possible ambiguity of such behaviour, will be discussed in the section on Social Play.)

THEORIES OF THE FUNCTION OF PLAY

As Mason (1965) points out, the problem of a comprehensive definition of play is certainly more difficult in theory than in practice, since there does exist so much agreement between observers familiar with a species as to when play is occurring. However, a problem more difficult in practice than in theory is that of the function of play.

The most generally accepted theory is to consider it as practice for adult activity (Pycraft, 1912; Mitchell, 1912; Groos, 1898). The main problem with this approach is that no clear distinction has been drawn between playful and non-playful behaviour in the young of any particular species. During the infancy and adolescence of many mammals it is possible to see immature forms of behaviour patterns which will appear in their complete form and appropriate context in adult life. In the young animal this kind of behaviour does not appear to serve the same biological ends that it does in maturity; and consequently observers have tended to regard as play all behaviour performed while the young animal's primary needs were being taken care of by its adult conspecifics. Often these immature forms of adult behaviour are performed in a characteristically playful manner, but equally often such behaviour is performed with the greatest possible degree of efficiency for whatever level of development the animal is at *at that point*. Thus, a kitten can and often does chase bits of paper and string in a playful manner; but it may equally well deal with them in a way that would have meant instant death to a mouse. In the same way, a fight between young chimpanzees may be quite as seriously intended as one between two adults, but the effects are obviously less drastic. It is a mistake, therefore, to regard as play all chase behaviour by a kitten, either because it is a kitten or because the object pursued is inedible; and the same holds for all young animals.

Nevertheless, it is a widely held view that the animal that plays – or practises – will become more expert, and thereby have a selective advantage over the animal that does not (Groos, 1898). None of this is to deny that practice or rehearsal of many forms of behaviour is likely to improve the efficiency with which they are performed in adult life. But it has yet to be shown that it is the *playful* execution of these particular patterns in infancy or childhood that is crucial to their later perfection, as opposed to their *non-playful* execution at whatever level of maturity the animal is then operating. Quite simply, it is not necessary to play in order to practise – there is no reason why the animal should not just practise. Certainly social interactions involving play within the peer group in rhesus monkeys have been shown by Harlow (1962) to be crucial for the full development of adult social behaviour. But the precise rôle of each of play's component parts – vision, smell, sound, physical contact, movement and any combination of these factors – has yet to be isolated and defined.

Furthermore, to regard play as practice for adult function does not account for the fact that adults as well as infants play in most mammalian species in which play occurs at all. They may certainly play less, but they still play. The fact that they play does not of course prevent the same behaviour in infants from serving as preparation of some kind; but as an explanation of function this must be enlarged to account for the persistence of, for example, play-wrestling in the adult cat.

The same objection is offered to the other major theory of the function of play: that it provides the animal with a constant stream of vital information about every feature of the environment with which it comes into contact. Again, it is simply not necessary to play in order to learn about the environment. The animal could explore as, in fact, some mammals that do not appear to play certainly do; for example, the rat. Of course it is inevitable that during play, or during any activity, an animal will be gaining additional knowledge about what or who it is playing with; but if this is the major function of play, one must wonder why the animal does not use a more economical way of getting hold of this information.

Brownlee's (1954) paper on play in domestic cattle suggests that play is the agency by which the somatic muscle groups involved in agonistic and reproductive behaviour are kept exercised in the young animal. These muscles are thus prevented from atrophying from disuse before the animal actually needs to engage in true agonistic or reproductive behaviour. More experimentation is necessary before this theory can be assessed; but again, it is unlikely that this is a sufficient explanation for the range and variety of all play behaviour.

THEORIES OF THE CAUSATION OF PLAY

The earliest attempt to account for the causation of play, especially as seen in the young, suggested that it represented the release or overflowing of abundant energy which had no other immediate outlet. A modern parallel to this theory is that of 'vacuum' activity, which occurs when a particular response has not been released for some time, and eventually occurs in the apparently total absence of any specific releasing stimulus. However, there are clear motoric differences between playful and vacuum activity, which led Lorenz (1956) to propose a distinction between them.

Tolman (1932) suggested that under certain conditions, men and lower animals have a need to redress a state of physiological imbalance produced by the presence of abundant energy by achieving a complementary state of mild fatigue. Beach (1945), however, considered this to be simply a 'variation . . . in modern dress' of the surplus energy hypothesis. He points out that interpretation of the energy expended in a particular action as surplus simply depends on whether you consider the behaviour under observation to be playfully or otherwise motivated. In any case, as Groos pointed out, young animals can be seen to lie panting and exhausted after a bout of play and suddenly to resume the game with apparently equal vigour. Although play is most likely to occur when the animal is not exhausted, so also is non-playful activity. So again, although energy may be a necessary condition for play, it has not been shown to be sufficient.

Play has sometimes been accounted for by stating that it is 'fun' (Bierens de Haan, 1952), 'expresses a joy of living' (Pycraft, 1912), 'is enjoyed purely for its own sake' (Tinkelpaugh, 1942), or, more recently, is 'self-rewarding activity' (Morris, 1962). Bolwig (1963b) describes the motivating condition for play as one of joy and goes on to describe joy as a condition which motivates increased activity. 'Postures of aggression and retreat never become complete, and non-aggressive movements are frequently exaggerated. In other words, joy is a condition which induces play actions.' All these statements may at the moment be the most useful shorthand way we have of describing the apparent affect accompanying certain motor patterns; but their use as an explanation of that same behaviour is unjustified and not at all useful.

The most recent of the theories dealing with the motivating conditions accompanying play behaviour is Mason's discussion of arousal, stemming from work developed by Bindra (1959), Duffy (1951), Hebb (1955) and others. Mason's results lead him to conclude that play is the preferred activity of the young chimpanzee. Play behaviour will occur at certain levels of arousal in the young animal, which in turn will depend upon the novelty of the situation and the animal's previous experience. Increases in novelty resulting from 'gross changes in the familiarity of the test situation' may depress play behaviour by increasing the level of arousal to the point where the animal prefers to cling to a familiar person. Mason considers that '. . . an orderly transition from one social pattern to another will occur as the result of progressive changes in arousal level'. In general, social play occurs in conditions of low or moderate arousal and is avoided when arousal is high, whereas the reverse is true of clinging.

The concept of arousal does not rely on any specific variable for its effect on social responsiveness. Thus the strangeness of the surroundings, unfamiliar sensory stimulation or separation from a cage-mate – all of them different aspects of novelty – may all affect the tendency to play. Mason emphasizes, however, that the 'presumed function of the level of arousal is to establish a predisposition to engage in certain patterns of behaviour rather that to serve as a precipitating cause'. Nevertheless, the work reported in this paper represents one of the only attempts to investigate experimentally the causation of play in chimpanzees, or in fact in any animal, and the concept of arousal has become popular among psychologists investigating responses to novel stimuli in primates.

PLAY IN PRIMATES

As Beach pointed out, '. . . the amount, duration and diversity of play behaviour in animals of a given species is related to their position on the phylogenetic scale'. In general, the higher the animal on the scale, the more frequent and varied is its play. That the relationship between phylogeny and play is not a direct one is made clear by Lorenz (1956) and Morris (1964) who propose a distinction between animals whose mode of survival is highly specialized, either structurally or behaviourally, and those who are 'opportunists' – that is, 'specialists in non-specialization'. The opportunists' stock-in-trade is their restless curiosity; what Morris (1964) has called their *neophilia*, or love of the new. Behaviourally, the neophilic animals share a tendency to play (although not all 'players' are neophilic), and in general maintain higher levels of activity than more specialized species. Morris ascribes this to the evolution in non-specialists of '. . . a nervous system that abhors inactivity', developed as protection against the hazards of non-specialization, since, '. . . having no special device, they cannot afford to miss any chance of a small reward'. Of such animals, the primates are probably the supreme examples; and of all the sub-human species the chimpanzee is probably the most playful. Certainly, most of the detailed observation of play patterns have been carried out on chimpanzees (*Pan troglodytes*) or rhesus monkeys (*Macaca mulatta*). With very few exceptions these have been captive animals. Goodall (1965) and Reynolds (1965) have also recorded observations of play in chimpanzees in the wild. Schaller (1965) has made a similar record of the play of wild mountain gorillas (*Gorilla gorilla beringei*). Harrison (1962) records some play in young captive orang-utans (*Pongo pygmaeus*), and Carpenter (1964d) on gibbons (*Hylobates lar*). Some of these observations will be referred to in more detail in later sections of this chapter.

ONTOGENY OF PLAY IN CHIMPANZEES

A condensation of various accounts of the ontogeny of play in two young captive chimpanzees will illustrate the range of play behaviour seen in this species. Both chimpanzees discussed were removed from their mothers shortly after birth and were reared by humans, several of whom they became very familiar with.

Jacobsen, Jacobsen and Yoshioka (1932) divided play into several categories, each of which appeared at a different level of development. 'Exploration, manipulation and simple play' made their appearance within the first two months of life. The infant's attention was directed outside its own body to the walls and floor of its crib, which it scratched or probed with its index finger. This activity increased up to the third month, after which it declined;

this was presumably the point at which locomotory forms of bodily activity predominated over stationary acrobatics. These authors do not give the grounds on which they distinguish exploratory activity from play. Most observers would classify the repetitive dropping and retrieving of a grapefruit seed with the lips as playful activity; but it is again emphasized that play is best defined by the quantitative change in the motor patterns involved in any activity, rather than by the apparent aimlessness of the activity itself. There are few, if any, behaviour patterns within the normal repertoire of the primate that cannot be performed in a playful manner. However, there are forms of some patterns of behaviour that will occur mainly in play (excessive repetition, for instance), and equally some patterns that will rarely occur in a playful context (excretion, for example, under normal conditions).

A general increase in exploration and manipulation was observed in Jacobsen's chimpanzee during the first twelve months of its life. It was especially noticeable with simple inanimate objects. These authors consider this behaviour as the basis for what they go on to call organized play, or play behaviour that has a logical sequence to it. Thus the animal was seen at six months to tear the pages from a magazine into pieces, lie down and roll over the torn paper, and later to adorn her head with the same pieces of paper. Similarly, repetitive activities involving climbing, standing and jumping in a regular sequence are seen as organized play. This type of activity was comparatively rare.

Bodily activity is the third general category of playful behaviour, in which '. . . the movement of the body was the essential characteristic'. This included all types of acrobatics, which were often repeated many times with only minor changes from instance to instance. They developed from stationary (arm-waving, kicking, etc.) to locomotory, at about the third month.

Finally the authors discuss social play, which is seen as activity which is primarily social in significance, although, as is pointed out, much of the above behaviour may occur in social situations. Play-threatening and attacking, which involves a social partner, was observed by the fifth month; and a month later threatening and swaggering postures were frequently present. (The development of aggressive behaviour was, the authors noted, accompanied by a marked increase in timidity. Hebb (1949) also comments on the relatively late development of fear responses to visual stimuli in chimpanzees. Upon being shown a chimpanzee head detached from the body, '. . . young infants showed no fear, increasing excitement was evident in the older (half-grown) animals, and those adults that were not frankly terrified were still considerably excited.' Pp. 243).) Also considered to be primarily social in significance was the interest shown by the three-month-old infant in the facial features, and in particular the nose and teeth, of her human keepers, although this behaviour does not appear to be greatly different from the characteristic exploration of similar aspects of inanimate objects. Welker (1956) has pointed out that up to a point, heterogeneity in a stimulus, including 'movability', provokes more exploration and play in the chimpanzee than does relative homogeneity.

Budd, Smith and Shelley (1943) have included in their account of the hand-rearing of an infant female chimpanzee some notes on the development of play. By five months, the animal '. . . showed a tendency to play with any object within reach, preferring a soft article to a hard one. Preliminary examination of the object was made by scratching it with her outstretched hand'. By six months play had become more extensive and included much acrobatic activity. By the seventh month 'toys which had caused Jacqueline much amusement during the preceding month were now practically ignored, more boisterous play being preferable. The infant would assume a threatening manner towards her keeper and then make a playful attack, thoroughly enjoying the rough handling which followed . . .' At the age of nine months a young gorilla was introduced, which increased the scope of the infant's acrobatic and social play.

These authors also comment on the chimpanzee smile, or playface, first seen during the twenty-second week of life.

Since both the chimpanzees on which these observations were made were raised in isolation from other members of their own species, these ontogenies lack any description of the development of social play with either adult or infant conspecifics. Nevertheless, it is clear that a relationship existed between the infants and their foster-mothers, which expressed itself in curiosity about individual features of their bodies or clothing, mock-aggressive behaviour towards them, or a positive inclination to romp, or be played with in a manner involving direct physical contact.

Two relatively distinct categories of play emerge from these descriptions and they will be treated separately, even though several behavioural features may be shared between them, and at some points they overlap. They are *non-social play*, which includes gymnastics and exploratory play; and *social play*. Both of them employ motor patterns that are as species-specific as are the patterns occurring in their originally motivated contexts. Although the majority of examples will be drawn from chimpanzees, examples in other primates will be quoted where there is available material. In general, both forms of play are most characteristic of younger rather than older primates.

NON-SOCIAL PLAY

(a) *Gymnastic or Acrobatic*

Bierens de Haan (1952) regards play in which the animal has 'no playfellow and no apparatus' as that which will be most characteristic both of the individual and the species. From his observations on a five-year-old male chimpanzee he lists more than a dozen distinct ways in which the animal showed playful activity entirely from the resources of his own body in combination with the floor, walls or ceiling of his otherwise empty cage. In the absence of playfellow or apparatus the animal is forced to be inventive: Bierens de Haan notices the large amount of repetitive behaviour that occurs, including rolling, swinging, jumping, banging and drumming. Having removed from the situation all possible sources of stimulation other than those which the animal can provide itself, he states that 'Animal play is without utility value: its only object lies in itself'. However, as Morris (1964) points out, this is precisely the kind of sterile situation in which the non-specialist animals may be forced to develop apparently non-utilitarian behaviour patterns in order to save themselves from '. . . the most dangerous state of all, namely gross inactivity'. Such a situation is of course exceptional. Normally the young primate lives in an environment full of opportunities for exploratory and gymnastic play and it engages in this activity very fully. Carpenter (1964b) mentions the characteristic gymnastic play patterns of young spider monkeys who run and jump from one branch to another. So also do young gibbons, whose greater agility enables them to engage in even more complicated aerial acrobatics and extreme variations of posture and locomotion. Carpenter's (1964c) observations on two young captive gorillas show, within the limitations of their environment, much the same sort of behaviour, including somersaulting, posturing, manipulation of hands and feet and spinning on the ropes and chains provided. Harrison (1962) reports of young captive orang-utans that they liked to hang from a rope while banging on a wire screen, that they spun on the rope and blew 'raspberries' through the wire. Gymnastic play in young primates seems to differ only in the capacity of each species to make use of the available features of the environment: patterns of swinging, jumping, spinning and somersaulting are common to them all.

It is difficult, particularly in the early stages of development, to draw a line between general activity and exercise and play. Possibly the arm-waving, kicking and manipulations of the very young primate are not usefully categorized as play. Later in ontogeny, however, the division between play and more random motor activity becomes clearer. Both in the wild (Goodall, 1965; Reynolds, 1965) and in captivity (Morris, 1964) chimpanzees will 'invent' play[1]; will repeat on or about available trees or pieces of equipment patterns of climbing, swinging and dropping. These may well be repeated many times. Certainly one of the functions that this particular form of play must serve is that of increasing familiarity with and consequently mastery over, the environment: the animal becomes acquainted at an early age with the limits both of its own body and of the features of the environment with which it is likely to come into contact. Thus at an age when such information is likely to be crucial it does not waste valuable and perhaps vital time jumping for branches which are out of its reach, or landing on creepers which are unable to bear its weight. It is noticeable that the amount of acrobatic play is maximal when the animal's weight and strength are changing most rapidly and that it decreases and may virtually stop altogether at the point where the animal's weight and strength become relatively stable; that is, in adulthood.

Possibly the individual differences to be seen in adult chimpanzees in acrobatic skill and inclination reflects the opportunities they had as infants for practice, but it is more likely that there are simply athletic and non-athletic chimpanzees just as there are

[1] It is probably useful to reserve the word 'game' for the highly organized and structured play of the language-using human species.

athletic and non-athletic human beings. The range of individual acrobatic skill is probably greater in the anthropoid apes than in the gibbons and monkeys, and greater still in human beings, for the obvious reason that it is no longer essential to survival and would not be subject to the same selection pressures.

(b) *Exploratory and Investigatory*

Berlyne (1960) and Welker (1961) have provided comprehensive reviews of the literature on curiosity and exploration in animals; and more recently Glickman and Sroges (1966) have undertaken a survey of curiosity in zoo animals as manifested in their reactivity to a standardized set of novel objects placed inside their cages. None of these authors, however, has considered it necessary, or perhaps even possible, to distinguish between investigation and investigatory play, either in terms of motivation or of the motor patterns involved. The following section will review briefly the work that seems to be more directly concerned with investigatory play, most of which has been done using the chimpanzee as subject. The chimpanzee's extreme responsiveness to novel stimulation makes it an ideal subject. Glickman and Sroges report that their chimpanzee subjects were '. . . all highly reactive and . . . produced a wide variety of contact responses'.

Novel objects produce in the chimpanzee varying amounts of curiosity and fear, perhaps better described in behavioural terms as tendencies towards approach and withdrawal. This will vary according to the type and amount of the animal's previous experience with similar or related situations (Menzel, 1962; Welker, 1956b). Novelty is, of course, always relative: the materials may be familiar but the form they take novel, or vice versa or any combination of these variables. At a certain point in the sequence of investigation of a novel stimulus may come a stage at which the motor patterns change subtly, and the behaviour takes on the characteristic appearance of play (p. 178).

The following extract from daily observations gives a fairly representative picture of the change from investigation to play, as approach tendencies outweigh those to withdraw. All observations in the record not considered to have been directly relevant have been omitted. The subject was a wild-born chimpanzee, aged approximately three years, which had been in captivity since approximately one year of age.

14TH JULY, 1965

I bounce a tennis ball in front of the cage several times so that she hears as well as sees it and place it inside on the floor. She backs away, watching ball fixedly – approaches with pouted lips, pats it – it rolls. She backs hurriedly to the wall. Hair erection . . . J. pokes at it from a distance, arm maximally extended, watching intently; looks at me; pokes ball and immediately sniffs finger . . . She dabs at ball and misses, sniffs finger; she backs away and circles ball from a distance of several feet, watching it intently. Sits and watches ball . . . (pause of several minutes) . . . walks around ball. J. walks past the ball again even closer but quite hurriedly. She lifts some of the woodwool in the cage to peer at the ball from a new angle, approaches ball by sliding forward on stomach with arms and legs tucked underneath her, so that protruded lips are very close to ball without actually touching it. Withdraws. Pokes a finger towards it and sniffs finger . . . Returns to ball, again slides forward on stomach with protruded lips without actually connecting. Pokes with extended forefinger, connects and it moves; she scurries backwards; more dabs at it with forefinger and it moves again (but not far because of the woodwool in that area of the cage). J. dabs, ball rolls and she follows, but jumps back in a hurry as it hits the far wall. She rolls the ball on the spot with her forefinger resting on it, then rolls it forward, watching intently the whole time. She dabs again – arm movement now more exaggerated, flung upwards at end of movement. Tries to pick ball up between thumb and forefinger very gingerly . . . fails. Rolls it *towards* her, sniffs with lowered head. Picks it up and places it in front of her – *just* touches it with lips – pushes it into straw with right forefinger – touches it with lower lip pushed out – pokes, flicking up hand at end of movement, but backs away as it rolls towards her. Bites at own thumb. Dabs at it with lips, pulls it towards her and backs away. Examines own lip, squinting down, where it touched ball. Picks at it with forefinger and covers ball as it rolls (walking on all fours, with head down to watch ball as it rolls along at a point approximately under her belly). Pushes with outside knuckles. Stamps on it, dabbing at it with foot. Sits on it, rolls it with foot; carries it gingerly in hand and puts it on the shelf, climbing up to sit beside it. It drops down – she holds it in one hand and pats it increasingly hard with the other. Holds it in right hand, picks at stripe on ball with her left. Rolls it between two hands. Rolls it between hand and shelf. Holds and pats; bangs it on shelf. Holds and *bites*, examining ball after each bite. Ball drops from shelf and she pats at it on ground with right hand. Lies on her back, balances ball on her feet, holding it there with hands; sits up, holds ball under chin and rolls it two or three times round back of neck and under chin. It rolls away and she chases it immediately and brings it back to shelf. Lies on back and holds it on feet. Presses it against teeth with her feet and bites – all fear appears to be gone – lies and bites at ball held in feet, hands. Rolls it in feet, hands. Climbs to ceiling, ball drops and she chases it at once, J. makes playface, rolls and tumbles with ball, around, over, under ball, bangs it; bites it, rolls it over her own body.

These notes were made over the course of a forty-five-minute period. What they unfortunately cannot convey is the increase in the tempo of investigation towards the end of the session, or the change in affect that took place as the animal gained in confidence in her dealings with the ball.[1]

Over the next thirty minutes the final patterns of response described above were repeated many times, each individual pattern such as dropping, chasing and retrieving being repeated many times in succession. Such repetitive behaviour is a marked feature of both acrobatic and investigatory play. However, there is more than one kind of behaviour in which repetition occurs.

1. The type of repetitive behaviour described in the protocol above is characterized by an inventive and non-rigidified freedom of movement and is highly typical of play. The chimpanzee may perform a movement that is 'self-rewarding' (see discussion on p. 186) in that the chances are high that it will be repeated immediately. Each repetition, however, is subtly different in that some features may be shed and others added each time round. It has none of the apparently 'driven' and inflexible quality of, for example, the ritual patrolling of the cage area seen in many zoo animals (Hediger, 1964; Holzapfel, 1939).

---

[1] It is interesting to compare the progress of exploratory behaviour in the chimpanzee with a description of similar behaviour in the raven, another 'non-specialist'. 'A young raven confronted with a new object . . . first reacts with escape responses. He will fly up to an elevated perch and, from this point of vantage, stare at the object literally for hours. After this he will begin to approach the object very gradually, maintaining all the while a maximum of caution and the expressive attitude of intense fear. He will cover the last distance from the object hopping sideways, with half-raised wings, in the utmost readiness to flee. At last, he will deliver a single fearful blow with his powerful beak at the object and forthwith fly back to his safe perch. If nothing happens he will repeat the same procedure in much quicker sequence and with more confidence. If the object is an animal that flees the raven loses all fear in the fraction of a second and will start in pursuit instantly. . . . With an inanimate object the raven will proceed to apply a number of further instinctive movements. He will grab it with one foot, peck at it, try to tear off pieces, insert his bill in any existing cleft and then pry apart his mandibles with considerable force. Finally, if the object is not too big, the raven will carry it away, push it into a convenient hole and cover it with some inconspicuous material.' (Lorenz, 1956.)

2. The second type of stereotyped response occurs in normally reared animals placed in a restricted environment. The restriction may consist of a reduction in living area, such as a cage, or of reduction in the variety within the environment; or of course both, as often happens to wild-born animals brought into captivity. Animals in these circumstances may develop stereotyped responses where thwarted intention movements (of escape, flight, etc.) may be repeated endlessly in the absence of guiding information from the environment (Morris, 1964).

Alternatively, these movements may arise in those non-specialist animals whose existence depends on constant curiosity about and investigation of their environment. Again, in the absence of external input, these animals may develop new motor patterns to provide them with the perceptual and physical stimulation they vitally need. In both cases a reduction in input increases the likelihood of repetitive, stereotyped behaviour.

3. The third, most damaging and least reversible of sources of stereotyping occurs in primates raised in restricted and, in particular, socially restricted circumstances.

One of the major areas of difference between feral and laboratory or isolation-reared chimpanzees is in the amount of stereotyped behaviour observable in the laboratory-reared animals (Menzel, Davenport and Rogers, 1961; Menzel, 1964). Stereotyped responses are those which Menzel defines as patterns '. . . involving rhythmical rocking, swaying or turning movements of the body, or repetitive or persistent acts involving parts of the body, such as thumb-sucking and eye-poking'. These developed during the rearing period of animals deprived of their mothers (Davenport and Menzel, 1963). These authors consider that stereotyping in the infant chimpanzee '. . . is related to the absence or insufficient amount of stimulation that the mother ordinarily provides her infant as she grasps, hugs, rocks and carries him during the early months of life'. Deprived of factors entering into the normal course of maternal care, artificially reared infants will stimulate themselves with self-produced rocking, swaying or turning movements. Such stereotyped behaviour is related to rearing variables, the developmental status of the animal, the immediate stimulus situation and various forms of ongoing activity such as eating or sleeping.

Amongst immediate stimulus factors affecting the quality and quantity of stereotyped behaviour are the presence or absence of play objects, and a combination of the size and degree of familiarity of the cage in which the animal is tested (Berkson, Mason and Saxon, 1963). Cage size was also found to affect the amount of stereotyped behaviour observed in rhesus monkeys (Draper and Bernstein, 1963). In general, the larger the cage the lower the incidence of stereotyped patterns of response. The authors suggest that vertical space would substantially reduce the amount of stereotyping, since the flight reaction in rhesus monkeys involves movement upward as well as outward.

In general, laboratory-reared chimpanzees were more easily aroused than ferals, and the level of stereotyping increased with the level of arousal. Interest in or contact with stimulus objects in the environment which were not sufficiently novel (i.e. arousing) to increase stereotyping actually served to lower the incidence of stereotyped response. Thus grasping and manipulation are reciprocally related to stereotyping (Menzel, 1963).

The only instance of stereotyped behaviour in the chimpanzee reported in the wild (Goodall, pers. comm.) is of an infant whose mother had tended to avoid social contact, and which had therefore had less experience with her peer group than other wild-born chimpanzees. This infant performed solitary stereotyped rocking and pirouetting movements, suggesting that it is not only stimulation received during the course of maternal care that is responsible for the absence of stereotyping. The rôle played by interaction with conspecifics in the development of adult behaviour will be discussed in the section on social play.

Factors Affecting the Response to Novel Stimuli in the Chimpanzee
Apart from the degree of stereotyped behaviour to be seen in isolation- or laboratory-reared chimpanzees, Menzel (1964b) considers the generalized patterns of response seem to be the same for these apes in general. The progression from withdrawal through caution, investigation, play and finally satiation is seen as characteristic of the chimpanzee as a species, regardless of his rearing conditions. What *is* affected by the manner in which the animal was raised is the difference in the level of arousal created by similar situations. Thus, the more experience a chimpanzee has had with a particular class of object, the less easily aroused he will be by the introduction of a further object of the same or a related class. There is of course a theoretically optimum amount of arousal,

which will probably differ from chimpanzee to chimpanzee (Haselrud, 1938), above which play behaviour may never develop.

Within the optimum range of 'novelty' there is some evidence that objects that produce most initial caution will be manipulated for longer than objects which provoke little or no fear. Up to a point, degrees of novelty enhance the tendency for investigatory play as measured by the total amount of contact with the test object (Menzel, Davenport and Rogers, 1961). Two young male chimpanzees, both of whom had been raised in isolation with accordingly very limited experience, were tested with a group of novel stimuli differing in one or more dimensions from a standard stimulus. The standard stimulus consisted of a small white cube, with which the two animals were totally familiar before the testing began. Test stimuli differed from the standard stimulus in size ($1\frac{1}{4}$ in. vs. $2\frac{1}{4}$ in.), colour (black vs. white) and shape (cube vs. triangle). All test objects elicited more contact than the standard stimulus, the amount increasing as the number of ways in which they differed from the standard stimulus also increased. Thus, a large black triangle elicited most contact of all. However, these scores were cumulative; in every case the initial reaction of the animal towards the novel object was caution. The authors do not discriminate between simple manipulation and play.

The second experiment performed on the same subjects was concerned with an analysis of the cumulative effects of experience with a given class of objects on initial responsiveness to other objects of the same class. Again, contact duration scores were lower on the first than on any subsequent trial; and extensive contact occurred only after several five-minute exposures to each group of objects when most of the caution, or fear, had gone.

These authors conclude that there is a generalized transition in the young chimpanzee faced with novel stimuli from low to high amounts of contact. They interpret this as indicating that '. . . experience, interacting with maturational factors, presumably operates to modify the subjects' general standard of "what is novel", and consequently lowers arousal from a relatively high (fear-producing) level to levels that are more and more optimal for contact and play'.

Amongst other features determining the total intensity of stimulation is that of movement. Haselrud (1938) tested groups of adults and infants on their willingness to reach for food placed near animate and inanimate versions of the following stimulus objects: tortoise, snake, alligator, fire and ball. Although their results showed large individual differences in what was and was not considered frightening, the patterns of fear for both groups were similar, appearing and disappearing quickly. The initial reaction of a chimpanzee to practically every new object is at least caution, but if the stimulus value is very strong, i.e. if the object moves, the reaction may include overt fear. The infants adapted rapidly to new objects that did not hurt or startle them, but the adults were '. . . more persistently prudent'. The greater caution of the adults towards the inanimate objects would indicate that, particularly in the absence of motion, '. . . an object has potentially more stimulating value or meaning for the adult'.

Menzel (1962) analysed the interaction between stimulus size, novelty and individual differences in five young chimpanzees. A piece of painted plywood varying in size from 1 sq. in. to 251 sq. in. was placed inside the subject's home cage. Results seemed to show an optimum size of object for varied manipulatory reactions: the smaller objects provoked little manipulation, the larger objects were handled tentatively or not at all.

'The precise size of the optimum stimulus varied for particular responses, for different subjects, and also for the same subject as a function of experience with the same or similar objects. Experience tended to increase the size of object required to produce either avoidance or vigorous contact activities; thus over a period of weeks the behaviour of cautious animals came to resemble the initial behaviour of bolder animals.'

Menzel emphasized the fundamental continuity between avoidance of and approach towards novel objects: large objects might provoke fear when placed inside the cage, but be preferred to smaller objects when placed outside the cage wire. Distance and the barrier of the cage wire reduce the intensity of stimulation, and a fear-producing stimulus becomes an interesting one.

The most comprehensive study of factors affecting play and exploratory behaviour in the chimpanzee has been carried out by Welker (1956a, b). Here, the effects of stimulus situation, novelty, repeated presentation, age and experience of the animal are all taken into account. Play and exploration are seen as 'variable responses to novel objects'. The influence of the novelty of the stimulus in eliciting play behaviour is shown by the fact that the introduction of a new set of 'play-objects' (differently shaped stimuli varying in any of a number of dimensions, including size, form, movability, texture, brightness, etc.) resulted in increased responsiveness, as measured by the number of manipulations made by an animal in a given period of time. A similar increase in responsiveness was made to the same set of objects after an interval during which the animal did not see them; and with increased exposure to the same objects, interest finally waned. Results with different age groups also points up the rôle of novelty in eliciting investigation and manipulation, since the older animals (seven to eight years old) had presumably had more experience with the environment than the younger subjects (three to four years old) and showed correspondingly less interest. However, this overall lack of interest only became apparent during the sixth minute of the test sessions and is more accurately described as a faster satiation rate.

Certain stimulus objects were preferred by both groups of animals. Within certain limits, 'more movable, larger, brighter, more heterogeneous and changing auditory and visual stimulus configurations' were preferred over those less extreme in any of these dimensions. Welker suggests that these preferences may indicate special cases of attraction to novelty (i.e. 'spatial or temporal changes'), but points out that preferences for qualitative stimulus characteristics also occurred. Curved objects were preferred to straight-edged objects; and Budd *et al.* reported that soft toys were preferred to hard ones.

Variability during each session and each minute of each session was a marked feature of the responses of Welker's chimpanzees. Those animals that were most responsive to the objects also showed a greater frequency of change in their response to any individual object. Animals who were less responsive in general were attentive for longer periods of time to individual objects, or features of each object.

Moreover, the chimpanzees never continued to respond to only one aspect of the experimental situation, but shifted their attention rapidly from one object to another and from feature to feature of the same object. Welker suggests that the short attention span of the chimpanzee faced with a novel stimulus is the result of two factors working concurrently: *attraction to novelty* and *satiation of interest with familiarity*. The rapidity with which satiation will occur is a function both of the stimulus situation and of the individual experience of the subject. Thus an infant chimpanzee (one year old) will remain interested in a simple block of wood for longer than a three-year-old, who in turn will manipulate it for longer than an adult. Welker, like Menzel *et al.*, found that the age, and hence experience of the animal, is a major factor in determining the response to novelty. Chimpanzees raised under conditions of almost total perceptual restriction will find even the simplest object so novel that exploratory activity is totally inhibited, and the animal will huddle fearfully in a corner, sometimes for as long as the novel object is present.

Curiosity as manifested in the exploratory behaviour of primates in the wild state differs in quantity from that observed in captive animals. Goodall (personal communication) reported that the chimpanzees of the Gombe Stream Game Reserve showed little curiosity in either her or other human beings, or in the many objects that formed a part of their camp, most of which must have been unfamiliar. Menzel (1966) comments on comparable behaviour in free-ranging Japanese monkeys (*M. fuscata*). This behaviour appeared to differ from the 'neophobia' shown by wild rats (Barnett, 1963) in that new objects were ignored rather than positively avoided, at any rate by the chimpanzees. Menzel, however, reports that it was often difficult to distinguish observationally between avoidance and indifference.

Several points can be made about the apparently greater amount of curiosity seen in captive primates.

1. Hediger (1955) has pointed out that the captive animal does not have to search for food and avoid predators, activities which would normally occupy the greater part of its day. It will also learn quite quickly that the objects it encounters in its cage, far from being harmful, are likely to prove rewarding. Both these factors would tend to increase above normal levels an already-present tendency to investigate.

2. Many authors have reported that more investigatory behaviour is shown by the young of a particular primate species than by the adults (Glickman and Sroges, 1966; Menzel, 1966; Beach, 1945; Carpenter, 1965). Moynihan (pers. comm.) has suggested that it may be that juvenile levels of investigatory behaviour persist in the captive adult through being maintained in a dependent, or juvenile, rôle by keepers, who act as substitute parents.

3. Perceptual deprivation – or boredom – in captive animals confined to constricted living quarters or deprived of normal amounts of social stimulation, may be a major cause of the difference. If this is so, the amount of investigatory behaviour seen in captive animals must be considered to be artificially high, rather than that in wild animals unnaturally low. As such, it may be a further example of the non-specialist animal's seizing on any available opportunity to maintain its level of sensory input above the minimum necessary level (Morris, 1964). Yet on the whole, the type of exploratory behaviour seen in any captive primate is markedly freer and more varied than the rigidly stereotyped activity seen in animals fighting off potential sensory deprivation.

4. In chimpanzees, it may simply be that in the wild they are under so little pressure from their environment – from climate, food supplies or predators – that they are simply not operating at optimal levels. It is clear that chimpanzees may be trained to perform many extremely involved tasks, ranging from opening puzzle boxes to playing an active part in manned space-flight programmes, where mistakes cannot be afforded.

Faced with situations in which the only way to obtain a desired reward is by manipulation of the environment, the chimpanzee is capable of working out on his own fairly elaborate chains of activity which will bring him the reward. Some work related to this type of behaviour is discussed in the following section.

The Relationship of Exploratory and Investigatory Play to Problem-solving

Kohler (1925) writes of the ability of his chimpanzees to solve food-getting problems by means of insightful manipulation of boxes and sticks into the appropriate positions.

Later this theory was put to more rigorous test by other experimenters (Birch, 1945; Schiller, 1957). Birch's and Schiller's animals were tested on problems involving the raking in of food with a stick both before and after the opportunity for varying amounts of free play with sticks. Few of the animals were able to solve the problem during the pre-play session unless it were by 'accident'; for example, in one case the animal perceived the food as being directly connected with the stick, or attached to it. The animals were then observed while manipulating the sticks in their home enclosure (Birch). Over three days of playing with the sticks, patterns of response evolved characterized by the increasing use of the stick as a functional extension of the arm. In this way the possibilities of the stick were explored, and in the post-play test session every one of Birch's four- to five-year-olds solved the problem within twenty seconds. Birch concludes that the perception of functional relations in a similar situation is dependent on the previous experience of the animal. Insightful behaviour, therefore, is more accurately described as the integration of motor components acquired during earlier experiences into new and appropriate behaviour patterns.

Schiller (1957) carried out approximately similar experiments using both younger and older chimpanzees. With the two year olds he found that even two weeks' free play with sticks did not improve their performance; he concluded that a maturational factor was present. After a year's opportunity to play with sticks in an open enclosure three of his four animals made spectacular improvements in test performance, able to connect the stick with the food even when the two were not in the same visual field. Even so, none of the young animals reached the level of the seven or eight year olds, the earliest age at which Schiller found quick success with stick-using problems. He too emphasized the mechanical rather than the insightful nature of problem-solving of this sort: in each case learning proceeded in clear steps, each of which originated in the accidental raking-in of the food with the stick, or fitting of one stick into another to make a single long stick.

Once a complex solution was mastered both by Birch's and by Schiller's chimpanzees, the animals continued to use this pattern even when a simpler solution was possible (Schiller) or in a totally new situation (Birch). Thus Birch's animals faced with a food-in-pipe situation all attempted to use the stick as a rake on a table where food had been earlier, even though no food was now visible. The animals had begun to develop stereotyped responses to problem situations; but they were capable of discarding these responses when it became apparent that they were getting nowhere. Schiller considers the persistence of an elaborate pattern to solve a problem when a simpler one will do to be playful: they '. . . do it for fun'. It is more likely that the chimpanzees are simply continuing to behave in a way for which they have been rewarded in the recent past. Even human beings have difficulty in changing a 'set'.

Goodall (1965) describes an interesting use of twigs in the wild, where chimpanzees use them for poking down narrow holes in order to 'fish' for termites. It seems more likely that this type of behaviour would have originated in free play with sticks and twigs, in which many patterns of poking and insertion are seen (Birch, 1945), rather than in insightful perception of cause and effect. Moreover, such behaviour with sticks appears to be more characteristic of the chimpanzee as a species than of other primates.

Lorenz (1956) points out the probable survival value of investigatory and exploratory patterns in those animals that are non-specialists. A generalized mode of approach, which has often been considered to be playful because of its apparent lack of direction, will quickly show whether or not the novel object is potentially edible, available as a sexual partner or otherwise useful. Lorenz uses the raven as an example, a bird which lives in such widely differing environments as the sea-bird colonies of the North, in the desert and in Europe. 'By treating each new situation as if it were biologically relevant – first as a potential enemy, than as a prey – the raven will discover sooner or later the relevant objects in very different habitats.'

It is this generalized investigation and exploration, resulting in rapid adaptation to a new environment which has undoubtedly contributed to the enormous growth in size of the rhesus population. 'No primate species other than the rhesus spans such a wide ecological continuum from complete domestic commensalism with man to a remote forest life.' (Southwick *et al.*, 1965.)

SOCIAL PLAY

The Play-face

There is a particular facial expression that occurs so often during play behaviour in the chimpanzee that it has come to be known as the 'play-face' (van Hooff, 1962). Although it occurs most frequently in social play, where it has its most obvious function, the play-face is also observed in both the acrobatic and the investigatory forms of solitary play (but possibly only when other chimpanzees are present). Briefly it is characterized by slightly lowered eyelids, a wide-open mouth in which the mouth corners are in approximately their normal position; and lips which cover the teeth or only allow the bottom row to show. Goodall (pers. comm.) has reported that the play-face in the chimpanzees of the Gombe Stream Reserve often involves the showing of both rows of teeth, and it is possible that this represents a higher intensity of play. However, it is also possible that in certain socially ambiguous situations the increased baring of the teeth (and particularly of the upper row) in the play-face, represents the introduction of an element of appeasement since the facial expression now bears more resemblance to the grin-face, which can have an appeasing function (van Hooff).

Occasionally it may be assumed that the play-face itself functions as an appeaser in certain situations. For some time (1964–5) there was only one male chimpanzee in the colony at the London Zoo. When an older, stronger male was introduced a certain amount of fighting took place within the colony during which it became clear that the new male had established dominance. Now when the sub-ordinate male approaches the dominant male it often does so showing the play-face and with an unmistakable play-gait. Since this manner of approach does not provoke attack, it may be assumed that one of its signal functions is that of appeasing the dominant male by informing him that his dominance is not about to be challenged.

Bolwig (1963b) considers the play-face to be a development of the ritualized play-bite, or of the preparation for the play-bite in which the lips are drawn tightly over the teeth, thus effectively covering them. Smiling in humans is seen as a further step in the ritualization of the play-bite; and the fact that both rows of teeth are often exposed in the human smile further suggests the element of appeasement since it now resembles the sub-human primate's fear-grin more closely that in does the play-face. The soft guttural staccato exhalations that often accompany the play-face are seen by Bolwig to be the origin of laughter.

As mentioned earlier, the major signal area in the primate is the face, although there exist certain bodily postures and movements which indicate a readiness to play. Rhesus mothers encourage their infants to play by turning their backs on them and regarding them through their hind-legs (Reynolds, 1961). Simonds (1965) records the play invitation in the male bonnet macaque (*Macaca radiata*) as consisting of '. . . a bouncy walk, head twisted to one side, reaching with the hand, and slightly open mouth'.

Patterns of Social Play

Play occupies part of the waking activity of the young primate. Most field workers have been impressed with the quantity of play observed and many have felt strongly that it was a major factor in the determination of later social behaviour. General statements about the function of play are typified by the following. 'The drives to play and explore bring about a diversified sampling of the environment that is probably of great importance in adaptation' (Washburn and Hamburg, 1965). 'As a context for learning to get along with other monkeys . . . experimentation and mistakes go without punishment or the threat of danger from other monkeys' (Jay, 1965). Mason (1965b) finds that many field workers feel that '. . . experience gained in play facilitates the development of communicative skills'. Southwick, Beg and Siddiqi (1965) state of rhesus monkeys that two important results of infant play are 'the development of positive social bonds between individuals, and the development of motor skills such as running, climbing, grasping and manipulation'. Schaller (1965) remarks of the gorilla that in social play infants have their first opportunity to interact closely with other youngsters in the group. Such interactions took place most often between only two animals at a time (81 per cent of the play-groups observed), but a few groups contained three or four young. The size of the play-group in young rhesus monkeys (Southwick *et al.*, 1965) increased with their age. Infants tended to play in twos and threes, but by the time the animals became juvenile the size of their play-groups had increased from four to ten individuals.

Carpenter (1964a), however, felt able to be more specific about the function of such youthful social play. As early as 1934, when his paper on howler monkeys was first published, he stated that 'The patterns of play which involved two or more individuals are forms of activity which relate to the process of conditioning or integrating an individual into the clan. Social relations with animal other than the mother first occur with play-partners.' He goes further, and suggests that social play enables the young animal to find its own place in the '. . . existent social form of the group', since play serves to establish a dominance scale among the young animals. Animals showing the greatest facility in play control the course of action to a greater extent than do the others. Similarly with gibbons: Carpenter (1964d) feels that 'Communal association during . . . play may . . . serve to reinforce social relations among individuals of a gibbon family'. Moreover, it is in play that the existing social relations are most typically and completely expressed.

(a) *Play between young:* The most noticeable feature of play between conspecifics of most species, and particularly the primates, is that the patterns used in social play are those otherwise seen in agonistic behaviour. Chasing and being chased, grappling, wrestling, pulling, pushing and inhibited biting are the most obvious features of social play in primates. Southwick *et al.* (1965) observed that in infant rhesus monkeys the most common forms of infant play are chasing, jumping and wrestling. Juvenile play involves more complicated patterns of chasing and wrestling on and around many features of the environment. The play of young gibbons (Carpenter, 1964d) also consists primarily of wrestling, grasping and biting. 'The behavior was very vigorous and involved stealthy movements, surprise attacks, slapping each other with swift movements of the hands and quick diverting sorties.' His observations of two young captive mountain gorillas (1964c) reveals the same sorts of activity, though naturally modified by the gorillas' more massive structure. Wrestling, chasing and teasing form the major part of social play. At times the young animals would wrestle intermittently for several hours, although the periods of struggle would last on average only fifteen to twenty seconds. Rather less social play was seen by Schaller (1965) in his observations of wild mountain gorillas. What there was also consisted of vigorous wrestling, but was relatively rare and lasted only briefly. Both Schaller and Carpenter observed displays of chest-beating occurring in play.

Similarly, in spider monkeys (Carpenter, 1964b) chasing, catching, biting and prolonged wrestling were seen; in howler monkeys (Carpenter, 1964a) the patterns were varied by the fact that the young animals were able to hang suspended by their tails and wrestle together with all four limbs engaged. Play is seen as an important part of the life of the young marmoset (Fitzgerald, 1935). It consists mainly of hide-and-seek, playful bites and slaps at each other and wild racing around. The 'utmost harmony' is seen to prevail in the play of young captive bush-babies, who '. . . pursue each other, roll about and playfully grapple. At times they will hang suspended from a horizontal stick the while beating at each other with their free arms like two inverted pugilists' (Lowther, 1940).

In monkeys, sexual elements are seen in play, including mounting and pelvic thrusting, but among chimpanzees, gorillas and gibbons they are rare and incomplete. Play appears to give way to other forms of social behaviour as adolescence is reached. Chimpanzees (Goodall, 1965) occupy the time increasingly with social grooming. Adults occasionally play, but they do so much less often. The relationship between play and grooming needs to be explored, but certainly in some instances they can be seen to serve similar rôles in the inhibition of aggression and maintenance of the social hierarchy.

(b) *Play between young and adult:* One type of situation in which adult primates may either initiate play, or respond when invited to play, will occur in interactions with their own or possibly other familiar infants. Schaller (1965) records one instance of a female gorilla's playing with an infant by holding it down until it struggled free. Goodall (1965) saw frequent playful interactions between mothers and infants, in which the mothers would push the infants as they swung on a branch, tickle or spar with them. Bingham (1927) reports similar behaviour between mother and infant in captive chimpanzees, in which chasing, mild sparring, rocking and tickling all occurred. He also quotes reports of more elaborate play patterns, perhaps to be expected under the restricted conditions of captivity. 'Loca invented a special play. She lay on her back and held with her foot a little leg of her young one, let him crawl over her body, and then quickly drew him back again "over hill and dale".' Chimpanzees in the London Zoo colony with very young infants have been observed to hold the infant aloft by its arms, sometimes while lying on their backs, allowing the infant to kick for several minutes at a time (Manley, pers. comm.).

Simonds (1965) records play between adult males and juveniles, or sub-adult males, in the bonnet macaque. Southwick *et al.* report of rhesus monkeys that 'On three occasions infants were seen playing with adult males, climbing over them and remaining in close association for several hours', but that these occurrences were rare. In general, playful interaction between males and infants in primates is far less common than between females and infants, and most play occurs in sub-adult groups.

Females have been observed to become involved in infant play that is apparently becoming too aggressive. Romanes' (1883) report of D'Osbonville's observations of this kind of behaviour probably reflects more accurately the attitude to child-rearing at that time than it does the real function of the interaction, but it is one of the earliest such observations. According to D'Osbonville, the following took place between certain monkeys that he observed in the wild state: '. . . after suckling and cleansing them, the mother used to sit down and watch the youngsters play. These would wrestle, throw and chase each other, etc.; but if any of them grew malicious, the dams would spring up, and, seizing their offspring by the tail with one hand, correct them severely with the other.' However, Koford's (1963) observation that the offspring of high-ranking rhesus monkey mothers attain high-ranking status in the band '. . . apparently because of protection by their mother during youth' suggests that it is more likely that the infant chastised during play would be the partner rather than her own son. In possible support of this suggestion is an instance reported by Reynolds and Reynolds (1965) of chimpanzees in which '. . . a female, with no evident provocation, chased away a juvenile-one which had been quietly playing with her infant-two; the victim fled and hung screaming from a neighbouring tree, but ten minutes later it had returned and was again playing with the infant.'

(c) *Play between adults:* Play between fully adult animals is a feature of the social behaviour of lorisoid primates (Manley, pers. comm.), but in the higher primates play between fully grown adults is rare. Play may be seen between sexually mature but not yet fully grown captive chimpanzees, but such episodes become increasingly rare as full adulthood is reached. Both in the wild and in captivity, grooming appears to occupy the time and the rôle once taken up by play. The same is not necessarily true for the monkey. Some of the male bonnet macaques observed by Simonds continued to play regularly. Their play-patterns were more restricted than those of infancy, however, in that play was confined to wrestling. Moreover, more care was taken to distinguish it from genuine aggression in that no threat gestures or vocalizations were used, as would have happened in infant play. If these occurred, play stopped immediately and genuine threat might follow.

Goodall has recorded on film an instance in which an old female enters momentarily into play with her two sons, one of whom is fully adult, as they follow each other around a large tree trunk. It is possible that the more rigid the social hierarchy in a primate species, the less likely it is that play will occur among the *adults* of that species. Thus, play between fully adult male baboons has never been recorded, whereas the chimpanzee, which has a more flexible social system, can afford interactions between the adult members of the group other than those of pure aggression. The relationship between the quantity and quality of sub-adult play and the social structure of the adult group offers many possibilities for research.

In summary, social play between the sub-adult members of primate societies occupies the greater proportion of the time not spent in eating or sleeping. Patterns of social play are largely derived from those of agonistic behaviour, consisting of chasing, wrestling, tumbling, biting, dragging and chewing. During play certain adjustments appear to take place which enable the young primate to function properly as an adult member of his species and to occupy a niche within the social organization of his particular group. Social play is generally agreed to contribute to the socialization process within each species. Young which have not had the opportunity for adequate play with conspecifics are, as Carpenter (1965) points out, faced with two options: '. . . to be "maladjusted" within the group or to be excluded from the group'.

THE EFFECTS OF SOCIAL PLAY

In some species of bird, the phenomenon of imprinting results in the chick's forming a strong attachment at a very early age to a figure which is usually that of the mother. This bond seems to determine the class of object upon which the young bird will later focus its social and sexual behaviour. Although mammals do not appear to imprint in the precise way in which many species of birds do, certain situations that occur only in artificial circumstances nevertheless point up analogous features in the social development of birds and mammals.

It occasionally happens that a young mammal is raised entirely in the company of a species other than its own. This most often happens when humans hand-rear young animals, but there are also instances of puppies being raised in a litter of kittens, and vice versa. The animal reared in this way appears in many respects to behave like a normal member of its own species until sexual maturity is reached, when it is unable to engage in normal reproductive behaviour and instead attempts to seek out as a sexual partner a member of the species with which it spent its youth. This has been known to happen in primates: chimpanzees raised singly by hand appear not to know 'which species they belong to' in that they regard human beings as potential partners in all social activities, regardless of whether or not there are other chimpanzees present.

Social learning in mammals, and in particular the primates, probably takes place over a longer period than the few hours over which avian imprinting occurs. It is suggested that (amongst other functions) the period of maximum social play occurring in sub-adult primates corresponds to the brief period of imprinting in birds. It is a period in which the young animal becomes familiar with the sight, sound, smell and (probably for mammals most important of all) touch of its conspecifics. In other words, during this period the animal is learning 'which species it belongs to'.

Relevant to this hypothesis are some studies concerned with imprinting in ducklings (Hess, 1959), in which the length of exposure to the imprinting-object was compared with the distance travelled by the duckling during the imprinting period. Results showed that it was the distance travelled, or '. . . the effort expended by the duckling in following the imprinting object' that affected the strength of the imprinting, rather than the length of exposure to the object.

The fact that it is the effort expended during exposure that is crucial in determining the strength of the imprinting suggests a further comparison with the hypothetical analogous period of social learning in primates. The distinguishing characteristic of the movements of play is exaggeration – or lack of economy. This very quality would ensure a maximum expenditure of energy and thus increase the strength of the 'imprinting' or social learning. The longer the period of dependency in the infant animal the more vulnerable it is to influences and interactions that can affect its social development adversely, and therefore the more essential it is that there should be a means of ensuring a continuous and corrective interaction with its own species. Play might well serve such a purpose.

Some light is shed on this by Harlow's (1962) now famous studies on the relative importance of the mother and the peer group in the social development of the young rhesus. The results of this work provide evidence (and it is the only experimental evidence so far) for the field workers' impressions that play within the peer group is important in the full development of social behaviour in the adult. Although these studies were not primarily concerned with defining the exact rôle of play in social development, they nevertheless indicate the direction that further laboratory studies might take.

Two groups of monkeys, one raised with their mothers and one with surrogate mothers, were allowed to play with each other in a large playpen for brief scheduled periods each day. The behaviour of the mothered infants '. . . evolved more rapidly through the sequence of increasingly complex play-patterns that reflects the maturation and learning of the infant monkey and is observed in a community of normal infants'. The older they grew and the more complex the play patterns became, the greater was the observable difference between the two groups. However, by the end of the second year the differences between the groups had virtually disappeared. The monkeys which had been raised with mothers developed more quickly in social relations, since from the third or fourth month of life onwards the mother began occasionally to reject the infant, thus encouraging it into closer contact with its peers.

However, another study showed that even normal mothering was not sufficient to ensure socially normal offspring. Two mother-infant pairs were raised in isolation for the first seven months of their lives and then brought together in the playpen. Relations between the infants were limited to '. . . an occasional exchange of tentative threats'. These two infants, even after separation from their mothers and two months of living together, were only slightly less retarded socially than infants which had been raised in total isolation.

One further study showed that even brief daily play-sessions between infants raised with surrogate mothers fully compensated for their lack of real mothers. At similar chronological ages, these infants developed as full a repertoire of infant-infant play relations and, later on, adult sexual relations, as did infants raised with their mothers in the playpen. Thus surrogate-raised infants allowed twenty minutes daily of play with their peer group were considerably better adjusted (as adults) than infants raised with their mothers alone. This work provides a laboratory parallel for the situation reported by Goodall (pers. comm.) in which a young chimpanzee whose mother had avoided social contacts displayed abnormal and stereotyped behaviour (p. 30). There is thus strong evidence for supposing that interactions with the peer group are both necessary and sufficient for the development of normal adult social behaviour.

Mason (1965b) considers that the distinction between social and non-social forms of play in primates is not particularly meaningful in terms of the stimulus characteristics that elicit play and the patterns of play in both social and solitary situations. In both cases, '. . . extreme departures from the familiar tend to depress play, whereas moderate degrees of novelty tend to enhance it'. Patterns of agonistic, grooming and sexual behaviour are also seen in response to inanimate objects. However, social stimuli are variable from moment to moment in appearance and, moreover, '. . . are not passive recipients of contacts, but may reciprocate, repel or withdraw from the activities of another animal'. Mason considers these to be differences of degree only.

However, differences of degree should not be dismissed as unimportant. The fact that conspecifics can actively both solicit and repel play can have far-reaching consequences on the social development of the young animal; inanimate objects cannot affect development in this way. A further potentially significant difference lies in the length of the relationship that can exist between the player and the play-object. Since inanimate objects possess no inherent capacity for change, they can eventually become familiar to the point of predictability. This will lead (Welker, 1956a) to the primate's loss of interest, at least on a short term basis. Social partners on the other hand, can never become so familiar that they are also fully predictable. Thus it is not uncommon for an animal to become so satiated with an object over many play sessions that it is eventually totally ignored, whereas relationships established during early play may be maintained into adulthood.

Certainly in some situations social and solitary forms of play appear to be reciprocally related, in that as social stimulation decreases, the amount of play with objects will increase; and in a perceptually restricted environment the length of bouts of social play becomes extended. Nevertheless, to concentrate on the unquestioned similarities between social and non-social play, rather than on their real differences, is possible to miss much that is important, including the whole question of function.

The questions outlined above present the ethologist with many problems. Studies of play now being undertaken by the author are concerned with precise comparison of the motor patterns involved in this behaviour with the same motor patterns occurring in their originally motivated contexts, in an attempt to isolate the qualities that make so much play instantly recognizable to casual and professional observers alike. The rôle of play in establishing and maintaining social relations is also being examined. Far from being a 'spare-time', superfluous activity in either human or sub-human primate, it may be that play at certain crucial early stages is necessary for the occurrence and success of all later social activity within one's own species.

ACKNOWLEDGMENTS

I am very grateful to many people at the London Zoo for stimulating discussions on the subject of play, and in particular to Dr Desmond Morris and Dr Gilbert Manley for invaluable help and criticism. I should also like to thank the Medical Research Council who supported this work with a grant for Training in Research Methods.

Parts of this chapter have already appeared in No. 18 of the Symposia of the Zoological Society of London.

REFERENCES

ALTMANN, S. A. (1962). 'Social behavior of anthropoid primates: analysis of recent concepts', *Roots of behavior*, ed. Bliss, New York, Harper and Bros.

BALLY, G. (1945). *Vom Ursprung von den Grenzen der Freiheit, eine Deutung des Spieles von Tier und Mensch*, Basel.

BARNETT, S. A. (1963). *A study in behaviour*, London, Methuen.

BEACH, A. F. (1945). 'Current concepts of play in animals', *Amer. Nat.* **79**, pp. 523–41.

BERKSON, G., W. A. MASON and S. V. SAXON (1963). 'Situation and stimulus effects on stereotyped behaviors of chimpanzees', *J. Comp. Physiol. Psychol.* **56**, pp. 786–92.

BERLYNE, D. E. (1960). *Conflict, arousal and curiosity*, New York, McGraw-Hill Book Company.

BIERENS DE HAAN, J. A. (1952). 'The play of a young solitary chimpanzee', *Behaviour* **4**, pp. 144–56.

BINDRA, D. (1959). *Motivation: a systematic reinterpretation*, New York, Ronald.

BINGHAM, H. C. (1927). 'Parental play of chimpanzees', *J. Mammal.* **8**, pp. 77–89.

BIRCH, H. G. (1945). 'The relation of previous experience to insightful problem-solving', *J. Comp. Physiol. Psychol.* **38**, pp. 367–83.

BOLWIG, N. (1963a). 'Bringing up a young monkey (*Erythrocebus patas*)', *Behaviour* **21**, pp. 300–30.

—— (1963b). 'Facial expression in primates', *Behaviour* **22**, pp. 167–92.

BREDER, C. M. (1932). 'On the habits and development of certain Atlantic Synentognathi', *Pap. Tortugas Lab.*, **28** (1), pp. 1–35.

—— (1932). 'Papers Tortugas Lab.', *Carnegie Inst. Wash.*, *Publ.* 435, **28**, pp. 8–9.

BROWNLEE, A. (1954). 'Play in domestic cattle in Britain: An analysis of its nature', *Brit. J. Vet.* **110**, pp. 46–8.

BUDD, A., L. G. SMITH and F. W. SHELLEY (1943). 'On the birth and upbringing of the female chimpanzee "Jacqueline" ', *Proc. Zool. Soc. Lond.* **113**, pp. 1–20.

CARPENTER, C. R. (1964a). 'Behavior and social relations of howling monkeys', *Naturalistic Behavior of Nonhuman Primates*, Penn. State Univ. Press.

—— (1964b). 'Behavior of red spider monkeys in Panama', *Naturalistic Behavior of Nonhuman Primates*, Penn. State Univ. Press.

—— (1964c). 'An observational study of two captive mountain gorillas (*Gorilla beringei*)', *Naturalistic Behavior of Nonhuman Primates*, Penn. State Univ. Press.

—— (1964d). 'A field study in Siam of the behavior and social relations of the gibbon (*Hylobates lar*)', *Naturalistic Behavior of Nonhuman Primates*, Penn. State Univ. Press.

—— (1965). 'The howlers of Barro Colorado island', *Primate Behavior*, ed. DeVore, New York, Holt, Rinehart and Winston.

DAVENPORT, R. K. Jr. and E. W. MENZEL Jr. (1963). 'Stereotyped behavior of the infant chimpanzee', *Arch. Gen. Psychiat.* **8**, pp. 99–104.

DRAPER, W. A. and I. S. BERNSTEIN (1963). 'Stereotyped behavior and cage size', *Percept. and Mot. Skills* **16**, pp. 231–4.

DUFFY, E. (1951). 'The concept of energy mobilization', *Psychol. Rev.* **58**, pp. 30–40.

FITZGERALD, A. (1935). 'Rearing marmosets in captivity', *J. Mammal.* **16**, pp. 181–8.

GLICKMAN, S. E. and R. W. SROGES (1966). 'Curiosity in zoo animals', *Behaviour* **26**, pp. 151–88.

GOODALL, J. (1965). 'Chimpanzees of the Combe stream reserve', *Primate behavior*, ed. DeVore, New York, Holt, Rinehart and Winston.

GROOS, K. (1898). *The play of animals* (translated by Elizabeth L. Baldwin), London, Chapman and Hall.

HARLOW, H. H. and M. K. HARLOW (1962). 'Social deprivation in monkeys', *Scientific American* 207, pp. 136–46.

HARRISON, B. (1962). *Orang-utan*, Collins, London.

HASELRUD, G. M. (1938). 'The effect of movement of stimulus objects upon avoidance reactions in chimpanzees', *J. Comp. Physiol. Psychol.* 25, pp. 507–28.

HAYES, K. J. and C. HAYES (1951). 'The intellectual development of a home-raised chimpanzee', *Proc. Amer. Phil. Soc.* 95, pp. 105–9.

HEBB, D. O. (1949). *The organization of behavior*, New York, John Wiley.

HEBB, D. O. (1955). 'Drives and the c. n. s. (conceptual nervous system)', *Psychol. Rev.* 62, pp. 243–54.

HEDIGER, H. (1955). *Studies of the psychology of captive animals in zoos and circuses* (translated by G. Sircom), New York, Criterion.

—— (1964). *Wild animals in captivity* (translated by G. Sircom), New York, Dover.

HESS, E. H. (1959). 'Imprinting', *Science* 130, pp. 133–41.

HOLZAPFEL, M. M. (1939). 'Die Entstehung einiger Bewegungs-oteriotypien bei gehaltenen Saugern und Vogeln', *Rev. suisse Zool.* 46, p. 18.

—— (1956). 'Uber die Bereitschaft zu Spiel und Instinkthandlungen', *Z. Tierpsychol.* 13, pp. 442–62.

JACOBSEN, C., M. JACOBSEN and J. YOSHIOKA (1932). 'Development of an infant chimpanzee during her first year', *Comp. Psychol. Monog.* 9, pp. 1–94.

JAY, P. (1965). 'The common langur of North India', *Primate behavior*, ed. DeVore, New York, Holt, Rinehart and Winston.

KOFORD, C. B. (1963). 'Group relations in an island colony of rhesus monkeys, *Primate social behavior*, ed. Southwick, New York, Insight Books, Van Nostrand.

KOHLER, W. (1925). *The mentality of apes*, Harcourt, Brace, New York (reprinted by Penguin, London, 1957).

LORENZ, K. Z. (1956). 'Plays and vacuum activities', *L'Instinct dans le Comportement des Animaux et de L'Homme*, ed. Autuori *et al.*, Paris, Fondation Singer-Polignac, Masson et Cie.

LOWTHER, F. DE L. (1940). 'A study of the activities of a pair of *Galago Senegalensus moholi* in captivity, including the birth and post-natal development of twins', *Zoologica* 25, pp. 433–62.

MASON, W. A. (1965a). 'Determinants of social behavior in young chimpanzees', *Behavior of Nonhuman Primates*, ed. Schrier, Harlow and Stollnitz, New York and London, Academic Press.

—— (1965b). 'The social development of monkeys and apes', *Primate behavior*, ed. DeVore, New York, Holt, Rinehart and Winston.

MENZEL, E. W. Jr. (1962). 'Individual differences in the responsiveness of young chimpanzees to stimulus size and novelty', *Percept. Mot. Skills* **15**, pp. 127–34.

—— (1963). 'The effects of cumulative experience on responses to novel objects in young isolation-reared chimpanzees', *Behaviour* **21**, pp. 1–12.

—— (1964a). 'Responsiveness to object movement in young chimpanzees', *Behaviour* **24**, pp. 147–60.

—— (1964b). 'Patterns of responsiveness in chimpanzees reared through infancy under conditions of environmental restriction', *Psychol. Forsch.* **27**, pp. 337–65.

—— (1966). 'Responsiveness of objects in free-ranging Japanese monkeys', *Behaviour* **26**, pp. 130–50.

MENZEL, E. W. Jr., R. K. DAVENPORT Jr. and C. M. ROGERS (1961). 'Some aspects of behavior toward novelty in young chimpanzees', *J. Comp. Physiol. Psychol.* **54**, pp. 16–19.

MITCHELL, P. C. (1912). *The childhood of animals*, New York, Fredk. A. Stokes.

MORRIS, D. (1956a). 'The function and causation of courtship ceremonies in animals (with special reference to fish)', *L'Instinct dans le Comportement des Animaux et de l'Homme*, ed. Autuori et al., Fondation Singer-Polignac, Masson et Cie., Paris.

—— (1956b). ' "Typical intensity" and its relation to the problem of ritualization', *Behaviour* **11**, pp. 1–12.

—— (1962). *The biology of art*, London, Methuen.

—— (1964). 'The response of animals to a restricted environment', *Symp. Zool. Soc. Lond.* **13**, pp. 99–118.

PYCRAFT, W. P. (1912). *The infancy of animals*, Lond., Hutchinson & Co.

REYNOLDS, V. (1961). 'The social life of a colony of rhesus monkeys (*Macaca mulatta*)', unpublished Ph.D. Thesis, Univ. of London.

REYNOLDS, V. and F. REYNOLDS (1965). 'Chimpanzees of the Budongo forest', *Primate behavior*, ed. DeVore, New York, Holt, Rinehart and Winston.

ROMANES, G. J. (1883). *Animal intelligence*, London, Kegan Paul, Trench and Co.

SCHALLER, G. B. (1965). 'The behavior of the mountain gorilla', *Primate behavior*, ed. DeVore, New York, Holt, Rinehart and Winston.

SCHILLER, P. H. (1957). 'Innate motor action as a basis of learning: manipulative patterns in the chimpanzee', *Instinctive behavior*, ed. C. Schiller, New York, International Universities Press.

SIMONDS, P. E. (1965). 'The bonnet macaque in South India', *Primate behavior*, ed. DeVore, New York, Holt, Rinehart and Winston.

SOUTHWICK, C. H., M. A. BEG and M. R. SIDDIQI (1965). 'Rhesus monkeys in North India', *Primate behavior*, ed. DeVore, New York, Holt, Rinehart and Winston.

TEMBROCK, G. (1960). 'Spielverhalten und vergleichende Ethologie. Beobachtungen zum Spiel von Alopex lagopus', *Z. Saugetierk.* **25**, pp. 1–14.

TINKELPAUGH, O. L. (1942). 'Social behavior of animals', *Comparative psychology*, 2nd edition, ed. Moss, New York, Prentice Hall.

TOLMAN, E. C. (1932). *Purposive behaviour in animals and men*, New York, Century.

VAN HOOFF, J. (1962). 'Facial expressions in higher primates', *Symp. Zool. Soc. Lond.* **8**, pp. 97–125.

WASHBURN, S. L. and D. A. HAMBURG (1965). 'The study of primate behavior', *Primate behavior*, ed. DeVore, New York, Holt, Rinehart and Winston.

WELKER, W. I. (1956a). 'Some determinants of play and exploration in chimpanzees'. *J. Comp. Physiol. Psychol.* **49**, pp. 84–9.

—— (1956b). 'Effects of age and experience on play and exploration of young chimpanzees', *J. Comp. Physiol. Psychol.* **49**, pp. 223–6.

—— (1961). 'An analysis of play and exploratory behavior in animals', *Functions of varied experience*, ed. Fiske and Maddi, Illinois, Dorsey Press.

Addendum

Since this chapter was written, further descriptions and analyses of play behaviour in various primate species have appeared in print. The reader is referred in particular to:

HALL, K. R. L. (1965). 'Behaviour and ecology of the wild Patas monkey, *Erythrocebus patas*, in Uganda', *J. Zool.* **148**, pp. 15–87.

HINDE, R. A. and Y. SPENCER-BOOTH (1967). 'The behaviour of socially living rhesus monkeys in their first two and a half years', *Anim. Behav.* **15**, pp. 169–96.

KAUFMAN, I. C. and L. A. ROSENBLUM (1967). 'The waning of the mother-infant bond in two species of macaque', *Determinants of infant behaviour*, Vol. IV, ed. Foss, London, Methuen. (*Macaca nemestrina* and *Macaca radiata*).

Chapter Six

Variability in the Social Organization of Primates

T. E. ROWELL

SOCIAL groups are maintained by communication between individuals, which is brought about by a series of largely innately determined gestures and responses. These units of communication are constant within a species, and are often specifically distinct. Because of this, *patterns* of social interactions seen in one group have also been accepted as typical of its species, and differences between groups as representing genetically based species differences. When discussing social organization, however, we are not concerned with the motor patterns of communication themselves but with their distribution: with their distribution in time, that is their frequency and pattern of occurrence; and with their distribution within the group, that is between which individuals and age/sex classes they occur.

The way in which the term 'social organization' is used here is a deliberate attempt to place it as an empirical subject, capable of being expressed in terms of quantitative observations. In the past, observations have often been predigested and presented in the form of subjective concepts like leadership, dominance, protection, friendliness or aggressiveness; but such descriptions must be based on the distribution of communicative motor patterns and responses, even if collated at the subconscious level by the observer. For the comparisons we need to make between species and between environments, however, quantitative presentations like that of Kummer (1957) are far more useful. Until more studies like his are available, a discussion of this subject must remain a mixture of conjecture and assumption. In the following discussion it is assumed that variations in any of the factors of social organization mentioned above will be correlated with differences in the resultant organization. To see what this asssumption would predict, let us consider the distribution of the motor pattern of threat in some imaginary social groups:

1. A group in which five threats are seen in an hour will have a different type of organization from one in which fifty-five threats are seen, even if the contexts in which they occur and the classes of animals involved are the same.

2. A group where threats occur at the higher rate only in a feeding situation will none the less have a type of organization which reflects this difference in other situations as well.

3. A group in which threats only occur between adult males will have a different type of organization from one in which they occur only between adult females, even though the frequency of threats, and their contexts, are the same.

This usage of 'social organization' would cover the interactions of any two or more animals, even if they were in the process of a fight to the death. This is acceptable, because we can learn from simplified and extreme social situations about mechanisms which also operate in more typical and complex ones. In naturally occurring groups, however, there are elements of cohesion and stability in the social organization, which allow us to think of the more self-destructive artificial groups, however instructive, as abnormal, even though it would be difficult to draw a line of definition through the continuum observed. 'Cohesion' and 'stability' are not necessarily subjective terms – the former could be defined and quantified in terms of the ratio of 'friendly' to agonistic interactions observed, the latter by comparison of representative samples of group social behaviour recorded at intervals.

At present, primate studies which take social organization into account are being undertaken at a rate which makes it difficult even for those of us working in this field to keep track of all new developments. This is no time for a review – much of the work is still available only in preliminary reports, in which statements are made which are impossible to evaluate until their quantitative bases are published; and more is still under way. Already, however, there are signs that we shall soon have to think again about some established generalizations of primate behaviour. Such generalizations were based on a very few pioneer studies, excellent in themselves, but too slender pillars to support the edifices built on them.

Only very recently comparative statements about behaviour could refer to Zuckerman (1932) on the baboon, Carpenter (1964) on the gibbon, the howler and the rhesus, Yerkes (1943) and Nissen (1931) on the chimpanzee and a few anecdotes. From these were made generalizations about *the* Old World monkey, *the* ape, *the* New World monkey, especially by those who wanted to make statements about the evolution of human behaviour.

Now the number of completely unknown genera is dwindling fast; more important, studies are accumulating that allow really useful comparisons to be made of populations of a species – or closely related subspecies – in different environments, and of different species in comparable environments. The picture which is emerging is one of diversity – not so much in the basic units of behaviour, the motor patterns of communication which seem to vary only in detail throughout the Catarrhina (for example) but in the pattern of living which is built out of these units. This diversity does not seem to be related to taxonomic divisions, in fact some of the most interesting findings are of diversity of habits within species. Here are some of the examples which prompted me to the following discussion:

Baboons (*Papio anubis*) in Kenya (studied by DeVore and Washburn, 1964, and Altman, 1965, in the Amboseli reserve) use about fifteen square miles of open grassland as home range. Baboons in Uganda (slightly different in appearance, but freely interbreeding) prefer to live in forest; in a mixed forest-grassland habitat troops of comparable size used a range of about two square miles (Rowell, 1966) and there were other differences in behaviour which will be mentioned later.

Vervets (*Cercopithecus aethiops*) in Amboseli (Struhsaker, 1966) are aggressively territorial and maintain a strict hierarchy in their troops by fighting. Vervets in Uganda (Gartlan, 1965) live peaceably with little or no fighting and no pronounced rank order; males there moved frequently between troops.

Langurs (*Presbytis entellus*) (Jay, 1964; Ripley, 1966) in N. India and Ceylon lived in friendly, non-hierarchical troops of both sexes and all ages; those *Presbytis* studied by Sugiyama (1966) in Southern India lived in troops of females and their young, with one adult male. All-male bands lived between the female troops, and occasionally fought and deposed the troop male and replaced him with one of their number.

Diversity of this order makes one extremely cautious about any generalization about, or comparison between, taxa on ecological or sociological bases.

There are two 'explanations' of the diversity that I want to consider: first, that it is a methodological illusion; and second, that environmental factors are formative of primate social structure to a hitherto unsuspected degree.

ILLUSIONS

Observer Bias

Monkey watchers are very rarely interested in the same things, and it is almost impossible not to be selective in recording behaviour simply because it is rapid and complex. Thus, one finds all too often that there is very little basis for comparison between two accounts because different elements were recorded. If comparisons are made in these circumstances their value is often comparable to that of the time-honoured dichotomy, beloved of textbooks, between birds and fishes as 'instinctive animals' and mammals as 'learning animals', which is based on the first choice of research animals by the founders of two schools of behaviour studies. Simply using different recording methods may make comparisons impossible: the man with the computer available tends to record in great detail, for short periods, small units of easily countable behaviour, whereas the pencil-and-paper man tends to follow long and involved interactions in more general terms (Jensen and Babbitt (1965) and Bolwig (1963) are two extremes in this dichotomy). Zuckerman (1963) suggested that among field workers the observer's own temperament and sex might be an important filter in determining, for example, the amount of agonistic behaviour observed and reported in groups of primates. It is to be hoped his remark is not to be taken too seriously, otherwise we must abandon all efforts at comparison except between studies by a single worker.

The Time Factor

No study of social behaviour has yet been long enough. This is especially true of field work but even artificially established or manipulated colonies have not been followed consistently, without drastic interference. The exception to this generalization will probably be the studies on unrestricted but artificially fed *Macaca fuscata* by Japanese workers; the Takasakiyama monkeys have now been followed for over ten years, a time span approaching the life expectancy of the species. The problem is one of human organization – it is difficult to get finance and staffing continuity for long-term projects, and the time scale for many primate societies are of the same order as for the human – even for man there are embarrassingly few longitudinal studies. A baboon can live for forty years in captivity, and there is nothing to indicate that this environment is any more favourable than the natural one. One regularly encounters elderly baboons in the wild whose teeth are worn level with the gums, but which are well able to follow the daily routine of the troop. Such animals must be well over twenty years old at a very conservative estimate. Environments at the present are changing very rapidly, and animals with this sort of life span could live through great changes in habitat. For example, in Uganda a largely unpopulated area might be colonized, deforested, and monkeys become crop-destroying vermin to be 'controlled'. Or a populated area may be cleared as trypanosomiasis invades, used as a hunting area, made into a National Park, and finally have a large hotel and a main road built in it. The field student coming into a new environment naturally tends to assume it is stable in its present state, and so finds difficulty in interpreting the behaviour of primate troops which are in part influenced by the past experiences of their older members.

The social organization of a natural group is based on long-term relationships. The Japanese workers have found that mothers in their *M. fuscata* groups retain a special relation with their daughters long after their daughters mature (Kawamuru, 1958). Members of a group of baboons know the mother of all juveniles at least up to two years, well after weaning. In fact there is no reason to suppose that relationships are not recognized amongst adults that are impossible for a new observer to determine – but no one has followed a wild group long enough to find out if inter-adult relationships are affected by year-class or family. (I shall return to this problem later.)

The group in which most primates live is fairly small and contains only a few breeding females. Since the sex of their babies is randomly determined, the composition of a troop can vary a good deal, even though the sex ratio of the population as a whole remains equal. Even in baboons, which occasionally live in very large groups of over a hundred, the typical troop probably has thirty to fifty members, with perhaps five to ten breeding females. In the population I studied (where the overall sex ratio was about equal) one troop of thirty had five adult males and five adult females. All the four to five year olds in this group – seven of them – were males. In the next younger age group there were two females and two or three males. The present adult sex ratio is equal; if all the juveniles survive (and there was no evidence of predation or disease in the area) the adult sex ratio will be roughly one female to two males in two or three years' time (allowing for the more rapid maturing of females). Perhaps the social structure of the troop is flexible enough to accommodate such a change in composition; or perhaps the proportions of adults will be maintained and the extra males will move to other troops, or even be driven out. Either way, the younger juveniles have a different sex ratio, so as they mature the present pattern will not be repeated. Whatever happens, an observer looking at that particular troop two years from now is likely to see a very different pattern of social organization from that which I have observed in the last two years.

On the whole, field workers have not seen enough mortality to keep the populations constant. Predation is rarely observed, remains are rarely found and often the population at the end of the study is larger than it was at the beginning. (For example, my baboons were increasing at roughly 10 per cent a year over two years, and only one death was known to occur. Gartlan's vervets and Jolly's lemurs (Jolly, 1967) showed similar trends.) There is evidence of sawtooth changes in numbers of New World monkeys governed by yellow fever epidemics: there at least, and probably in other areas as well, social organization must be flexible enough to accommodate a wide range of population densities in a relatively short period of time. Since monkey populations all seem to be organized on a group basis, a change in population density means a change in troop size or in number of troops, either of which requires big changes in organization. Splitting a troop requires big changes in the ranging and social behaviour of its members, and is likely to be a process much longer than the average field study. Looked at at different times in the process, the same population would give a very different picture of inter- and intra-group organization. Sugiyama (1960) described the division of a troop of Japanese macaques. This was a relatively rapid process lasting less than a year, and he was able to follow the big changes in organization of both the bud troop and the main, parent troop during and after the division. These animals had a remarkably stable hierarchical social organization, but it was loosened and partly inverted during the division. A baboon population in Uganda also perhaps illustrates this process: the troops were spread along a river, so that each had only one neighbour either side. The best-known troop had quite different relationships with each of its two neighbours. With its upstream neighbour it occasionally joined forces to go to some rather distant fruit trees, and they would sleep in trees only a few hundred yards apart. Older juvenile males and the occasional adult male sometimes moved between the troops. The downstream neighbours rarely met the middle troop: when they did each side stared at the other intently and then one or other withdrew. The downstream troop had a closer relationship with *its* downstream neighbour and sometimes slept in the same or neighbouring trees. It was a large troop (over fifty), and occasionally split into sub-groups which followed different routes for all or part of a day, but as far as I could tell always rejoined at night. There was no appreciable change in this set-up during two years. One interpretation would be that the first two troops had divided relatively recently, and that the other was in the very early stages of division.

There has been some argument about the definition of a primate troop, and the degree of cohesion and permanence to be expected of the breeding unit – the problem is, of course, of interest to the taxonomist and the geneticist as well as the sociologist. The argument has been largely theoretical, in the field one is usually fairly confident to the entity of a troop. The problem resolves itself if one takes into account the time dimension in the structure of the population. The field student is in the position of someone taking a cross-section of a bush: he has no difficulty in recognizing the branch as an entity, but finds a great variety of spatial relation between branches, and at some places his section goes through branching points at various levels. As it stands, the cross-section is meaningless – if he could see the vertical elevation of the plant as well, its structure would be clear. In such a scheme, to argue whether the smaller units or the larger unit should be called a troop where two groups of animals are sometimes seen separately and sometimes together is a waste of time.

The above examples have been given to show how we might gain a false impression both of the variety and of the limitations of primate social organization by taking fleeting glimpses of these long-lived and long-memoried animals. Now I want to consider such evidence as there is for flexibility of social organization in the face of environmental pressures.

ENVIRONMENTAL FACTORS

Before we can say whether the frequency of social interactions, as well as their form, is genetically determined for each species, we obviously need comparisons between groups of one species in different habitats, and between different species in the same habitat. (The first difficulty here is in assessing the validity of 'species', at least among the Old World monkeys. For example, the baboons of East, Central and South Africa, divided at one time into several sub-genera, have also been described as geographical races of the same species. (The latter interpretation seems more useful on behavioural grounds.) Similarly, the macaques are frequently given species rank on the basis of single superficial characters like tail or pelage length, but many of them interbreed so freely and are so similar in other characters, their specific status seems open to doubt.)

Very few field studies have considered the effect of different habitats on a species. Hall (1963) on baboons and Southwick *et al.* (1961) on rhesus macaques concentrated mainly on troop size and composition, and diet, rather than details of social behaviour, nevertheless some interesting points emerge. In some areas the rhesus suffer severe predation on a limited age group as small juveniles are trapped for sale to laboratories – the pressure is sufficient to alter the composition of these troops drastically compared with unmolested ones, and this surely must alter the social behaviour of the troops. Of particular interest are Hall's observations on baboons trapped on a newly formed island by the rising water of the Kariba dam in S. Rhodesia, and in poor condition, short of food. In contrast to the more usual baboon troop organization seen on the adjacent mainland, these animals were seen foraging in small groups or even singly. Hall remarked on the extremely small amount of social interaction to be seen on the island: juveniles rarely played, females were not in breeding condition, so there was no sexual behaviour. One is tempted to suggest that where there is no time for anything but foraging, interactions become so few that social organization disintegrates.

Fortunately there is no reason to confine interest in this general problem to field studies. In fact, since we do not yet know the critical factors which constitute captivity, there can be no clear line between 'field' and 'captive' environments. It is doubtful if any 'field' environment today is not manipulated in some way by man, and several primate populations that have been studied have been only slightly controlled or interfered with – by artificial feeding of originally undisturbed troops, for example. Even the more extremely manipulated environments provided by captivity can quite legitimately be included in the range a species will tolerate, providing the animals survive and breed in them. Indeed, it could be argued that the only way to make a valid comparison between species is to keep them in identical captive environments, equally foreign to both. A comparison of pigtail and bonnet macaques under these conditions is being made by Kaufman and Rosenblum (1965). They clearly show a constellation of specific differences related to the amount of contact – especially huddling – between adults of the groups: *M. nemestrina* adults usually sit alone, *M. radiata* sit in clumps. This difference radically affects, for example, the environment in which the newborn is reared.

In the wild, two species that occupy the same area utilize its resources differently, to an extent which is usually difficult to assess. The captive environment will inevitably be far simpler, and therefore easier to describe. The simplification, of course, should not be carried too far; unless the cages provide sufficient space and scope, the comparison will be made within a very limited range of the animal's behaviour; and a diet which is adequate for survival and breeding may still be far from adequate for the animal to show its full behavioural capacity. This latter is a statement of the obvious regarding one species, man, but is infrequently extended to other primates. In assessing the cage environment and the limitation imposed, it is enormously helpful, almost essential, to have some knowledge of the animal in the wild. As a trivial example, baboons and rhesus monkeys carry their babies on their backs in the wild. They do not do so in cages where they frequently go through low doors.

One of the most interesting aspects of the problem is to discover

which parameters of a habitat might be relevant to social organization. At present we have mainly speculation with some empirical
findings from zoo management and a few scraps of evidence from
field studies. Some of the likely factors are considered below.

Space: What, for example, is a large cage: the term has been
used for cages four feet cube to cages twenty-five yards across for
the rhesus monkey (adult female weighs about sixteen pounds). In
fact probably more important than the total space available is its
arrangement – the amount of 'three-dimensional movement' possible, and the extent to which the area is broken up into sections
relatively separate from each other. One of the most striking differences between caged and wild baboons, for example, is that a caged
group occasionally appears to 'get trapped' in a cycle of agonistic
interactions which goes on and on until the animals inflict serious
damage on each other. In the wild these interactions do not go on
too long because one individual goes out of sight behind a bush
until the other has forgotten about the incident. Where a cage is
subdivided such potentially dangerous interactions can usually also
be broken by one animal 'escaping' to a different section; but if the
same total area is left as a single unit and the animals are always in
each other's sight, bad fights may be more frequent. In the amount
of cover available, then, we have a factor which can potentially alter
the frequency of any type of interaction, and so the whole type of
organization achieved. It could also be effective in the wild – the
baboon for example lives in a variety of habitats from thick secondary forest and dense bush to short open grassland. Studies made
where cover was plentiful (Rowell; Hall, 1962) have laid less stress
on agonistic than on friendly behaviour, whereas those carried out
in open grassland have tended to have the reverse emphasis
(DeVore and Washburn, 1964; J. Altman, 1966). Of course this is
very circumstantial, many other factors were operating.

Food: On the richness of the environment depends the amount
of time needed for foraging, and hence the time left over for social
interaction. As we saw in Hall's observations, if conditions are bad
enough, social behaviour almost disappears. But differences can be
seen before this extreme is reached. In a forest habitat baboons can
usually fill their stomachs in half an hour or an hour, and then
spend an equal amount of time sitting about, grooming, playing, dozing before walking a short distance to the next fruit
tree. Baboons in dry grassland on the other hand forage nearly

continuously, and though the young play early in the morning, later in the day all their energy is needed to keep up with the adults, and the number of interactions seen is small. Probably related to this is another difference in organization – the troops in the richer environment never left a straggler behind: if an individual lagged everyone sat and waited till it caught up. The savannah baboons (Altman, pers. comm.) ignored stragglers, and those which could not keep up were lost. At the other extreme, foraging time cannot be reduced beyond a minimum, apparently: even in a cage where food is abundant and needs no searching out, a group spends a lot of time turning over and inspecting the debris in the cage with movements used for foraging in the wild, bringing the time spent in this sort of activity to about the same level as that in a rich wild habitat.

Distribution of food is also important. Chalmers (pers. comm.) has shown that mangabeys (*Cercocebus albigena*) have more agonistic interactions when they are feeding on large fruit growing singly, than when the food is small fruit evenly dispersed on the trees. In the forest baboons mentioned above agonistic interactions were extremely rare, and most of their food sources were well spread out and abundant. But exchanges of threats occurred over the possession of piles of elephant dung at the stage of maturity when it contains sprouting seedlings much prized as food. It is significant here that recent studies of primate societies which have found clear-cut dominance hierarchies maintained by fighting have been made on free-ranging groups which come to feeding stations for the main part of their diet, e.g. Sade (1966). Partly equivalent to this situation is that of animals frequently fed titbits by visitors, as in the Nairobi National Park baboons (Kenya). For these animals, much of their food is obtained under the stressful conditions which occasionally obtain for the mangabeys and baboons mentioned above. Sade, on the Cayo Santiago rhesus colony, said that most agonistic interactions he saw occurred round the food hoppers, and the Japanese macaques of Takasakiyama are also figured fighting over food. An association between a high frequency of agonistic behaviour and competition at a restricted food source is of course, not confined to monkeys (see for example, Marler (1955) on chaffinches). In a typical regime of a caged monkey group, food is provided not only in one place, but at rare intervals. The animals become very excited at feeding times, and the period just before food is due is the one in which serious fights most usually break out. When food is given, it is taken in strict succession, alpha male first and omega animal last, a pattern which is best regarded as an extreme modification of social organization in an extreme environment.

The captivity situation also provides extreme conditions of group composition. As we have seen, the composition of wild troops may be subject to quite wide variation within one population over a short time span, and can presumably adapt to these changes. An artificially constructed group can lack whole sections of the typical composition, and provide highly eccentric proportions of others. But the traumatic experience with hamadryas baboons in London Zoo reported by Zuckerman (1933) shows that there are limits to this adaptation: that colony was started with all adults, a heavily male-weighted sex-ratio, and a very large number of mutually strange individuals. Even then the animals began to build the rudiments of an organization similar to that found in the wild by Kummer and Kurt (1963) but could not achieve stability.

The cage situation may be more difficult for some types of inter-action than for others, so that unlikely (by comparison with wild group) compositions may be perpetuated. Thus the mother-infant interaction in its early stages does not seem to be greatly affected, but there are fairly frequent reports of mothers – usually alone in cages with their offspring – who deprive them of food or are excessively aggressive from the age when the baby would normally begin to be independent (e.g. Lashley and Watson's (1913) report of a rhesus macaque and her baby). Adult males of many species rarely establish a stable social relationship with each other in cages, although in the wild the same species are known to live in multi-male troops. Thus it is a matter of practical colony management to start a new caged group with only one adult male, and often his sons are removed as they reach maturity to reduce fighting damage. Yet adult male baboons for example, observed in the wild, interact continually with each other, both at the purely 'social' level, and in the troop's disposition in relation to the environment. Every move-ment of a wild troop male seems to be made in relation to the positions and activities of his peers. The solitary adult male of this species in a caged group finds himself in a curiously isolated position with a much reduced level of social activity limited largely to sexual and grooming interactions with females.

Eccentric population composition, and environmental stresses like lack of cover and feeding competition are fairly obvious factors in the extreme conditions of captivity that lead to unusual or unstable social organization. Less often recognized, but possibly equally important, is a common feature of the social environment of captivity: caged groups are often made up for breeding purposes of a collection of adult animals hitherto strange to each other. In contrast wild monkeys interact almost exclusively with animals they have known from birth. It is not too far-fetched, I think, to suggest that for later inter-adult behaviour to be on a friendly and stable basis, two individuals must establish their communication pattern when one or both of them is an infant. Of course strange monkeys thrown together will, like most animals, establish some sort of crude hierarchical organization in a very short time (e.g. Bernstein and Mason, 1963). But experience in establishing breeding groups suggests that this is often not the final pattern, and the group is still in the process of settling and adjusting at least a year after it is established. Even after this there may still be unexpected flare-ups of serious fighting – perhaps occasioned by some slight temporary disability in one animal – so possibly the founder members of such a group can never be expected to become fully integrated into a social unit. The most striking difference between a caged and a wild group of baboons is the vastly greater amount of agonistic interaction between adult females among the caged animals (Rowell, 1967). (I suspect this is also true for macaques and vervets at least, but I have no field experience of them.) Most of the serious wounds are inflicted by females on each other, some by the adult male when breaking up a fight between females. In the wild population the very rare bouts of threatening between females were seen to lead even to contact on only two occasions, and there was no biting. This observation can be interestingly juxtaposed to two others: whereas there is accumulating evidence that males of several species move in and out of troops in the wild, adult females, as far as is known, remain in the troop in which they were born, with their immediate female relations. Secondly, adult females caught in the wild are notoriously difficult to acclimatize successfully in captivity, and their high mortality rate makes dealers unwilling to handle them. Males and young animals in contrast seem to adapt reasonably quickly to captivity and new social environments. It seems possible that the female is peculiarly sensitive to changes in her social environment, and in some way depends to a great extent on stability of social organization based on individual relationships established with other females during infancy. Wrenched out of this organization she 'goes all to pieces', showing atypical behaviour or becoming extra susceptible to disease. This is a good example of the way extreme situations provided by captivity might throw light on methods of social organization which would not be apparent in the wild however extensive the observations.

There is only one species of primate for which we have some idea of the range of naturally occurring social organization, and that is man. Comparison of human with other primate societies can be very misleading, but it would be unreasonable to avoid them altogether since the availability of funds for research on non-human societies depends largely on the assumption that the findings are peculiarly relevant to human needs. The main source of confusion seems to be that there is no non-human equivalent to what is broadly thought of as 'society' in industrialized countries: monkey social groups can be compared only with human more or less extended family groups, perhaps in some cases with the isolated farming hamlet. However, it surely requires only an elementary degree of sophistication to avoid arguing from the hierarchical structure of a caged baboon troop to the naturalness and inevitability of fascism, while on the other hand we have a lot to learn at this stage of our studies from anthropology. Disregarding the reasons people give for their behaviour and taking into account only behaviour that can be observed, there is still a very wide range of types of social organization to be seen in this single species at any one time, and the societies also change their structure quite rapidly in response to changing external conditions. In many cases the structure of a society can be related directly to dominant features of the present environment and is regarded as adaptive; elements of the structure of other societies may seem highly unadaptive, but are maintained by tradition in the face of environmental changes, and may have arisen in response to earlier environmental pressure, or even as a result of a false hypothesis about the environment.

It is unlikely that we shall find variety of social organization in other primates to approach that of man, if only because our species is so much more numerous than any other. However, it should not be surprising to find the same *sort* of variability in the similar social organizations of related species. It is important that we should recognize at an early stage in the accumulation of comparative studies of social organization that there *may* be no such thing as a 'normal social structure' for a given species, and that a description of social organization is only useful if accompanied by a description of the environment in which it occurs – and further, that we are still only guessing about which features of the environment will be essential in such a description.

SUMMARY

Social organization can be described in quantitative terms. It depends on three variables – the frequency of interactions, their pattern of occurrence in time, and between which individuals and age/sex classes they occur. 'Cohesion' and 'stability', two other concepts which are often used in describing social organization are also capable of quantitative description. Since the basic motor patterns of communication are relatively similar throughout the primates, quantitative descriptions of social behaviour are essential for useful inter- or intra-specific comparisons.

New field studies are bringing to light an unsuspected variability in the social organization of primates. It is possible that this variability is only apparent, an artefact of observer biasses, and the short time span of most studies relative to the longevity of primates – some possible effects of this are discussed. An alternative is that environment has quite radical effects on organization. Environments in captivity provide often extreme conditions which may give clues about smaller differences in natural habitats. Space, food and group composition are considered as affecting social organization.

Only for one species, man, is the range of variability in social organization fairly well known, but it is suggested that variability of the same order might reasonably be expected in other primates.

REFERENCES

ALTMAN, J. (1966). 'Agonistic interactions in baboons', *Social communication among primates*, ed. Altman, University of Chicago Press.

BERNSTEIN, I. S. and W. A. MASON (1963). 'Group formation by Rhesus monkeys', *J. Anim. Behav.* **11**, pp. 28–31.

BOLWIG, N. (1963). 'Bringing up a young monkey', *Behaviour* **21**, pp. 300–30.

CARPENTER, C. R. (1964). *Naturalistic behavior of non-human primates*, Penn. State Univ. Press.

CHALMERS, N. (1967). 'A field study of the mangabey (*Cercocebus albigena*)', Ph.D. Thesis, Cambridge.

DEVORE, I. (1963). *A comparison of the ecology and behavior of monkeys and apes*, Viking Fund Publications in Anthropology II, 37.

DEVORE, I. and S. L. WASHBURN (1964). 'Baboon ecology and human evolution', *African ecology and human evolution*, ed. F. C. Howell, New York.

GARTLAN, J. S. (1965). 'Dominance in East African monkeys', *Proc. E. Afr. Acad.* **2**, pp. 75–9.

HALL, K. R. L. (1962). 'Sexual, agonistic and derived social behaviour pattern of the wild Chacuna baboon', *Proc. Zool. Soc. Lond.* **139**, pp. 283–327.

—— (1963). 'Variations in the ecology of the Chacuna baboon', *Symp. Zool. Soc. Lond.* **10**, pp. 1–28.

JAY, P. (1964). 'Mother-infant relations in Langurs', *Maternal behavior in mammals*, ed. Rheingold, New York, Wiley.

JENSEN, G. D. and R. A. BABBITT (1965). 'An observational methodology and preliminary studies of mother-infant interaction in monkeys', *Determinants of infant behaviour*, III, London, Methuen.

JOLLY, A. (1967). *Lemur Behaviour*, Univ. of Chicago Press.

KAUFMAN, I. C. (1965). *Determinants of infant behaviour*, III, London, Methuen.

KAWAMURU, S. (1958). 'The matriarchal social order in an *M. fuscata* group', *Primates* I.

KOECHWSKY (1937). 'Brain mechanism and variability, I, II, III', *J. Comp. Psychol.* **23**, pp. 121–38, 139–64, 351–64.

KUMMER, H. (1957). 'Soziales verhalten einer Mantelpaviangruppe', *Beihalt zur Schweizerischen Zeitschrift fur Psychologie und ihre anwendungen*, nr. 33.

KUMMER, H. and F. KURT (1963). 'Social units of a free-living population of hamadryas baboons', *Folia primat.* **1**, pp. 4–19.

LASHLEY, K. S. and J. B. WATSON (1913). 'Notes on the development of a young monkey', *J. Anim. Behav.* **3**, pp. 114–39.

MARLER, P. (1955). 'Studies of fighting in chaffinches. I. Behaviour in relation to the social hierarchy', *J. Anim. Behav.* **3**, pp. 111–17.

NISSEN, H. W. (1931). 'A field study of the chimpanzee. Observations of chimpanzee behaviour and environment in Western French Guinea', *Comp. Psych. Monogr.* **8**, pp. 1–105.

RIPLEY, S. (1966). 'Intertroop encounters among Ceylon grey Langurs', *Social communication among primates*, Univ. of Chicago Press (in press).

ROSENBLUM, L. A. (1966). 'Mother-infant relations in pigtail and bonnet macaques', *Social communication among primates*, Univ. of Chicago Press.

ROWELL, T. E. (1966). 'Forest-living baboons in Uganda', *J. Zool. Lond.* **149**, pp. 344–64.

—— (1967). 'A quantitative comparison of the behaviour of a wild and a caged baboon group', *Anim. Behav.*

SADE, D. S. (1966). 'Dominance in a group of free-ranging Rhesus monkeys', *Social communication among Primates*, Univ. of Chicago Press (in press).

SOUTHWICK, C. H., M. A. BEG and M. R. SIDDIQUI (1961). 'A population survey of rhesus monkeys in Northern India. Pts. I and II', *Ecology* **42**, pp. 538–47 and 698–710.

STRUHSAKEV, T. (1966). 'Vocal communication by vervet monkeys', *Social communication among primates*, Univ. of Chicago Press.

SUGIYAMA, Y. (1960). 'On the division of a Natural troop of Japanese monkeys at Takasakiyama', *Primates* **2**, pp. 109–48.

—— (1966). 'Social organisation and its change in Honuman Langurs', *Social communication among Primates*, Univ. of Chicago Press.

YERKES, R. M. (1943). *Chimpanzees,*

ZUCKERMAN, S. (1932). *The social life of monkeys and apes*, London, Kegan Paul.

—— (1931). 'The menstrual cycle of the primates, Part 3. The alleged breeding season of the primates, with special reference to the chacma baboon', *Proc. Zool. Soc. Lond.*, pp. 325–43.

—— (1963). 'The Primates. Part I. Concluding remarks of chairman', *Symp. Zool. Soc. Lond.* **10**, pp. 119–23.

Chapter Seven

Comparative Aspects of Communication in New World Primates

M. MOYNIHAN

THIS paper is a tentative, preliminary attempt to survey and summarize some features of the social signal systems of platyrrhine monkeys.

During the last seven years, I have been able to study three species in considerable detail, both in the laboratory and in the wild. These are *Aotus trivirgatus* (described in Moynihan, 1964), *Callicebus moloch* (Moynihan, 1966), and *Saguinus geoffroyi*. Other species of *Saguinus* and *Callicebus* and at least one species of all the other genera commonly recognized (except '*Brachyteles*') have been observed more briefly. Although there are published accounts of the behaviour of several species by other students, all references to behaviour patterns throughout the following pages are based upon personal observation unless specifically stated otherwise.

The generic names used throughout this paper follow Hershkovitz (1958). The specific names of most of the forms occurring in Panama and Colombia follow Hershkovitz (1949 and 1963), and those of other species follow Fiedler (1963) unless specifically stated otherwise.

The New World primates are diverse in both ecology and social organization. All the species are tropical and inhabit areas of forest and more or less forest-like scrub (savannah faunas are poorly developed in the New World tropics). Within these limits, they show a great range of habitat preferences. For convenience, however, they can be divided into two main groups: (1) forms which usually live *within* tall, mature or nearly mature forests (e.g. *Aotus, Alouatta, Cebus, Ateles*); and (2) forms which usually occur along the edges (including the upper layer or top surface) of old forest and/or in relatively low, bushy, second growth (e.g. *Saimiri* and all or most species of *Saguinus*). The edge and second growth species are comparatively small. All the large forms usually live within tall forest (but not all the species in this habitat are large). The diet of many or all of the large species includes a relatively large amount of vegetable matter. Most of the small species are primarily insect-ivorous (usually prefer to eat insects whenever available). At least one species, *Cebus capucinus*, comes down to the ground with some appreciable frequency. Others, e.g. *Saguinus geoffroyi*, come down occasionally. Still others, e.g. *Alouatta* spp., seem to be exclusively arboreal. One species, *Aotus trivirgatus* (the only species of the genus), is nocturnal. The rest are diurnal.

None of the New World primates seems to be as nearly com-pletely solitary as some lemurs (see Petter, 1962); but some of them are only slightly gregarious. *Aotus trivirgatus* in central Panama, for instance, apparently are never found in organized social groups of more than two or three individuals, i.e. mated pairs, sometimes with their (single) young of the year. Many *Callicebus moloch* occur in similar groups. Other species, e.g. *Saguinus geoffroyi* and *S. midas* (Thorington, in press), often form larger groups of five to seven individuals. These may be composed of adult pairs with the young of several years. Still other species tend to associate in even larger troops or bands, in which the sex ratio may be unequal and in which special 'family' groups (apart from mother-infant relationships) are difficult to detect. Among these more gregarious species are *Alouatta palliata* (Carpenter, 1934), *Ateles 'geoffroyi'*[1] (Carpenter, 1935), *Saimiri sciureus*[2] (Sanderson, 1957), *Lagothrix lagotricha* (Fooden, 1963), *Cebus apella* (see references in Nolte, 1958, and Hill, 1960), *C. albifrons* (Hill, 1960), *C. nigrivittatus* (Hill, 1960), and *C. capucinus*.

Any behaviour pattern which often conveys information from one individual to another may be considered a social signal, even if it has many other functions as well. The most conspicuous

---

[1] It may be doubted if there is really more than one species among all the forms assigned to *Ateles* by most authors (i.e. all the typical spider monkeys, excluding the Woolly Spider Monkey '*Brachyteles*'). The oldest valid specific name for the group as a whole seems to be *paniscus* (Kellogg and Goldman, 1944).

[2] Sanderson does not give the specific name of the individuals he observed; but his observations were made in Guiana and *sciureus* is supposed to be the only form of the genus in that region.

(although not necessarily most important) social signals are 'displays'. The term 'display' may be applied to any pattern which has become 'ritualized', i.e. specialized in form or frequency as an adaptation expressly to convey information.

Most of the patterns which will be discussed below are displays in this sense.

The signal patterns of primates and other mammals can be divided into four main types, according to the senses by which they are perceived. These types may be called tactile, olfactory, visual, and acoustic.

TACTILE SIGNALS

These are not particularly varied among Platyrrhini. None of the species studied has a very large repertory of such patterns. Some species and groups of species have one or more tactile signals which are peculiar to themselves alone (e.g. the Tail-twining[1] of *Aotus* and *Callicebus* moloch). Most of these patterns of limited distribution seem to play a relatively minor rôle in the social life of the species. The only tactile signal which is both very widespread and obviously a significant factor in many social interactions is 'Allogrooming', the grooming of one individual by another. Probably mothers groom their infants in all or most species of New World primates, as in most other mammals. More interesting from a comparative point of view (and better studied) is mutual grooming among adults and juvenile individuals which have already developed most of the adult signal patterns. This type of Allogrooming has been observed in the following species: *Callimico goeldii, Callithrix jacchus* (Fitzgerald, 1935), *Saguinus geoffroyi, Aotus trivirgatus, Callicebus moloch, Alouatta palliata* (Carpenter, 1935), *Pithecia pithecia, P. monacha, Cacajao rubicundus, Saimiri sciureus* (Ploog, Blitz and Ploog, 1963), *Cebus apella, C. capucinus, Ateles 'geoffroyi' panamensis* and *A. 'fusciceps' robustus*. To my knowledge, the only species of New World primate which has been studied at some appreciable length but which has not been seen to perform adult Allogrooming is *Cebuella pygmaea*. (This apparent absence may not be very significant. Extreme alarm seems to inhibit the expression of Allogrooming tendencies in all species of Platyrrhini. *C. pygmaea* is a particularly timid species, perhaps the

[1] Throughout this paper, the initial letters of all certainly or probably ritualized patterns will be capitalized.

Fig. 1 Tail-twining, a tactile display, by adult *Callicebus moloch*
(from Moynihan, 1966)

shyest in the whole group, and has been observed only in captivity. Perhaps the individuals observed were continually alarmed by the unnatural environments to which they were exposed. Thus, it would be premature to state that adult Allogrooming definitely is not part of the normal repertory of the species.)

Allogrooming must subserve several functions. It must help to maintain the skin and hair of the groomed animals in good condition. As an adult social signal, however, its primary function probably is to reduce or remove certain social 'barriers' between individuals. Like most other vertebrates of comparable age, all adult and sub-adult New World primates usually show some reluctance to allow themselves to be touched by other individuals. This reluctance must be overcome if certain social reactions are to be carried out successfully. Allogrooming apparently helps to do this because it (frequently or occasionally) 'pleases' or 'satisfies' the individuals being groomed (this is obvious from their behaviour before, during, and after the grooming process). The satisfaction provided by physical contacts during Allogrooming probably facilitates other reactions involving contact.

The form of Allogrooming is much the same in all the New World primates which exhibit this behaviour; but it may occur in different ranges of social circumstances, and its frequency varies enormously from species to species. *Aotus trivirgatus* illustrates one extreme. Among adults of this species, Allogrooming occurs very frequently in close association with copulation, but is rare or absent in other circumstances. The corresponding behaviour of *Saguinus geoffroyi* is less common, and somewhat less closely associated with overt copulation, but nevertheless seems to be purely sexual in motivation. In *Callicebus moloch*, by contrast, adult Allogrooming is very common in a great variety of circumstances and definitely is not purely sexual. In this species, it seems to be a 'general social' or 'gregarious' pattern, stimulating 'friendly' relations between members of the same social group (non-mated individuals as well as mates) throughout the year (during the non-breeding as well as the breeding season). The Allogrooming of adult *Pithecia* and *Cacajao* also is relatively common and certainly is a general social pattern, if nothing else (the sexual behaviour of these forms has not been studied). The Allogrooming of adult *Cebus* certainly is general social and apparently not sexual (see Hill, 1960) but it is performed less frequently than the corresponding behaviour of either

Callicebus or *Pithecia*. In most of the other species, Allogrooming among adults may be more or less intermediate in quality, partly sexual and partly general social, but even rarer than in *Cebus*. It seems to be least frequent in *Alouatta palliata* and *Saimiri sciureus*.

This list indicates that there is no widespread, general correlation between type or frequency of adult Allogrooming and type of vegetation inhabited or usual size of social groups.

Marler (1965) has suggested that there may be a correlation between frequency of adult Allogrooming and 'harshness' or 'rigidity', i.e. extreme development, of dominance relationships in different species of Old World monkeys and apes. There is little or no evidence of this in New World primates. None of the New World primates exhibits very conspicuous dominance relationships under natural conditions. Even the degree of aggressiveness, which certainly influences the development of dominance, does not seem to be very relevant in this connection. *Aotus trivirgatus*, *Saguinus geoffroyi* and *Callicebus moloch* are all comparatively aggressive. So, at least in captivity, are *Cebus* spp. and *Saimiri*.[1]

There also is very little indication of any general correlation between Allogrooming and morphology or anatomy. Thus, for instance, the hands of *Alouatta* are poorly constructed for delicate manipulation (the first two fingers oppose the last three, to form a chamaeleon-like grasping organ), but the hands of *Pithecia* are essentially similar to those of *Alouatta*, while the hands of *Saimiri* are well adapted to manipulation (they are at least as good, in this respect, as those of *Callicebus* or *Aotus*). The type of pelage may be more significant – but only in some cases. *Callicebus* has comparatively long and dense fur, *Aotus* has short and dense fur, and *Pithecia* has rather sparse but long hair; but the pelage of *Saimiri* also is thick and that of male *Alouatta* is approximately as long as that of *Cacajao* and longer than that of *Cebus* or *Saguinus*.

Finally, it should be mentioned that the distribution of Allogrooming is different from that of 'auto-' or 'self-grooming'. Thus, for instance, *Saimiri sciureus* performs autogrooming with some appreciable frequency (although *Alouatta palliata* does not).

[1] Ploog, Blitz and Ploog (*op. cit.*) state that actual fighting is rare among *Saimiri* in captivity. Nevertheless, they did observe an appreciable amount of aggression and what they call 'rage' in the group they studied. My own observations suggest that aggressive hostile vocalizations are among the least uncommon displays of captive *Saimiri* individuals near other individuals with whom they are not very familiar (see also below).

This apparent lack of general correlations is puzzling. It may be useful, therefore, to consider the problem of the adaptive significance of adult Allogrooming in Platyrrhini at somewhat greater length. In view of the fact that adult Allogrooming obviously is advantageous to some New World primates (it was selected for during their evolution), and that theoretical considerations would suggest that it should provide some advantage(s) for all species, and that it is very highly developed in many Old World monkeys and lemuroids, the crux of the problem would seem to be to explain why it is rare and/or restricted to a limited range of circumstances in certain particular species. Several possibilities come to mind.

It may not be coincidental that both the species which are known to perform Allogrooming as a sexual pattern but (certainly or probably) not as a general social reaction (i.e. *Aotus trivirgatus* and *Saguinus geoffroyi*) also sleep in holes in trees. In both species, there is evidence that all the members of the same family or social group usually sleep (or prefer to sleep) in the same hole. Perhaps the physical contacts in restricted sleeping quarters provide most of the same social advantages that other species obtain by general social Allogrooming.

The hypothesis that sleeping in holes does not favour the development of extensive Allogrooming is supported by comparative data.

Among the other primates which sleep in holes are *Cebuella pygmaea*, the tree shrew *Tupaia glis*, and the lemur *Microcebus*. The extreme rarity or absence of adult Allogrooming in *C. pygmaea* has already been mentioned. It certainly is performed by both *Tupaia glis* (Vandenbergh, 1963) and *Microcebus* (Andrew, 1964), but only very infrequently.

The habit of sleeping in holes seems to be directly correlated with size. All very small primates, both diurnal and nocturnal, seem to have acquired this habit. Presumably it gives them added protection against predators. Small species probably are vulnerable to a larger range of predators than are large species. (Also, of course, there probably are more small holes than large holes in most forests.)

Alouatta palliata individuals certainly do not sleep in holes; and the sleeping habits of *Saimiri sciureus* in the wild seem to be unknown. The two species are very different in morphology, and are not very closely related. They do, however, resemble one another in several other aspects of behaviour, in addition to the slight development of Allogrooming. Both are comparatively highly gregarious, and both have developed special, apparently unique, pre-copulatory displays, i.e. 'Tongue-pumping' in the case of *Alouatta* (Carpenter, 1934) and 'Penile Erection' in the case of *Saimiri* (Ploog and Maclean, 1963). Individuals of *Saimiri sciureus* apparently form the largest groups known among Platyrrhini (Sanderson, *op. cit.*), and individuals of the same group tend to stay very close together (at least in captivity). Groups of *Alouatta palliata* usually are somewhat smaller, on the average, but again individuals of the same group tend to stay close together. Both *Saimiri* and *Alouatta* groups seem to be more consistently cohesive than groups of other highly gregarious forms such as *Cebus* and *Ateles*. (*Alouatta* individuals also are comparatively slow moving. This may reduce the tendency to scatter.) Possibly the mere fact that individuals of these species usually are in very close proximity to one another obviates the 'need' for general social Allogrooming among adults. Similarly, the performance of special pre-copulatory displays may obviate the 'need' for sexual Allogrooming.

If the performance of adult Allogrooming in a narrow range of circumstances is partly or indirectly correlated with the habit of sleeping in holes, and if the latter is positively correlated with small body size, then it seems quite likely that Allogrooming of the type shown by *Saguinus geoffroyi* is primitive in Platyrrhini. *Saguinus* itself is obviously specialized in some ways, and its small size may be the result of secondary reduction; but most of the known fossil prosimians which may be related to the ancestors of the Platyrrhini are equally small or smaller (see summary in Simons, 1963).

It might be added, in this connection, that the adult Allogrooming known to occur in other orders of mammals seems to be largely or exclusively sexual in motivation (Andrew, 1964).

If this suggestion is correct, then Allogrooming must have evolved along divergent lines in different groups of New World primates. In some groups, the originally sexual Allogrooming was 'diffused' to become a general friendly reaction. In other groups, it declined and was largely or completely replaced by species-specific and more highly specialized forms of sexual display. Both developments probably occurred independently in several different groups.

There is also another possibility to be considered. It has been suggested that *Callicebus* has retained a larger number of primitive morphological characters than any other living genus of New World primates (see references in Moynihan, 1965). It may also be rather primitive in some aspects of habitus or ecology. This might suggest that general social Allogrooming among adults is primitive. It does not, however, contradict the hypothesis that adult Allogrooming has been reduced during the evolution of some, perhaps many, of the more specialized groups of modern Platyrrhini – especially the more highly gregarious types.

In the present state of our knowledge, the subject might be summarized as follows: Allogrooming among adults must be advantageous in some ways. But it probably has corresponding disadvantages. These disadvantages, like some or all of the advantages, probably are consequences of the physical contact involved. Physical contacts are always 'risky'. An individual approached or touched may become frightened, and retreat, or become irritated and start fighting. (Even in those species which sleep in holes, disputes may occur immediately before or during entry into the sleeping quarters.) Thus, there would seem to have been selection pressure against adult Allogrooming whenever suitable substitutes could be developed in its stead. Selection has favoured the replacement of Allogrooming by patterns not involving physical contact in species whose social organization and other habits are such that non-contact patterns can usually be perceived without difficulty.

OLFACTORY SIGNALS

Olfactory signals have been studied much less thoroughly than other forms of communication in New World primates.

It is obvious, however, that they really *are* very important in the social life of many species. The sense of smell may be as important to many platyrrhines as to many lemurs. (Perhaps the most conspicuous indication of its significance is the fact that the initial response of individuals of all or most species to a strange object or animal, of their own or another species, is to sniff at it.)

The most widespread olfactory display of New World primates is a type of 'marking' which may be called Anal Rubbing, i.e. repeated rubbing of the genito-anal region along a branch. (This region is richly provided with skin glands in all or most species;

and there are additional specialized glandular areas on adjacent parts of the body, such as the base of the tail, in some species.) The species known to perfom Anal Rubbing include *Saguinus geoffroyi* and some other species of the same genus (belonging to the group of 'white-faced' tamarins which Hill, 1957, calls '*Tamarinus*'), *Aotus trivirgatus*, and *Alouatta palliata*. *Saimiri sciureus* and *Callicebus moloch* have been seen to perform movements which look like the Anal Rubbing of other species, but only relatively rarely.[1] *Callicebus moloch* also performs Chest-rubbing, which seems to spread an odoriferous secretion from a large gland in the centre of the chest. *Cebus apella* has been seen to perform similar movements, although it is not known to have a conspicuous gland on the chest or any adjacent area. All the Rubbing patterns are performed in social circumstances which are at least partly hostile. Some or all of them may be employed in territorial defence, at least occasionally.

Rubbing patterns almost certainly are absent in *Cebuella pygmaea*, *Pithecia monacha*, *Cacajao rubicundus*, *Cebus capucinus*, and *Ateles*. It is possible, however, that some of these species have other patterns which serve some of the same functions.

Cebuella pygmaea has an elaborate 'presentation' display, in which the intricately marked genito-anal region is shown to 'opponents' (captive individuals frequently display to human beings). Possibly this display provides olfactory as well as visual stimuli during intraspecific encounters. The circumstances in which it is performed by captive individuals would suggest that it may be used in close-range territorial defence under natural conditions.

Cebus apella (Nolte, *op. cit.*), *C. capucinus*, *Aotus trivirgatus* (Hill, 1957), and *Saimiri sciureus* (Andrew, 1964), have been observed to 'wash' their hands and/or feet with urine in captivity. It is not known how important this type of behaviour may be in the wild; but it certainly could fun :tion as territory marking.

The only New World primates which have been observed at some length but which have not been seen to perform any sort of pattern which could provide olfactory stimuli for territorial defence are *Pithecia monacha*, *Cacajao rubicundus*, and *Ateles*.

---

[1] Anal Rubbing movements by *C. moloch* are not described in Moynihan (1966); but they have recently been seen to be performed by an adult male while carrying its infant young.

Ateles certainly is not closely related to the other two genera; but they all share one character in common: a tendency to assume upright or semi-upright postures. *Ateles* is well known to be a brachiator or 'semi-brachiator' (see, for instance, Erikson, 1963, and Napier, 1963). *Pithecia* and *Cacajao* not only brachiate (pers. obs.) but also occasionally walk on their legs alone (Sanderson, *op. cit.*). All three genera have rather long legs.[1] Maintaining proper balance may be a more difficult and/or more complex process for them than for all or most other Platyrrhini. These factors might help to explain why such species may have lost territory marking patterns in the course of evolution. They probably could not perform Rubbing movements as conveniently or rapidly as other species with shorter legs and/or more consistently secure balance.

Another factor which may be significant in this connection is mobility. *Ateles* individuals are very rapid in their movements, and range over wide areas every day. The behaviour of *Pithecia* and *Cacajao* in captivity indicates that they may be equally active. Marler (*op. cit.*) has already suggested that great mobility reduces the value of olfactory marking of the environment. Presumably, individuals of very mobile species are too often far away from their markings for the latter to be of much use to them. It seems likely, at least among primates, that olfactory markings are very effective only when frequently reinforced or supplemented by visual and/or acoustic stimuli provided by the near presence of the marker.[2]

---

[1] The peculiar locomotory habits of *Pithecia* and *Cacajao* have not been widely recognized because they are not reflected by the (relatively few) measurements used by anatomists interested in locomotion.

This is not the place to discuss the problem in detail. But the distinctive features of the two genera may be crudely summarized as follows. They have developed moderately long legs, presumably as an adaptation to occasional bipedal walking. They also have developed moderately long arms, as an adaptation to occasional brachiation or semi-brachiation. At the same time, however, they have retained the long trunk which is undoubtedly primitive in Platyrrhini. The *relative* proportions of their limbs and bodies may not be very different from those of such species as *Cebus* and *Saimiri* (see Erikson, *op. cit.*), but they frequently move in very different ways.

[2] Marler (*op. cit.*) cites Eisenberg and Kuehn (unpublished) to the effect that *Ateles* individuals have throat glands which may, conceivably, produce special secretions as olfactory signals. My own observations would suggest that, if these glands do produce secretions, they are used in 'greeting' displays and/or to facilitate individual recognition.

VISUAL SIGNALS

Most of the visual displays of New World primates can be divided into three groups, whose evolutionary history obviously has been quite different.

Special postures and 'gross' movements involving the whole head and body, or the head alone, or large parts of the body, or the limbs, or the tail.

All species have a few patterns of this type. *Cebuella pygmaea* has more than any other species whose behaviour has been studied. Even in *Cebuella*, however, they are less common than some other displays.

A few special postures and gross movements are very widespread among New World primates, occurring in a great variety of species. Two alarm patterns, Swaying movements and Head-down postures, and one aggressive (probably threat) pattern, Tail-lashing, may be cited as examples. They certainly are ritualized in some species, although possibly not in all. It seems likely that they were included, in one form or another, in the repertory of common ancestor of all the existing Platyrrhini.

They illustrate a phenomenon which is characteristic of many other groups of animals, in addition to the primates. Alarm patterns (in which a tendency to escape is greatly preponderant) and threat or other very aggressive patterns (in which a tendency to attack is greatly preponderant) often are much more conservative during evolution than more ambivalent patterns. The reasons for this are obvious. Both alarm and very aggressive patterns tend to occur at 'crucial' moments. It usually is vital for the individuals perceiving such patterns to react to them instantly. It also is highly advantageous to the individuals performing such patterns to have them reacted to instantly. Any sudden evolutionary change in a pattern of this type, no matter how minor, must increase the chances that the signal meaning of the pattern will be misinterpreted, or reacted to slowly. Thus, it probably will be selected against, unless it has some really appreciable compensatory advantage.

There is no need to suppose that any recondite 'group selection' is involved here. Rapid and (therefore) efficient responses to alarm patterns may be advantageous to the individual performing the patterns in any one or all of several different ways. They may

Fig. 2 Several variants of the Arch Posture, a hostile display, by adult *Aotus trivirgatus* (from Moynihan, 1964). The Arch is the most conspicuous of the few ritualized visual signals of the species. Note that it is not accompanied by any special facial expression or Pilo-erection

Fig. 3 Visual displays of adult *Callicebus moloch* (from Moynihan, 1966). An extreme Arch Posture is shown at bottom right. This certainly is homologous with the Arch of *Aotus*; but it is accompanied by both Pilo-erection and a special facial expression. A slightly different form of the same expression is shown at top left

help to save the individual's own offspring from predators or opponents. They also may induce 'mobbing' responses by the individual's companions, and thus help to save the individual itself. Similarly, an individual whose threat patterns are particularly effective probably will not have to fight, and thus will not run the risk of suffering physical injuries.

As might be expected, the nocturnal *Aotus trivirgatus* has a visual signal system which is rather different from that of all other New World primates. It has few or no visual displays except postures and gross movements – and not even many of these. Most of its visual displays also are less highly ritualized, i.e. less exaggerated in form, than the homologous patterns of other species. Possibly it performs unritualized patterns which function as visual signals more frequently than displays; but the actual frequency of such patterns probably is not great, and they probably do not play a very important rôle in most social encounters, simply because individuals usually cannot see one another very well in the dark.

Facial Expressions

Some of the facial expressions of some species of New World primates have been described and discussed by van Hooff (1962) and Andrew (1963). I will not attempt to repeat, or add to, their descriptions here.

Most species of Platyrrhini have a small number of facial expressions. *Callicebus moloch*, for instance, may protrude its lips, or bare its teeth, or partly close its eyes; but it does little or nothing else to alter the appearance of its features. *Pithecia monacha, Cacajao rubicundus, C. calvus, Alouatta palliata*, and *A. seniculus* probably have repertories of comparable size, although differing in detail. *Saimiri* and all or most of the tamarins and marmosets (*Callimico, Leontideus, Saguinus, Callithrix*, and *Cebuella*) seem to have even fewer expressions. Some of them are almost as 'poker-faced' as *Aotus*.

The only genera which are definitely known to have a great variety of facial expressions (including some very complex patterns) are *Ateles, Lagothrix*, and *Cebus*. *Cebus* has a larger repertory of such patterns than the other two genera.

It will be noted that there is some correlation between size and variety of facial expressions. All the diurnal forms with compara-

Fig. 4 Ritualized facial expressions of different species. Top: a juvenile *Pithecia monacha*. Centre: an adult *Ateles* (cf. *belzebuth*). Bottom: an infant *Alouatta palliata*

tively few expressions are small. Presumably their faces are so small that many types of expressions would be useless because they could not be seen at a distance. All the forms with a great variety of expressions are large. But not all the large forms have a great variety of expressions. *Alouatta* is anomalous. It is the heaviest of New World primates, but it has notably fewer expressions than the other relatively large species.

Pilo-erection

Obviously ritualized raising of hair has been seen to be performed by *Callicebus* and most of the tamarins and marmosets. *Callicebus moloch* has one general Pilo-erection display, in which all the hairs on the whole body and tail are raised. All or most of the tamarins and marmosets have several different types of more localized Pilo-erection. They can raise different 'tracts' of hair independently as different signals. Some species have developed special, elongated, erectile tufts and ruffs of hair around the face. These tufts and ruffs may be white or brightly coloured. Rather surprisingly, however, the species with the largest number of different Pilo-erection displays is the comparatively dull coloured *Cebuella pygmaea*.[1]

It will be noted that all the species with Pilo-erection displays are moderately to very small, and have little variety of facial expressions. Possibly some Pilo-erection displays are functionally equivalent to some facial expressions of larger species. Presumably raising long hair is visible over a longer distance than a change in facial expression.

It will also be noted that the smallest species of New World primate, *Cebuella pygmaea*, has the largest number of *all sorts* of visual displays, except facial expressions.

There is only one moderately small species which is definitely known not to have any special Pilo-erection displays. This is *Samirii sciureus*. It is anomalous among the small species in much the same way that *Alouatta* spp. are anomalous among the large forms.

Both *Saimiri* and *Alouatta* have comparatively few ritualized visual signals; and most of their few ritualized signals are postures

[1] There is no obvious correlation between brightness of coloration and habitat preference among New World primates. Both brightly coloured and dull species are included among both the forms living inside forest and those preferring edge habitats. But there does seem to be a relatively larger number of species with intricate or complex colour patterns in edge than inside forest.

and gross movements of the whole body, limbs, etc. In these respects, they recall *Aotus*; but they differ in several other ways. Their visual display repertories are not quite as impoverished as that of *Aotus* (note the pre-copulatory displays mentioned above). More important, both genera seem to employ, and respond to, unritualized visual signal patterns with some appreciable frequency.

Fig. 5 Ritualized facial expressions and Pilo-erection patterns of different species. Top left: an adult *Leontideus rosalia*. Top right: an adult *Saguinus leucopus*. Centre left: an adult *Cebus capucinus*. Bottom right: a juvenile *Saguinus geoffroyi*

It seems likely that the similar visual signal systems of *Saimiri* and *Alouatta*, like some of the other resemblances between the two genera, are correlated with their extreme gregariousness. As both genera are diurnal, and form very closely knit social groups, individuals usually can see their social companions very clearly. They seem to pay constant attention to many of the unritualized postures and movements of their companions, especially the pre-locomotory

Fig. 6 Pilo-erection patterns of adult *Cebuella pygmaea*. The bottom drawing shows one type of Pilo-erection combined with a conspicuous and possibly ritualized posture (during a special form of locomotion) and one of the few distinct facial expressions of the species

'intention' patterns; and seem to be able to interpret the meaning of such patterns (i.e. predict what their companions are going to do) quite promptly and without difficulty most of the time, in spite of the fact that the patterns are not exaggerated or stereotyped.

These differences and resemblances among *Aotus*, *Saimiri*, and *Alouatta* could be summarized as follows. All three genera have visual repertories which are relatively simple in form. In the case of *Aotus*, this probably is because a great variety of elaborate visual signals would be useless. In the cases of *Saimiri* and *Alouatta*, it probably is because a great variety of elaborate visual signals is unnecessary.

The fact that *Saimiri* and *Alouatta* have few facial expressions and no special Pilo-erection displays might also suggest that such patterns are not more efficient than postures or movements of the limbs for short-distance communication.

ACOUSTIC SIGNALS

Some species of Platyrrhini have a few non-vocal acoustic displays. The most widespread pattern of this type is 'displacement' Sneezing. In none of the species, however, are the non-vocal acoustic patterns as common or as important as vocalizations. All species have a great variety of vocalizations and use them more or less frequently.

It is possible to recognize a schematic 'typical platyrrhine' vocal repertory. This is characteristic of many species of different genera, including *Aotus trivirgatus*, *Callicebus moloch*, *C. torquatus*, *Pithecia monacha*, *Cacajao rubicundus*, *Saimiri sciureus*, *Cebus capucinus*, *C. albifrons*, *C. apella*, *Ateles*, and (almost certainly) *Lagothrix lagotricha*. The vocalizations of these species include several groups of patterns.

High-pitched Sounds
These can be divided into two sub-groups.

The first includes Squeaks, Whistles, and Trills. Infants of at least some species, e.g. *Aotus trivirgatus*, utter such sounds as 'generalized' not very intense, distress signals. Adults of all[1] species

---

[1] Throughout the following account of typical platyrrhine vocal repertories, such terms as 'all' or 'most species' refer only to the species which have repertories of this type.

utter the same or very similar sounds; but they do so relatively much less frequently than do infants, under natural conditions.[1] Among adults, the signal function(s) of such sounds seem to be comparatively narrow or restricted. They may indicate low intensity hostility (e.g. in *Callicebus*) and/or some combination of hostility and friendly or sexual tendencies (*Aotus*). In all or most species, Trills are produced when the escape tendencies are relatively stronger than when Squeaks and/or Whistles are produced.

The second type of high-pitched sounds includes Screams. These are expressions of high intensity distress in infants of at least one species (*Aotus*), and of very high intensity hostility, including strong escape tendencies, in juveniles and adults of all species.

Low-pitched Sounds

The most common patterns of this group are short, rather harsh notes which can be described as Grunts in some species, as Barks in others. Some species have a considerable variety of slightly different Grunts or Barks; but all or almost all seem to be purely hostile and relatively more aggressive than all or most high-pitched sounds. In some species, e.g. *Callicebus moloch* (W. A. Mason, pers. comm.) and *Pithecia monacha*, the Grunts or Barks develop into Roars at the very highest intensities of motivation.

Other low-pitched sounds may be described as Moans. These are similar to typical Grunts or Barks, but definitely plaintive in tone (to human ears). They usually or always are produced by some combination of hostile and non-hostile tendencies.

Many species also utter sharp alarm or warning notes which do not fit very well into either the high-pitched or low-pitched groups cited above. In some species, e.g. *Callicebus moloch*, these sharp notes sound intermediate in pitch, to human ears. In other species, e.g. *Cebus albifrons*, the equivalent and apparently homologous patterns are 'ultra-sonic' and completely inaudible to human ears.

Some of the species which utter the full range of typical platyr-

---

[1] Individuals raised in captivity often retain essentially infantile or juvenile repertories for unusual lengths of time, until well after they have become morphologically adult. Thus, accounts which are based upon studies of captive individuals (e.g. Andrew, 1963) may convey a rather misleading impression of the importance and frequency of high-pitched vocalizations among adults.

rhine sounds have also developed elaborate, compound 'Song' phrases, long series of many notes arranged in specialized sequence. The most conspicuous song patterns are largely or completely composed of low-pitched notes, more or less resonant modifications of ordinary Grunt or Bark notes.

The basic components of this typical repertory are really quite similar in form in all species. They would appear to have been more conservative in evolution than many anatomical (osteological and dental) characters. (It is known that ancestors of the *Pithecia*, *Cacajao* and *Saimiri* groups were already distinct, morphologically, in the Miocene – see Stirton, 1951.)

This is remarkable because the species which possess the typical repertory are very diverse in ecology and social structure (almost as diverse at the Platyrrhini in as whole).

They also *use* their homologous vocalizations in different ways. Many of the vocalizations of adult *Aotus*, for example, seem to be precise signals. They usually induce the same response by all other individuals of similar age, sex, physiological condition and social status, irrespective of the other external circumstances in which they may occur. The vocalizations of adult *Callicebus moloch*, by contrast, seem to be much less precise as 'releasers', *per se*. They express motivation very clearly; but they frequently produce very different social effects, i.e. induce different responses by similar individuals in different circumstances. They certainly do so much more frequently than do the homologous patterns of *Aotus*. (This subject is discussed in more detail elsewhere (Moynihan, 1966).)

Thus, the typical platyrrhine vocal repertory would appear to be extremely adaptable, as well as conservative (and the two features certainly are causally related). Both the typical vocal repertory and Allogrooming may be more adaptable than the visual display system. In all the species of Platyrrhini which have been studied, the same visual display apparently tends to induce similar responses by similar individuals in all circumstances; and the homologous visual displays of different species usually subserve similar functions.

The principal physical differences between the vocalizations of different species possessing the typical platyrrhine repertory are matters of frequency. Two types of differences are particularly obvious:

Some species are much more consistently noisy, i.e. utter all or most vocalizations much more frequently, than others. This appears to be directly correlated with vulnerability to predators. Sounds probably are much more likely to attract the attention of predators than any other type of social signal. In a forested environment, they should be noticeable at greater distances than visual signals (and intra-specific tactile signals are not distinguishable from visual signals from a predator's point of view). They also should indicate the current location of the signalling animal(s) more precisely than olfactory signals such as marking. Thus, it is not surprising that more vulnerable species are less frequently vocal than less vulnerable species.

Vulnerability, in turn, usually is correlated with such factors as size (see above) and gregariousness. Most small species are less frequently vocal, on the average, than most large species. And most highly gregarious species are less vocal than less gregarious species of similar size. (Obviously, a large group of animals must be more conspicuous, on the average, and therefore more easily detected by predators, than a small group of animals, if individuals of both groups utter similar sounds equally frequently. Conversely, large groups may be able to detect predators more efficiently than small groups, and may even be able to defend themselves against predators more efficiently; but neither factor seems to completely outweigh the disadvantages of increased conspicuousness in most species of highly gregarious New World primates.)

Callicebus spp. seem to be the most frequently vocal species possessing typical platyrrhine repertories. They also are the largest diurnal species which are not highly gregarious. *C. moloch* is the only species of the genus which has been studied in the wild. It seems to be preyed upon relatively very infrequently (Mason, pers. comm.). This is confirmed by other behavioural characters. *C. moloch* individuals pay less attention to new stimuli in their environment, and the appearance of other species which certainly are potentially capable of acting as predators (e.g. human beings and large hawks), than any other New World primate with which I am familiar.

Saimiri sciureus, the smallest highly gregarious platyrrhine, probably is the least frequently vocal of the species which have been observed in the wild. It also seems to be shyer than the other species possessing typical platyrrhine repertories.

In at least one species (*Aotus*), the major vocal patterns of adults are quite discrete, i.e. intermediate patterns are relatively rare. In other species (e.g. *Callicebus moloch*), intermediates are relatively very common and many of the adult patterns seem to intergrade almost perfectly. These differences probably are correlated with the need for clarity of messages. It has already been suggested that 'There may be a general rule . . . that species or classes of individuals that are largely dependent upon auditory signals for the regulation of their social behaviour tend to have discrete, sharply delimited vocal patterns, while species or classes of individuals that are less dependent upon auditory signals tend to have intergrading vocal patterns' (Moynihan, 1964). In other words, only species and individuals whose vocalizations usually are accompanied by non-acoustic information can 'afford' to 'take the risk' of uttering sounds which are not discrete – simply because intergrading or intermediate sounds, when heard by themselves alone, are more likely to be misinterpreted and/or reacted to less rapidly than more sharply differentiated signals.

There are only two groups of New World primates which have very distinctive vocal repertories.

One of these is *Alouatta*. All the vocalizations of *A. palliata* are more or less low in pitch (see descriptions in Carpenter, 1934, and Altmann, 1959). This probably is characteristic of the genus as a whole. *A. seniculus* certainly utters many of the same sounds as *palliata*, and has not been heard to utter any high-pitched notes. Brief and anecdotal accounts of travellers in South America (see, for instance, the citations in Cabrera and Yepes, 1940) would suggest that other species also have similar voices.

The other aberrant group includes all the tamarins and marmosets. They are peculiar in just the opposite way from *Alouatta*. All their sounds are more or less high in pitch. (Some of the sounds of some species include ultra-sonic components; and some sounds of at least one species, *Cebuella pygmaea*, are purely ultra-sonic.)

The distinctive quality of the repertory of *Alouatta palliata* seems to be due to the complete absence of homologues of many or all of the high-pitched patterns of other species which have typical platyrrhine repertories. This probably is the result of actual loss during the evolution of *Alouatta*. The closest relatives of *Alouatta* among living New World primates probably are *Aotus* and *Callicebus*. As both the latter genera utter high-pitched notes, and similar notes are uttered by so many other species of such diverse habitat preferences and social organizations, it seems likely that at least some high-pitched sounds were included in the repertory of the ancestors of *Alouatta* (see also below).[1]

Thus, the vocal repertory of *Alouatta* may be said to be comparatively simple insofar as it includes relatively few basic components. (*A. palliata*, however, is another species in which intergradation between components is frequent. It is possible, therefore, to recognize a great many intermediate patterns in its repertory. Many of these intermediates are listed, and given separate names or labels, in Altmann, *op. cit.*, but they obviously are nothing more than variations on a few major themes.)

The distinctive quality of the repertories of marmosets and tamarins may be the result of more complex evolutionary changes. Some of the high-pitched patterns of these species seem to be strictly homologous with patterns that are low-pitched in the repertories of other species. Thus, either originally low-pitched patterns were 'transposed into a higher key' during the evolution of the tamarins and marmosets or originally high-pitched patterns were transposed into a lower key during the evolution of other Platyrrhini. In any case, all or most of the distinctive features of the tamarin and marmoset repertories do not seem to be due to loss of patterns.

It seems likely that the consistent high pitch of the vocalizations of tamarins and marmosets is correlated with their small size, and consequent vulnerability to predators. (It certainly is not due to inevitable physical limitations of a small vocal apparatus. Young

[1] It would be very useful to know which (if any) platyrrhine vocal patterns are strictly homologous with vocalizations of other groups of primates. Unfortunately, there are not enough detailed descriptions of the right kind available, yet, to permit very reliable or profound comparisons of individual patterns of species of different families or sub-orders. Certainly, many vocalizations of New World primates are more or less similar, in form and function, to patterns of such Old World monkeys as *Macaca mulatta* (Rowell, 1962; Rowell and Hinde, 1962), *Papio* (Hall and DeVore, 1965), *Erythrocebus patas* (Hall, Boelkins, and Goswell, 1965), and *Presbytis entellus* (Jay, 1965). But one would expect that patterns conveying similar information in similar situations would come to resemble one another, as a result of convergent evolution, even when they are completely unrelated to one another phylogenetically (see also comments in Marler, *op. cit.*). Thus, for instance, some characters which are shared by many species of primates of different families (e.g. the harshness of aggressive sounds and the use of short, sharp notes as alarm signals) are also common in many other groups of mammals and even birds.

Callicebus, which are smaller than adult tamarins, are perfectly capable of uttering low-pitched sounds.) Even in an open environment, high-pitched sounds do not carry as far as low-pitched sounds. (The energy of a sound is lost in heating the air through which it moves; and sounds of higher pitch are used up more rapidly than those of lower pitch.) This difference is exaggerated in forest environments. The relatively short waves of high sounds may be scattered by obstructions (leaves and branches) which the longer waves of lower sounds simply flow around. Thus, all other factors being equal, a monkey uttering high-pitched sounds is less likely to attract the attention of a predator than a monkey uttering low-pitched sounds.

The apparent causal relationship between small size and high pitch would support the hypothesis that *Alouatta* has lost high-pitched sounds in the course of evolution (its ancestors certainly were small – see above).

One would expect that all high-pitched patterns, of all species, would be relatively short-range signals. This expectation seems to be confirmed by observation of some species which possess full typical platyrrhine repertories. The high-pitched patterns of both *Aotus trivirgatus* and *Callicebus moloch* usually are uttered by individuals in actual contact with others and/or in situations in which the performers know, or have reason to believe, that other individuals probably are nearby.

There may be a general rule among all New World primates, with the possible or probable exception of *Alouatta* and *Callicebus*, that vocalizations are not adapted to carry any further than is usually absolutely necessary. In all species, the principal disadvantages of too penetrating calls probably is the danger of attracting predators. In some species, however, it also may be disadvantageous when a sound is heard by too many other individuals of the same species. Wide broadcasting of certain types of sexual, greeting, and appeasement patterns might cause social confusion, especially in some of the more highly gregarious species.

The consistent low pitch of the repertory of *Alouatta* may be explained by a combination of factors. Like *Callicebus*, and unlike most other highly gregarious forms, *Alouatta* spp. probably are not preyed upon frequently. (This certainly seems to be true of *A. palliata* – see Carpenter, 1934. And individuals of both *A. palliata* and *A. seniculus* seem to pay less attention to human observers and other 'potential predators' than *Cebus* spp. and *Saimiri sciureus* in the same or adjacent environments.) Perhaps minor mistakes in signalling, i.e. the reception of signals by individuals for whom they are not intended, seldom have serious consequences in *Alouatta* spp. because their social organization is characterized by both minimal hostility between the members of the same band and the absence of pair bonds. There may also be a physical factor involved in this case (unlike that of the tamarins and marmosets). The vocal apparatus of *Alouatta* is very highly specialized (see descriptions in Hershkovitz, 1949). This may be an adaptation to produce long-sustained Roars (the other species which utter Roars have less specialized vocal mechanisms and apparently can sustain the sounds for only a few seconds at a time). Perhaps the structure of the *Alouatta* vocal apparatus is such that the production of high-pitched sounds is physically impossible.

Alouatta spp., of course, are famous for the volume of some of the noises they utter. Their only rival in this respect is *Callicebus*. They also utter sounds comparatively very frequently, almost as frequently as *Callicebus*, and more frequently than all or most other highly gregarious forms. Both *Alouatta* and *Callicebus* have many more different types of vocal patterns than other kinds of display. Thus, both genera may be said to be 'highly vocal', in much the same ways (see also above), in spite of the fact that an appreciable number of their vocal patterns are quite different in form. This general resemblance or parallelism between the two genera which seem to be least vulnerable to predators probably is not coincidental. It would suggest that vocalization is the most effective method of communicating most of the information that New World primates need to give one another – whenever it is not too dangerous.

Some published accounts state that vocalization is much less important in other groups of primates. Thus, for instance: 'These findings' (referring to Hinde and Rowell, 1962, and Rowell and Hinde, *op. cit.*) 'and our observations lead us to infer that for monkeys vocalization is used in communication in a secondary and supportive rôle to add emphasis to posture and expression and is rarely a sole or primary means' (Cole, 1963). Statements of this type seem to be based entirely on studies of *Macaca* and related terrestrial or semi-terrestrial Old World monkeys. They may well be correct as descriptions of *Macaca* and related forms; but they probably do not apply to all Old World monkeys, much less all non-human primates.

There are, in fact, so many other published references to frequent and conspicuous calling by all sorts of different kinds of primates, from lemurs (e.g. Petter, 1965) to apes (e.g. Carpenter, 1934), as to suggest that vocalization is the principal means of communication in the order as a whole. It probably is actually or potentially the most efficient method of communication between individuals of all species. It probably has been selected against, in some species, only because it is equally or more efficient in communicating information to the 'wrong' individuals (of the same or other species) as to the 'right' individuals.

APPENDIX

Several papers on tree shrews have appeared very recently.

Van Valen (*Evolution*, 1965, vol. 19, no. 2, pp. 137–51) has suggested that the phylogenetic relationship between tree shrews and (other) primates is remote. This does not lessen the comparative interest of tree shrew behaviour. At the very least, the tree shrews and the (other) primates have been derived from similar ancestral groups and have evolved along partly parallel lines.

Sorenson and Conoway (*Folia Primatologica*, 1966, vol. 4, no. 2, pp. 124–45) have described some behaviour patterns of *Tupaia minor*, *T. gracilis*, *T. chinensis*, *T. longipes*, and *Tana tana* in captivity. All the species of *Tupaia* were observed to perform Allogrooming. In addition, female *longipes* were seen to groom other females of the same species quite frequently, and one male *longipes* was seen to groom another male occasionally. It is possible that some or all of the latter performances were essentially 'homosexual', produced by sexual rather than general social motivation, as sexual behaviour may be 'misdirected' in the abnormal conditions of captivity.

ACKNOWLEDGMENTS

I am indebted to the Zoological Society of London for permission to reproduce Figs. 1 and 3, and to the Smithsonian Institution for permission to reproduce Fig. 2.

REFERENCES

ALTMANN, S. A. (1959). 'Field observations on a howling monkey society', *J. Mammal.* **40**, pp. 317–30.

ANDREW, R. J. (1963). 'The origin and evolution of the calls and facial expressions of the primates', *Behaviour* **20**, pp. 1–109.

—— (1964). 'The displays of the primates', *Evolutionary and genetic biology of primates*, vol. II, ed. J. Buettner-Janusch, New York and London, Academic Press.

CABRERA, A. and J. YEPES (1940). *Historia natural Ediar; mamiferos sud-americanos*, Buenos Aires.

CARPENTER, C. R. (1934). 'A field study of the behavior and social relations of howling monkeys (*Alouatta palliata*)', *Comp. Psych. Monogr.* **10**, 2, pp. 1–168.

—— (1935). 'Behavior of red spider monkeys in Panama', *J. Mammal.* **16**, pp. 171–80.

—— (1940). 'A field study in Siam of the behavior and social relations of the gibbon (*Hylobates lar*)', *Comp. Psych. Monogr.* **16**, 5, pp. 1–212.

COLE, J. (1963). '*Macaca nemestrina* studied in captivity', *Symp. Zool. Soc. Lond.* **10**, pp. 105–14.

ERIKSON, G. E. (1963). 'Brachiation in the New World monkeys', *Symp. Zool. Soc. Lond.* **10**, pp. 135–64.

FIEDLER, W. (1956). 'Ubersicht uber das System der Primates', *Primatologia*, vol. I, ed. H. Hofer, A. H. Schultz, and D. Stark, New York, S. Karger.

FITZGERALD, A. (1935). 'Rearing marmosets in captivity', *J. Mammal.* **16**, pp. 181–8.

FOODEN, J. (1963). 'A revision of the woolly monkeys (genus *Lagothrix*)', *J. Mammal.* **44**, pp. 213–47.

HALL, K. R. L., R. C. BOELKINS and M. J. GOSWELL (1965). 'Behaviour of Patas Monkeys (*Erythrocebus patas*) in captivity, with notes on the natural habitat', *Folia Primat.* **3**, pp. 22–49.

HALL, K. R. L. and I. DEVORE, (1965). 'Baboon social behavior', *Primate behavior*, ed. I. DeVore, New York, Holt, Rinehart and Winston.

HERSHKOVITZ, P. (1949). 'Mammals of northern Colombia. Preliminary report no. 4: monkeys (Primates), with taxonomic revisions of some forms', *Proc. U.S. Nat. Mus.* **98**, pp. 323–427.

—— (1958). 'A geographic classification of neotropical mammals', *Fieldiana: zoology* **36**, 6, pp. 581–620.

—— (1963). 'A systematic and zoogeographic account of the monkeys of the genus *Callicebus* (Cebidae) of the Amazonas and Orinoco river basins', *Mammalia* 27, pp. 1–79.

HILL, W. C. O. (1957). *Primates*, III, Pithecoidea, Platyrrhini, Edinburgh Univ. Press.

—— (1960). *Primates*, IV, Cebidae, Part A, Edinburgh Univ. Press.

HINDE, R. A. and T. E. ROWELL (1962). 'Communication by postures and facial expression in the rhesus monkey (*Macaca mulatta*)', *Proc. Zool. Soc. Lond.* 138, pp. 1–21.

JAY, P. (1965). 'The common langur of north India', *Primate behavior*, ed. I. DeVore, New York, Holt, Rinehart and Winston.

KELLOGG, R. and E. A. GOLDMAN (1944). 'Review of the spider monkeys', *Proc. U.S. Nat. Mus.* 96, pp. 1–45.

MARLER, P. (1965). 'Communication in monkeys and apes', *Primate behavior*, ed. I. DeVore, New York, Holt, Rinehart and Winston,

MOYNIHAN, M. (1964). 'Some behavior patterns of platyrrhine monkeys. I. The Night Monkey (*Aotus trivirgatus*)', *Smithson. Misc. Coll.* 146, 5, pp. 1–84.

—— (1966). 'Communication in *Callicebus*', *J. Zool. Lond.* 150, pp. 77–127.

NAPIER, J. (1963). 'Brachiation and brachiators', *Symp. Zool. Soc. Lond.* 10, pp. 183–94.

NOLTE, A. (1958). 'Beobachtungen uber das Instinktverhalten von Kapuzineraffen (*Cebus apella* L.) in der Gefangenschaft', *Behaviour* 12, pp. 182–207.

PETTER, J. J. (1962). 'Recherches sur l'Ecologie et l'Ethologie des Lemuriens Malgaches', *Mem. Mus. Nat. Hist. Natur.*, Serie A (Zoologie), 27, 1, pp. 1–146.

—— (1965). 'The lemurs of Madagascar', *Primate behavior*, ed. I. DeVore. New York, Holt, Rinehart and Winston.

PLOOG, D. W., J. BLITZ and F. PLOOG (1963). 'Studies on social and sexual behavior of the Squirrel Monkey (*Saimiri sciureus*)', *Folia Primat.* 1, pp. 29–66.

PLOOG, D. W., J. BLITZ, F. PLOOG and P. D. MACLEAN (1963). 'Display of penile erection in Squirrel Monkey (*Saimiri sciureus*)', *J. Anim. Behav.* 11, pp. 32–9.

ROWELL, T. E. (1962). 'Agonistic noises of the rhesus monkey (*Macaca mulatta*)', *Symp. Zool. Soc. Lond.* 8, pp. 91–6.

ROWELL, T. E. and R. A. HINDE (1962). 'Vocal communication by the rhesus monkey (*Macaca mulatta*)', *Proc. Zool. Soc. Lond.* 138, pp. 279–94.

SANDERSON, I. T. (1957). *The monkey kingdom: an introduction to the primates*, New York, Hanover House.

SIMONS, E. L. (1963). 'A critical reappraisal of Tertiary primates', *Evolutionary and genetic biology of primates*, vol. I, ed. J. Buettner-Janusch, New York and London, Academic Press.

STIRTON, R. A. (1951). 'Ceboid monkeys from the Miocene of Colombia', *Univ. Calif. Publ., Bull. Dept. Geol. Sci.*, **28**, pp. 315–56.

THORINGTON, R. W. Jr. (in press).

VAN HOOFF, J. A. R. (1962). 'Facial expressions in higher primates', *Symp. Zool. Soc. Lond.* **8**, pp. 97–125.

VANDENBERGH, J. G. (1963). 'Feeding, activity, and social behaviour of the tree shrew (*Tupaia glis*) in a large outdoor enclosure', *Folia Primat.* **1**, pp. 199–207.

Chapter Eight

The Effect of Social Companions on Mother-Infant Relations in Rhesus Monkeys

R. A. HINDE AND Y. SPENCER-BOOTH

INTRODUCTION

EARLY experience plays a fundamental rôle in shaping the adult behaviour of all higher animals, e.g. Harlow and Harlow (1962). The most important element in the environment of a young primate is its mother. Her body not only provides the major part of the baby's physical environment during the period after birth, but her behaviour determines its relations with the physical world and with other animals. In particular, mothers vary in the extent to which they permit or restrict the amount of interaction between the infant and the other members of its social group (Plate 1). Reciprocally, however, the interaction between mother and infant is itself influenced by the group: this chapter summarizes the results of an attempt to assess the extent and nature of this influence amongst captive rhesus monkeys.

The monkeys in question live in outdoor cages (18 ft × 8 ft × 8 ft) each communicating with an indoor room (6 ft × 4 ft 6 in. × 7 ft 6 in.; for further details see Hinde and Rowell, 1962). For the most part each such unit is occupied by a group of animals consisting of a male, 3 or 4 females, and their young. In the course of collecting data on mother-infant interaction under these conditions, it became apparent that the mother's relations with her infant were being much influenced by the 'aunts' present in the group – 'aunts' being the term we use for adult or adolescent females other than the infant's own mother, with no implication of blood relationship (Rowell, Hinde and Spencer-Booth, 1964).

The importance of the aunts was first brought to our attention by the behaviour of an adolescent female who was living in a group with a male and 2 adult females, each of which had a young infant (Nik and Andy). This adolescent female tried, often with success, to groom, hold, cuddle and play with the infants. She was apparently the male's favourite at the time, and he came to her support on many occasions when disputes arose with one of the mothers over possession of her baby. The mothers were thus unable to dominate the adolescent aunt, and became very restrictive with their infants. The consequences of this appeared in many aspects of the data we were obtaining in routine watches on these and other infants. For example, 7 other infants also living in groups but with less interference from aunts spent on average 37 per cent (range 13–63 per cent) of the half-minutes of our morning watches wholly more than 2 ft from their mothers when 17/18 weeks old: Nik and Andy were hardly ever seen to stay away from their mothers for a whole half-minute up to this age. Again, whereas all other infants were first seen to climb 4 ft up the netting sides of the cage when between 17 and 42 days old, Nik and Andy were first seen to do so only when 106 and 152 days old respectively (Hinde, Rowell and Spencer-Booth, 1964).

Furthermore, the situation was clearly a harassing one for the mothers. During a 3-months' period when one of these mothers was repeatedly being pestered by the adolescent she lost condition and became weak. There were times in which she often swayed and seemed in danger of falling from branches, and her gait became abnormal, with a pronounced 'goose-stepping' action of the forelimbs. She was unable to put her mouth accurately to the spout of the milk bottle, and when offered small pieces of food would persistently try to take them 1–2 in. from where they actually were. During the period of illness such particularly acute deteriorations in the mother's condition were noted 9 times and in at least 7 of them the baby was known to have been stolen within the previous 24 hours. It seems clear therefore that these were effects of the aunt's behaviour. When the baby was about 22 weeks old, and became able to escape from the adolescent, the mother's condition improved and the symptoms disappeared.

The behaviour of this adolescent female was adventitious in the sense that it had an important influence on the mother-infant interaction we were studying, but one which we have not planned or controlled beforehand. Furthermore, her effect on the behaviour of the infants seemed to be extreme. We therefore thought it necessary to see whether such influences were operating in all our group-living mother-infant pairs, even where their effects were less dramatic. Accordingly, 4 mother-infant pairs were kept alone in cages, similar to those used for the groups, from shortly before parturition until their infants were a year old. For the rest of this chapter we shall be concerned with a comparison between the mother-infant interaction in these 'isolates' (so-termed, although they could see and hear other monkeys) and mother-infant interaction in 'group-living' animals.

The latter control group consisted of 9 infants, including Nik and Andy. The mother of one of them died when it was *c*. eight months old: in this case data on mother-infant interaction were included only up to the end of the twenty-sixth week. The others comprised 3 females and 5 males. Of the isolates, 3 were female and 1 male. The group-living infants were watched for 6 hours during each weekly period until they were 26 weeks old, and thereafter for 6 hours in successive fortnightly periods; the isolate mother-infant pairs were watched for 6 hours a week for the first 8 weeks and thereafter for 6 hours per fortnight; all watches were carried out between 0900 and 1300 hours (GMT or BST). The data were recorded on check sheets graduated in half-minutes, with simultaneous qualitative observations. For the analysis, the data were lumped into weekly periods up to week 6, fortnightly periods between weeks 7/8 and 25/26, and monthly periods thereafter: this gave 23 observation periods in all. Further details on the methods of observation and data on the group-living animals are given by Hinde, Rowell and Spencer-Booth (1964) and Hinde and Spencer-Booth (1967).

QUANTITATIVE ASPECTS OF RELATIONSHIP TO MOTHER

We may first consider some comparisons between the group-reared and isolate infants in the proportion of time which they spent on, off, or at a distance from their mothers. The data considered here were obtained from the check sheets, and are summarized in Figs. 1–10. For assessing the differences between the groups, two tests were used. First the medians of the two groups were compared over all observation periods for which data was available (usually twenty-three), the significance of the difference being assessed with a Wilcoxon matched pairs signed-ranks test (one-tailed). In some cases inspection suggested that it would be profitable to apply a similar test to only a limited sequence of observation periods. Second, the data from the animals in the isolate group ($n_1 = 4$) were compared with those from the group-reared animals ($n_2 = 8$ or 9) in each observation period with a Mann-Whitney U test.

In addition, since the sex ratio of the group-living animals was in favour of males, while the isolates consisted of only one male and three females, we also compared the median scores obtained by the three females in each group for each observation period. The differences between these medians over the whole period of study were assessed with a Wilcoxon matched pairs test, and the probability values are given below ('♀♀ only'). With group sizes of only three it was not possible to assess the significance of differences within particular observation periods for the females.

Eyes closed (Fig. 1). During the morning hours in which our observations were made, the infants slept only while on their mothers. Over the whole sequence of observation periods the medians for isolate and group-reared animals did not differ significantly. During weeks 1–12, however, the isolate animals spent more time with their eyes closed ($p < 0.01$; ♀♀ only, $p < 0.01$) though the only single period in which the difference was significant at $p < 0.05$ was week 6. This difference during the early weeks can perhaps be ascribed to the smaller degree of disturbance suffered by the isolate mothers.

Eyes open on nipple. Rhesus infants spend much time attached to the nipple but not sucking, so that it was not possible for us to assess the proportion of time spent feeding from the mother. The proportion of time that the infants were on the nipple with their eyes open is shown in Fig. 2. The median for the isolates was lower than that for the group-reared animals in every observation period ($p < 0.005$; ♀♀ only, $p < 0.005$), the differences within individual periods being significant at $p < 0.01$ in weeks 13/14, 25/26, 35/38 and 39/42, and at $p < 0.05$ in weeks 6, 7/8, 19/20, 23/24, 27/30, 31/34 and 43/47. The differences were thus rather more clear-cut in the second half-year.

On mother, off nipple. Rhesus infants spend a relatively small proportion of their time on their mothers but not attached to the nipple (Fig. 3). The median for the isolates was less than or equal to that for the normals in every period after week 3 (p<0·005: ♀♀ only p<0·005). The data for individual periods differed with p<0·01 in weeks 11/12, 13/14, 19/20, 25/26, 35/38 and 43/46, and with p<0·05 in weeks 4, 6, 9/10, 27/30 and 39/42.

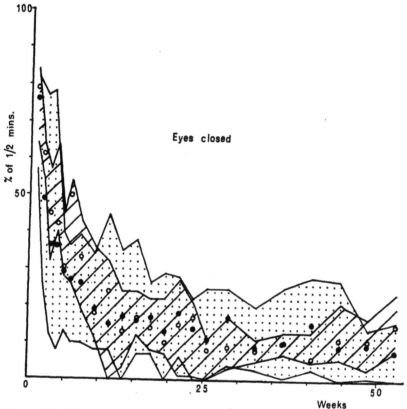

Fig. 1 Number of half-minute periods in which infants were on the nipple with their eyes closed, expressed as a percentage of the number of half-minute periods for which the mother and infant were watched

Shaded area shows the range for the isolate infants
Dotted area shows the range for group-reared infants
O = median for isolate infants
● = median for group-reared infants

Total time off mother. Turning now to the time spent off the mother, the proportions of half-minute periods during which the infants were recorded off their mothers are summarized in Fig. 4.

The median for the isolates was greater than or equal to that of the normals in all observation periods (p<0·005; ♀♀ only p<0·005). Within individual periods the difference was significant at p<0·01

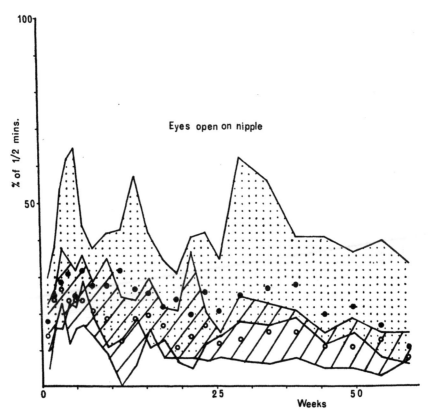

Fig. 2 Number of half minutes in which infants were recorded on the nipple with their eyes open, expressed as a percentage of number of half-minute periods for which mother and infant were watched

Conventions as Fig. 1

in weeks 13/14, 23/24, 25/26, 31/34, 35/38 and 39/42, and at p<0·05 in weeks 1, 5, 11/12 and 27/30.

Time more than 2 ft. from mother. On our check sheets we discriminated between the time spent by the infant at a distance of more than 2 ft from its mother, and the time spent off her but within this distance. Two feet was chosen as being roughly the distance within which the mother can pick up the baby quickly in an emergency. Fig. 5 shows the number of half-minute periods during which the infants went more than 2 ft from their mothers

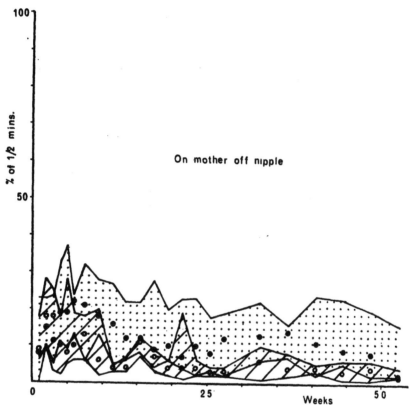

Fig. 3 Number of half minutes in which infants were on their mothers but were not attached to the nipple, expressed as a percentage of the number of half-minute periods for which they were watched

Conventions as Fig. 1

expressed as a percentage of the number of periods in which they were off her at all. The medians for the isolates are higher than those for the group-living animals in sixteen out of twenty-one periods for which data is available (p<0·005; ♀♀ only, p<p·005). The difference is significant at p≤0·01 in weeks 3, 4, 5, 6, 7/8 and 35/38, and with p≤0·05 in weeks 11/12 and 31/34. The difference

was therefore most marked in the early weeks, when many of the group-reared infants rarely went far from their mothers. Thus the isolates were not only off their mothers in a higher proportion of

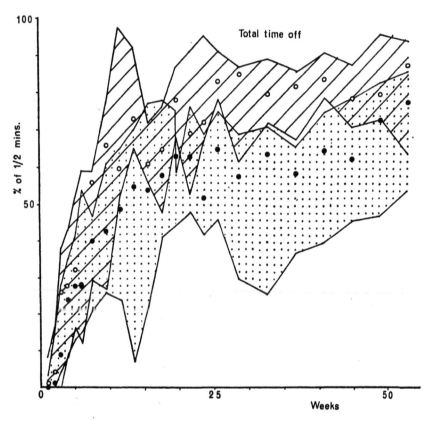

Fig. 4 Number of half minutes in which the infants were recorded off their mothers, expressed as a percentage of number of half-minute periods for which they were watched

Conventions as Fig. 1

the half-minute periods, they were also more frequently more than 2 ft from her.

Initiative in leaving or approaching mother. A higher proportion of time being spent at a distance of more than 2 ft from the mother could be due to the behaviour of either the mother or the baby. Whenever the distance between mother and baby changed from less than 2 ft to more, or vice versa, we recorded which of the two animals was responsible for the change.

Fig. 6 shows the percentage of times in which the distance was

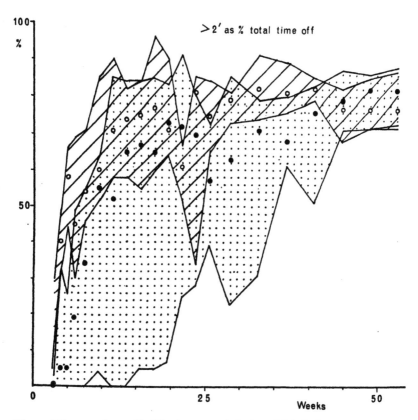

Fig. 5 The number of half-minute periods in which infants were recorded more than two feet from their mothers expressed as a percentage of the number of periods in which they were off her at all

Conventions as Fig. 1

increased to more than 2 ft which were due to the infant. The medians were smaller for the isolates than for the group-reared infants (p<0·005; ♀♀ only, p<0·005), though the difference was significant only in week 43/46 (p<0·05). Put in another way, the isolate mothers tended to walk off and leave their infants more than did the group living-mothers.

Fig. 7 shows that the percentage of times in which the distance was decreased to less than 2 ft by the infant was higher for the isolates, especially during the first six months ($p < 0.005$; ♀♀ only,

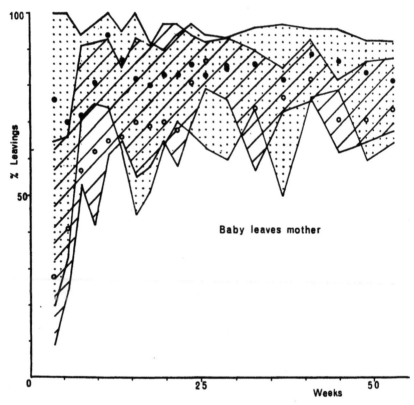

Fig. 6 The percentage of times in which distance between mothers and their infants increased from less than two feet to more than two feet which were due to the infants

Conventions as Fig. 1

$p < 0.005$). The difference is significant ($p < 0.05$) only in weeks 9/10 and 31/34. This we interpret as being due to the frequency with which group-living mothers approach their infants to protect them from other animals. We acknowledge that, had the difference turned out to be in the opposite direction, it would have been possible to attribute it to greater independence on the part of the

276

isolate infants. However, it seems reasonable to suppose that the member of the pair which appears more indifferent about leaving would also be more independent with respect to approaching, and

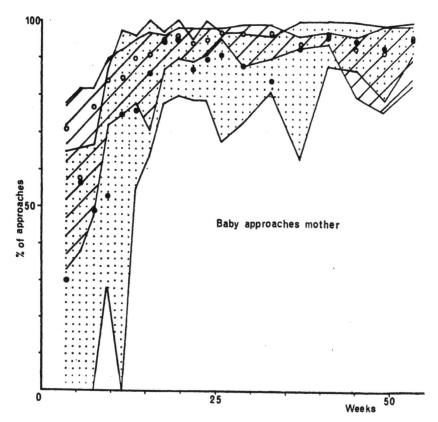

Fig. 7 The percentage of times in which distance between mothers and their infants decreased from more than two feet to less than two feet which were due to the infants

Conventions as Fig. 1

our interpretation of the data on decreasing the distance fits in with that on its increase.

Length of bouts off mother. A higher proportion of time spent off the mother could be due to longer or to more numerous bouts off her. An index of the length of these bouts is provided by the number of half-minute periods during which the infants were recorded as being both on and off their mother. The more numerous such periods, relative to the number of periods in which the infant was off, the shorter (on average) must the bouts off the

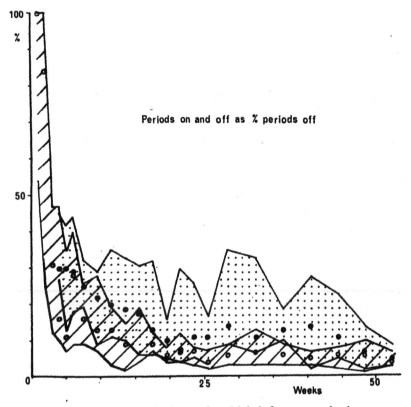

Fig. 8 The number of half minutes in which infants were both on and off their mothers expressed as a percentage of the number of half minutes in which they were off their mothers

Conventions as Fig. 1

mother have been. In practice the data (Fig. 8) show that the median proportion for the isolates was smaller than or equal to that for the normals in all periods ($p < 0.005$; ♀♀ only, $p < 0.005$). In individual periods the difference was significant with $p < 0.01$ in weeks 35/38 and with $p \leqslant 0.05$ in weeks 11/12, 19/20, 25/26, 27/30, 39/42 and 43/46. Thus the isolates tended to spend longer bouts off their mothers.

Plate I An infant rhesus monkey being restrained by its mother

Proportion of half-minute periods spent totally more than 2 ft from mother. Although when off their mothers the isolate infants went to a distance more than 2 ft from her more frequently than did the

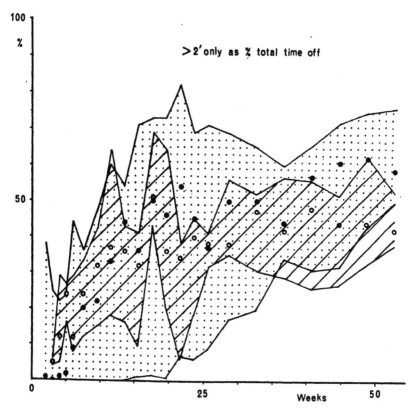

Fig. 9 The number of half minutes throughout which infants were more than two feet from their mothers expressed as a percentage of the number of half minutes in which they were off their mothers

Conventions as Fig. 1

normals, there was no significant overall difference in the proportion of half-minute periods spent off the mother in which the infant was continuously more than 2 ft from her (Fig. 9). It is, however, profitable to consider separately the first ten weeks and the rest of the first year. During the earlier period the proportion was higher for the isolates (p < 0·05; ♀♀ only not significant): during this time, as we have seen, some of the group-reared infants rarely went to a distance from their mothers. The difference was significant at $p < 0.01$ in week 5 and at $p \leqslant 0.05$ in week 3. After week 11 it was the group-reared infants who spent a higher proportion of half-minute periods wholly more than 2 ft from their mothers ($p < 0.005$; ♀♀ only, $p < 0.005$), though the difference in individual periods was significant only in weeks 47/50 ($p < 0.05$) and 51/54 ($p < 0.01$). During this latter period, of course, the group-living infants played with other infants and had more distractions to keep them at a distance from their mothers.

The same difference between isolate and group-living animals is reflected in the data in another way – namely in the frequency with which the 2-ft-radius circle round the mother was crossed (see Figs. 6 and 7). On a comparison of medians this was consistently greater for the isolates ($p < 0.005$).

Grooming. The proportion of half minutes during which the mothers groomed their infants is shown in Fig. 10. The median for the isolates was consistently higher than that for the group-living animals ($p < 0.005$), though the difference was significant only in two individual periods (weeks 6 and 9/10, $p < 0.05$).

Grooming of the mother by the baby is relatively rare during the first year, but the median frequency was higher in the isolate infants in eight out of ten periods after weeks 15/16; in the other three periods the scores were the same.

Acceptances/rejections. When a baby went on to or was picked up by its mother we recorded whose initiative this was, and we also recorded the number of times an infant attempted to go on its mother and was rejected by her. We have calculated for each period the ratio of the number of rejections to the sum of the number of acceptances and the number of times that the mother picked up the baby on her own initiative. Our data were incomplete for the early weeks of the group-living animals, but in ten out of the twelve periods from weeks 17/18 onwards there was a higher ratio of rejections for the isolates than for the group-living animals ($p < 0.04$, two-tailed sign test). The difference in individual periods was significant only in weeks 17/18 ($p < 0.05$).

Interactions with mother – summary. To summarize the data presented so far, the measures fall into three groups.

First, the time spent with eyes closed. During the first twelve weeks this was greater for the isolates than for the group-living animals: less disturbance to the mothers may have accounted for this. Thereafter there was no difference: it seems reasonable that the proportion of time spent sleeping should depend on the age of the infant rather than on details of the environment.

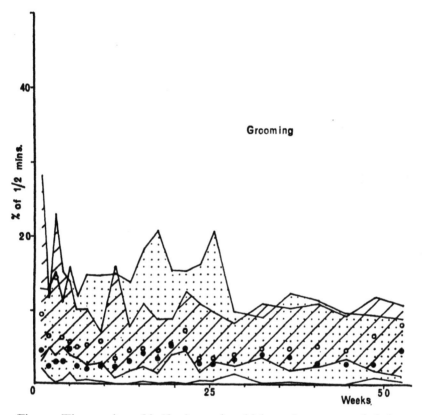

Fig. 10 The number of half minutes in which mothers groomed their infants expressed as a percentage of number of half minutes for which they were watched

Conventions as Fig. 1

Second, the isolates were on their mothers (either on or off the nipple) while awake less, and off their mothers more, and for longer bouts, than the group-living animals. It thus seems that the effect of the aunts was to keep the infants more on their mothers during their waking phases than they would otherwise have been. In conjunction with this, the presence of the aunts also resulted in the infants being more than 2 ft from the mother during fewer half-minute periods than the isolates. However, this effect was through the behaviour of the mothers rather than that of the infants, for the approaching/leaving data indicate that the mothers were going away and the infants following them more than was the case with the group-living infants.

Third, before ten weeks the isolates spent more half minutes wholly at a distance from their mothers, presumably because the aunts were exerting an influence on the restrictiveness of the group-living mother as discussed above. Later, however, the isolates spent a smaller proportion of half-minute periods more than 2 ft from their mothers than did the group-living animals. This we ascribe to the reduced carrying and cuddling of babies by aunts after 10–13 weeks (Rowell, Hinde and Spencer-Booth, 1964) permitting another factor to appear. It is to be expected that, in the absence of playmates, these social animals should keep returning to the only other animal available. The greater proportion of rejections to acceptances shown by the isolate mothers, and the greater frequency with which the isolate infants groomed their mothers, are probably also consequences of the same factor. However, the greater grooming scores of the isolate mothers suggest that they also were influenced by the absence of social companions.

SOME DATA ON ACHIEVEMENTS

During the routine watches records were made of the first appearance of various criteria of sensori-motor co-ordination, locomotor achievement and independence of the mother (Table I), and also of various noises and gestures (Table II). Although few of the differences are significant, these data are on the whole in harmony with the conclusions reached in the previous section. Thus the isolate infants were off their mothers at an earlier age than the group-living ones ((a), Table I $p < 0.05$, one-tailed), and the mothers were earlier to leave them in one part of their quarters while they were in the other ((b), not sig.). Both these findings are in harmony with the greater restrictiveness of the group-living mothers. On the other hand, various indications of the babies' ability or initiative in leaving the mother ((c) to (f), Table I, not sig.) occurred later in the isolates: this may be related to the absence of attraction from social companions.

TABLE I

Median age (days) of first appearance of various criteria of locomotor achievement, sensori-motor co-ordination and independence from mother. Range of ages given in brackets. n= number of individuals for which data available.

			Group-reared		Isolate
		n		n	
(a)	Off mother first time	9	8 (4–15)	4	5 (2–7)
(b)	Mother goes inside leaving baby outside, or outside leaving baby inside	9	133 (31–312)	4	106 (33–245)
(c)	Walks from mother	9	8 (5–23)	4	17 (10–33)
(d)	Climbs 1' up netting	9	14 (7–28)	4	26 (17–31)
(e)	Climbs 4' up netting	9	25 (17–152)	4	34 (26–39)
(f)	Climbs to top of netting	7	37 (31–205)	4	45 (30–66)
(g)	First seen climbing down tail first	9	50 (25–205)	4	30 (24–31)
(h)	Last seen attempting to climb down head first	9	62 (46–205)	4	128 (46–193)
(i)	Mother last seen to rescue or support on wire	6	41 (30–44)	4	60 (52–81)
(j)	Cavorts	9	39 (22–82)	4	38 (18–91)
(k)	Grasps object after fixation	8	17 (12–24)	4	32 (22–39)
(l)	Mouth to object after fixation	5	11 (7–14)	4	20 (7–31)
(m)	Object in hand to mouth or mouth to object in hand	9	18 (10–36)	3	29 (12–31)

TABLE II

Median age of first appearance of various noises and gestures. Range of ages given in brackets. d = days, w = weeks, n = number of individuals for which data available.

			Group-reared		Isolates
		n		n	
(a)	Geckering	6	d3 (d1–w4)	4	d16 (d1–w4)
(b)	Screaming	6	w7 (w4–w10)	4	w10 (w4–w14)
(c)	Squeak	5	d7 (d1–w2)	4	d2 (d1–w2)
(d)	Lipsmack	8	w9 (d3–w15)	4	w11 (w5–w16)
(e)	Tantrum	5	w19 (w12–never)	4	w14 (w7–w22)
(f)	Shakes	6	w21 (w13–w23)	4	w14 (w11–w16)

Infant rhesus only gradually start to come down vertical wire-netting backwards: at first they try to come down head first. The isolates started to come down tail first earlier than the group-living infants ((g), p<0·02, two-tailed), but persisted with head-first attempts later ((h), not sig.). That the isolate mothers were seen to rescue their infants on the wire-netting later than the group-living ones is perhaps surprising ((i), p<0·02, two-tailed), but may be a further indication of the isolate infants being slower to become competent at getting down the wire.

Cavorting, a form of solitary locomotor play, appeared at about the same age in both groups ((j)).

Three aspects of hand-eye-mouth co-ordination appeared later in the isolates ((k), p<0·016, two-tailed, (l) and (m), not sig.). This we cannot explain.

Turning to the expressive movements in Table II, two of the three calls given in frightening or frustrating situations ((a), (b) and (c), not sig.) were first heard later in the isolates, presumably because the appropriate situations were less likely to arise. The same factor probably underlies the later appearance of lip-smacking ((d)), a predominantly social gesture.

We saw earlier, the isolates showed a higher ratio of rejections/acceptances by the mother: in harmony with this, tantrums appeared earlier ((e)).

Wire- or branch-shaking in first-year infants first appears in the course of locomotor play; it appeared earlier in the isolates ((f), p<0·07, two-tailed).

QUALITATIVE NOTES

During our watches we made descriptive notes of the infants' behaviour. On the whole the behaviour of the isolate babies differed little from that of the group-living ones, but three points are worth mentioning. First, the only male in the isolated groups masturbated (by holding his penis) quite frequently from week 22 onwards. We have never observed this in a group-living baby. The second point illustrates the lack of play-companions felt by the isolates. Three of them frequently seemed to 'play' with infants in adjoining pens, running along the wire watching the other infant and keeping opposite it.

Finally, two of the mothers habitually carried their babies on their backs. Whilst this is often seen briefly in our group-living animals, these isolate ones differed in that the babies would more often jump on to their mother's backs than go on to her belly when going to her, and indeed were often deliberately directed or pushed there by the mother. They were frequently carried thus, after days 16 and 25, for the whole length of the run or from the ground to the perches, and perhaps then pulled round to the belly to be nursed. The two mothers concerned were wild-caught as adults and were almost certainly not primiparous (which the other two isolate mothers were); it is difficult to know whether isolation was a factor in causing this behaviour. The group-living infants with permissive mothers began to ride on their mothers' backs at the same age as the isolates (days 17–29) but much sooner than Nik and Andy (days 72 and 100 respectively).

CONCLUSION

These data demonstrate unequivocally differences in the nature of the mother-infant interactions between the four isolated mother-infant pairs and the group-living ones. Since the mothers of the isolate group were selected at random from the monkeys coming to the laboratory, the differences are unlikely to be due to genetic effects, and it is reasonable to ascribe them to the different social conditions under which the animals were living.

The differences in mother-infant relations can be ascribed to two main environmental differences. First, in the absence of aunts the isolated mothers were less restrictive and their infants ranged to a distance from their mothers more freely. In this respect they differed from the majority of the group-living animals in the opposite direction from the two (Nik and Andy) with restrictive mothers. Second, in the absence of play-companions, during the second six months, the isolated infants returned more often to the only other animal in their pen, their mother, and thus spent shorter periods at a distance from her. This had further repercussions, leading for instance to a greater proportion of rejections by the isolate mothers. By this time Nik and Andy, who had play-companions like the rest of the group-living animals, no longer differed from them in such measures.

Thus during the first six months the effect of isolation seems to have been primarily due to absence of aunts, and in the second six months primarily to lack of play-companions. The importance of play-companions in infant rhesus monkeys has also been shown by Harlow (1962).

The long-term effect of these differences on the behavioural characteristics of the growing young remains to be assessed: we hope to report on these in a later paper.

REFERENCES

HARLOW, H. F. (1962). 'Development of the second and third affectional systems in Macaque monkeys', *Research approaches to psychiatric problems*, ed. T. T. Tourlentes, S. L. Pollack and H. E. Himwich.

HARLOW, H. F. and M. K. HARLOW (1962). 'The effect of rearing conditions on behaviour', *Bulletin of the Menninger Clinic*, **26**, pp. 213–24.

HINDE, R. A. and T. E. ROWELL (1962). 'Communication by postures and facial expressions in the rhesus monkey', *Proc. Zool. Soc. Lond.* **138**, pp. 1–21.

HINDE, R. A., T. E. ROWELL and Y. SPENCER-BOOTH (1964). 'Behaviour of socially living rhesus monkeys in their first six months', *Proc. Zool. Soc. Lond.* **143**, pp. 609–49.

HINDE, R. A. and Y. SPENCER-BOOTH (1967). 'The behaviour of socially living rhesus monkeys in their first two and a half years', *Anim. Behav.* **15**, pp. 169–96.

ROWELL, T. E., R. A. HINDE and Y. SPENCER-BOOTH (1964). ' "Aunt"-infant interaction in captive rhesus monkeys', *J. Anim. Behav.* **12**, pp. 219–26.

Chapter Nine

Mother-Offspring Relationships in Free-ranging Chimpanzees

JANE VAN LAWICK-GOODALL

SECTION I – INTRODUCTION, MATERIAL AND METHODS

THE data on which this chapter is based was obtained during a four-year field study of free-ranging chimpanzees, *Pan satyrus* (*or troglodytes*) *schweinfurthi*, at the Gombe Stream Chimpanzee Reserve, Tanzania, East Africa.

The Gombe Stream Reserve consists of a narrow mountainous strip of country stretching for some ten miles along the east shore of Lake Tanganyika just north of Kigoma, and running inland about three miles to the peaks of the mountains of the Rift escarpment. These mountains rise steeply from the lake (2,334 ft) to heights of about 5,000 ft. Many steep-sided valleys and ravines intersect the mountains, a number of which support permanent streams. The thick gallery forests of the valleys and lower slopes give place to more open deciduous woodland on the upper slopes and many of the peaks and ridges are covered only by grass.

The field study was commenced in 1960 since when I have spent 45 months in the field, and at least 4,000 hours in direct observation of the chimpanzees (data on mother-offspring relationships were obtained during approximately 2,000 hours of observation). After investigating all areas of the Reserve a main study area of some 15 square miles was established. During the first year most observations were from one or other of the open peaks from where, with the aid of 8 times 30 binoculars, it was possible to survey a wide area of country. Gradually the chimpanzees became habituated to my presence: at first they ran off if they saw me 500 yds away: after

287

10 months it was possible to walk to within 100 ft of many individuals and after 14 months they carried on their normal activities
when I was only 30 ft away (provided they were up in a tree or in
thick undergrowth, and not in the open).

During 1963 an artificial feeding area was set up at the base
camp and it became possible, for the first time, to make fairly
regular observations on the different individuals. The number of
chimpanzees visiting this area gradually increased and by the
middle of 1964 a total of forty-five individuals, comprising all age
groups of both sexes, had become frequent visitors. These
chimpanzees were all habituated to the presence of human
observers and many observations were made when the subjects
were within 10 to 15 ft.

When I left the Reserve in March 1965, Miss Edna Koning
remained behind to continue routine observations. She was with
me for ten months prior to this, during which time she became
familiar with my observation and recording techniques. I have
included Miss Koning's observations where they fill gaps in the
data presented in this chapter – on locomotor development and
maternal play for the infants Goblin and Cindy from the ages of
six and three months old respectively, and suckling data for all
infants subsequent to March 1965. I have also included relevant
observations on the infant Pom.

Table I lists the individual mothers and their offspring which
visited the feeding area, and from observations of which the main
bulk of data for this chapter were obtained. The relationships
between the mothers and their adolescent offspring, in all cases but
one,[1] were traced, at very irregular intervals, from the early part of
the study when the offspring concerned were still juveniles.

Table I also lists the number of observation periods when each

---

[1] The genealogical relationship between the mother Flo and her presumed
adolescent son, Faben, only became clear during 1964. Evidence in support of
their relationship is as follows:

    (1) No mother was observed to groom an adolescent other than her own off-
spring for longer than 10 minutes: Flo frequently groomed Faben for over
one hour.

    (2) No mother was observed to play with an adolescent other than own off-
spring: Flo played vigorously with Faben on several occasions.

    (3) Flo protected her infant Flint from all individuals other than her own off-
spring during his first few months: Faben was permitted to touch and
groom Flint at all times.

From the above I have concluded that Faben was, in fact, an offspring of the
female Flo, and shall refer to him as such in this chapter.

offspring was observed with its mother. An 'observation period' is used to describe the time when it was possible to observe the individuals concerned for more than ten minutes at a time. Most, but not all, of the observation periods listed refer to occasions when the individuals were at or near the feeding area. These periods were sometimes as long as six hours – the average mean length was calculated for three individuals (Flo, Melissa and Mandy) and, in each instance, was between forty-eight and fifty-three minutes.

Detailed observations on the development of mother-infant relationships during the first six months of the offspring's life were mostly obtained from four infants. Table II lists the total number of hours per week that it was possible to observe each of these infants and it will be seen that the infant Flint was observed for longer than any of the others. It should be emphasized that at no time was mother-offspring behaviour studied to the exclusion of other forms of behaviour.

Before discussing mother-offspring relationships it is necessary to describe briefly the social structure of the chimpanzee community (Goodall, 1965). This consists of a number of small groups which may be composed of any combination of age and sex. In fact, each group (apart from a mother and her younger offspring) is merely a temporary association of individuals which may remain stable for a few hours or a few days. Individuals are continually leaving one group and joining up with another, and sometimes a number of groups move about together forming a temporary association of up to thirty individuals or so. It is common for mature or adolescent males to move about alone, and sometimes females are also seen on their own. The only unit which remains stable over a period of several years is a mother with her infant and young juvenile offspring. Such a mother is joined from time to time by her adolescent offspring. A mother-infant group often moves about on its own, but may join temporarily with other groups.

SECTION II – MATERNAL CARE

The young chimpanzee has a long period of complete dependence on its mother when it relies on her for food, transport and protection. The stages in the development of wild chimpanzees are summarized in Table IV and it will be seen that even after the child has been weaned, at about three years of age, and has achieved complete locomotor independence, it continues to associate closely with its mother during the next few years.

A wild chimpanzee mother under normal conditions may give birth every two and a half to four years (see also Reynolds and Reynolds, 1965; Kortlandt, 1962; Goodall, 1965). Observations so far suggest that there may be a high infant mortality rate in this chimpanzee community, although only one infant was seen dead (Jane – see Table I). Three females (Olly, Madam Bee and Passion – see Table I) were seen with infants which subsequently disappeared. A year after the presumed death of these babies all three mothers gave birth again. In the case of Olly there was then an age gap of five to six years between her two surviving offspring. Age gaps of five years or more between the offspring of two other mothers (Sophie and Marina, Table I) suggest that they also may have lost infants. In addition another female (Circe, Table I) was well advanced into maturity when she gave birth in 1965 and was almost certainly not a primiparous mother; another female, accompanied only by a five- to six-year-old juvenile, also probably lost an infant. The above observations, together with an examination of Table I, suggest that at least 35 per cent of the infants born between 1960 and March 1965 died. Certain aspects of mother-infant behaviour which may contribute to this death rate will be discussed in Section IV.

The behaviour to be described in this section is mainly that relating to the rôle played by the mother which helps to ensure the successful physical development of her offspring.

1. *Mother's Behaviour during Pregnancy*

The gestation period, for captive chimpanzees, is about eight lunar months: during the first three to four months there is normally periodic recurrence of swelling of the anogenital area, but menstrual bleeding does not occur (Yerkes, 1943). In the wild, four of the five females observed regularly during their pregnancies showed recurrences of sexual swelling, and during such periods became attractive and receptive to males. Copulations occurred, with full penetration and ejaculation.

The final two to two and a half months of pregnancy are characterized by pronounced swelling, first of the abdomen and subsequently of the breasts. During these months, captive females behave with extreme caution and try to avoid strenuous exercise (Yerkes, 1943), but such self-protective behaviour is not obvious in the wild. It is possible, however, that females go off by themselves during the final days of pregnancy, and thus avoid strenuous forms of group activity. Evidence in support of this is slight: firstly, two females, obviously pregnant, were seen on their own; secondly, the pregnant females mentioned above all ceased to visit the feeding area for periods of up to a week before the birth of their infants.

2. Behaviour of Mother with Newborn Infant
Parturition was not observed in the wild, but one mother, Melissa, was seen at 6 p.m. with a baby that had undoubtedly been born

Fig. 1 A mother walking with the 'hunched gait' and thus supporting her small infant with her thighs. (Drawing from photograph)

earlier that day. During the thirty minutes that she was under observation she supported the infant continually. When she walked she kept one hand under his back and shoulders, and moved with the 'hunched gait' (Fig. 1) rounding her back and taking short steps with partially flexed legs so that he was also supported by her thighs. She sat down at least once in every fifteen steps (I followed her for approximately half a mile). When sitting she cradled the baby closely, keeping her knees bent up, one arm or hand behind the infant's shoulders, and her feet close together under his rump. He was thus almost completely encircled by her body, legs, arms and feet (Plate I). On the following day Melissa's behaviour was similar. Once she momentarily removed her hand from the infant's back – he at once lost his grip with both hands and would have fallen had not Melissa caught him. Another female, Passion, was seen when her infant was two days old. Her behaviour was less efficient despite the fact that she had been seen with an infant the previous year. When she sat she allowed the baby to lie with its back on the ground for several minutes on three occasions. This infant, too, lost its hand grips when the mother removed her supporting hand.

For at least thirty-six hours after the birth of Melissa's infant the placenta remained attached to the baby and trailed along the ground as the mother walked. Once it became caught in some twigs and nearly jerked Goblin from his mother's grasp. Melissa immediately stopped, reached back to free the cord, and continued. (An African assistant who observed an experienced mother, Flo, with her newborn, described how she draped the umbilical cord around her neck when she moved off.) Melissa immediately gathered up the placenta and cradled it with her baby when a juvenile reached out towards it. Captive rhesus monkey mothers have been observed to protect the afterbirth in a similar way (Hinde, Rowell and Spencer-Booth, 1964).

The behaviour of captive chimpanzee mothers towards the afterbirth varies. Yerkes (1943) reports that nearly all mothers ate the placenta soon after its delivery. He also describes one mother biting through the cord and discarding the placenta. One mother in the London Zoo ignored the placenta which was cut away by a keeper; another gave birth from a high platform and the cord broke as the infant fell to the ground (Budd and Smith, 1943). Similar discrepancies of behaviour occur in various other primates in captivity and more data on parturition in the wilds are badly needed.

Melissa was probably a primiparous mother, yet her handling of the infant was efficient. In captivity, inexperienced chimpanzee mothers are often afraid of their first-borns, refusing to touch them or to allow them to cling to them (Yerkes, 1943). Similar behaviour is recorded for gorilla mothers – indeed one, after pulling her baby from the vaginal opening, bit off first a hand, then a foot, and finally punctured its skull with her teeth (van den Berghe, 1959). Such abnormal behaviour suggests that some experience normally gained by wild females before the birth of their first infant is sometimes lacking in captive individuals. This hypothesis is strengthened by the fact that rhesus monkey females that had been deprived, during the first years of their lives, of social contact with their mothers and others of their kind, either ignored or violently abused their own infants (Harlow, Harlow and Hansen, 1963).

In the wild, although the degree of maternal efficiency may

increase with experience, it seems likely that young females normally care for their first-borns. A wild chimpanzee female, by the time she has her first baby, has not only enjoyed normal relations with her mother and other individuals, but has had the opportunity to watch other mothers with small babies and to play with and carry around infants.

3. *Support and Transport*

(a) *Cradling.* An infant, during its first three months or so, is almost invariably supported by the mother whilst she is sitting or lying. The typical cradling position of the mother has been described (Section II, 2). At times, a mother may bend up one or other of her feet to give added support to the back or head of

Fig. 2 The mother Flo cradling her two-month-old infant whilst she is lying. (Drawing from photograph)

her infant (Plate 2). Whilst thus cradled the infant may sit up straight on the mother's lap, its feet and hands holding hair on either side of her body between her armpits and groins. Or it may slump over to the side with its head behind her right or left knee. When a mother remains for some time in one position the infant may lie stretched flat out across her lap.

Whilst lying, mothers normally hold their infants close with one arm, or cradle them in their groins. Flo, when lying with Flint in her groin, invariably flexed one or both thighs and held one or both of her own feet with one hand in order to maintain the cradling position (Fig. 2).

Cradling behaviour permits the infant to relax whilst sleeping. Thus although infants frequently retained strands of the mother's hair between their relaxed fingers while sleeping, they often ceased to hold on at all. On four occasions when mothers leaned forward to reach some object and parted their knees without supporting their sleeping infants with either a hand or foot, the babies concerned immediately fell back and away from their mothers. When the mother of a sleeping infant stands up, she often keeps her hand under its back for several yards until it grips tightly again.

As the infant grows older the mother gradually relaxes her cradling positions and the infant itself must climb to a comfortable place on or beside its mother.

(b) Transport of infant

(i) For the first six to nine months the infant is transported from place to place in the ventro-ventral position, gripping the mother's hair between flexed fingers and toes. Chimpanzee infants, unlike many other primate species (e.g. baboon (DeVore, 1963), rhesus (Hinde, Rowell and Spencer-Booth, 1964), and langurs (Jay, 1963)), do not use nipple attachment as an additional means of support. During the first few days, infants were not able to cling on for more than a few minutes unaided, and their mothers gave them extra support, either pressing them close with one hand, themselves proceeding on three limbs, or walking with the hunched gait (Fig. 1) and thus supporting them with their thighs.

During the first few weeks the mother gradually supports her infant less frequently. Thus Goblin was supported continuously (during observation) for his first two days. He was not seen on his third day. On the fourth and fifth days he was supported less often by his mother's hand but almost continuously by her thighs. During his second week he was seen clinging, unsupported, for at least forty yards, and during his third week Melissa only supported him for the first few yards after moving off. If he lost a hand or foot hold, however, she immediately pressed him back into place. The data for the infants Cindy and Pom follow the same general pattern.

Flint, however, was supported at least once every forty yards during his first two months, and at frequent intervals even during his third. This was probably due to the fact that he was never able to grip properly with his right foot, and when Flo removed her hand from beneath him he invariably lost first one and then the other foot hold. By the time he was three months he was often able to maintain the ventral position by pressing the right foot against

Plate IV Mother (Melissa) pressing her infant into the ventral position with her hand and foot as she moves off. The infant is about four months old. © National Geographic Society, 1967, courtesy National Geographic Magazine

Plate III Mother (Mandy) supporting her two-month-old infant as she moves along a branch. © National Geographic Society, 1967, courtesy National Geographic Magazine

Plate V The infant Flint, at six months old, clinging to his mother's
back. © National Geographic Society, 1967, courtesy National
Geographic Magazine

Plate VI Infant Goblin at twenty-four weeks old has climbed a few inches up a sapling. This was the first occasion on which climbing was observed. © National Geographic Society, 1967, courtesy National Geographic Magazine

Plate VII The mother Flo tickling her daughter and thus distracting the child's attention from the infant Flint (here about three months old). © National Geographic Society, 1967, courtesy National Geographic Magazine

Plate VIII Mother (Circe) gently pushing away a one-year-old infant with the back of her hand. The one year old is trying to look more closely at Circe's three-week-old baby. © National Geographic Society, 1967, courtesy National Geographic Magazine

Plate IX Mother (Olly) leading her two-year-old female infant (Gilka) away from a group of mature individuals. After being led away Gilka returned on three successive occasions and was fetched each time. (The group is eating salt from the ground.) © National Geographic Society, 1967, courtesy National Geographic Magazine

Plate X Mother (Marina) crouching over her two-year-old infant
during a low intensity attack by a mature male. © National Geographic
Society, 1967, courtesy National Geographic Magazine

Plate XI The mother Flo surrounded by all her known offspring. She is grooming the elder of her adolescent sons – the other is stretched out in the foreground. Her juvenile daughter is playing with the infant Flint (here about three and a half months old).
© National Geographic Society, 1967, courtesy National Geographic Magazine

Plate XII The mother (Flo) is tilting up her infant's chin and gazing into his face. She did this for at least half a minute. The infant is ten months old.
© National Geographic Society, 1967, courtesy National Geographic Magazine

Plate XIII Mother (Flo) tickling her infant Flint when he was five
months old. Flint shows the typical playface. © National Geographic
Society, 1967, courtesy National Geographic Magazine

Plate I Mother (Melissa) cradling her one-day-old infant. The umbilical cord and placenta are still attached to the infant, and can be seen hanging beside the mother's left leg. © National Geographic Society, 1967, courtesy National Geographic Magazine

Plate II Mother (Mandy) supporting the head of her sleeping infant with one foot. The infant is about two months old. © National Geographic Society, 1967, courtesy National Geographic Magazine

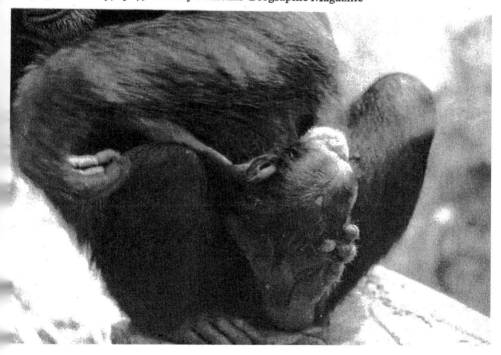

Plate XIV The mother (Flo) sits whilst her elder offspring chase each
other around her. The elder adolescent male (Faben) is in the lead and
the juvenile (Fifi) brings up the rear. © National Geographic Society,
1967, courtesy National Geographic Magazine

Flo's flank whilst gripping hair in the normal way with the other. Mandy also supported her infant at frequent intervals for three months, but it was not certain whether this was due to weakness on the part of the infant, or over-solicitude on the part of the mother.

Most mothers, whilst moving about in trees, carefully support their small infants, either with one hand or, more usually, by exaggerated flexion of one thigh (Plate 3). During brachiation both thighs may be drawn up towards the body.

The mother of a small infant may get up and move off without warning her baby if it is gripping on tightly. If it has relaxed its grip, however, the mother presses it close as she moves off (Plate 4). When the infant is older, and begins to crawl about on or near the mother, it normally responds to a light touch on its back by clinging in the ventral position. A similar touch serves as a fairly specific signal for the infant to cling on in a number of other primate species such as gorillas (Schaller, 1963), baboons (DeVore, 1963), and rhesus (Hinde, Rowell and Spencer-Booth, 1964).

The normal response of an infant chimpanzee to a sudden movement of its mother is to tighten its grip, but this is not always effective in preventing it from falling. Mothers usually pressed their infants close when they made sudden movements; on six occasions when they leapt forward without supporting their infants the babies lost either their hand or foot holds and would have fallen had not their mothers caught them; and once Flint did fall when he was eight weeks old. Such mishaps may occur when the infant does not grip fast enough, or when it makes an 'erroneous' grip. Once, for example, when Flo gave a sudden jerky movement Flint at once gripped tightly to Flo's body with one hand, but to his own head (which he happened to be scratching) with the other.

The infant, even after it has commenced to ride on its mother's back, frequently reverts to the ventral position if the mother is moving fast, or if she is travelling through thick undergrowth or swinging through the trees. Infants of two years or more do not normally grip hair with their feet, but press the sides of their ankles against the mother's flanks and turn their feet inwards over her rump.

(ii) *Dorsal position.* From about five to seven months of age the infant commences to ride regularly on its mother's back. At first it lies with its head in the region of her neck or shoulders and grips on to hair with its hands and feet wherever it can. Later it sits up in the 'jockey' position either with its legs gripping the mother's sides (Fig. 3c) or with its knees drawn up and feet resting on her back. In either case hair is normally grasped with the hands only.

The change over from the ventral to dorsal position was observed in two infants only and in both cases was initiated by the mother. When Flint was fifteen weeks old Flo was first seen to try and push him towards her back with one hand as she moved off. She pushed him in this way several times during the next week, but Flint invariably slithered down at once into his usual ventral position, or dangled down, clutching hair in the region of Flo's shoulder with his hands. The following week Flo several times picked him up by one arm and pushed him over her shoulder from the front, but always he slid down immediately. However, when he was seventeen weeks old Flint was observed riding on Flo's back for over forty yards, and from then onwards was seen on her back more and more frequently. By the time he was twenty-eight weeks old he nearly always rode in the dorsal position, often clambering up from the ventral position while Flo was moving, and without prompting.

Melissa was not observed to push Goblin towards her back from the ventral position until he was twenty-one weeks old, and he slithered down immediately. The following day Melissa pushed him, head first, over her shoulder whilst sitting – as soon as she let go Goblin slid to the ground. She next stood up, again pushed him over her shoulder, and then held him until he turned to face the right direction. As she walked forward he slipped and twice she tried to pull him into the ventral position. Finally, after walking some thirty yards, she sat down and gathered him on to her lap. On two occasions, later in the week, she pulled him over her head when he slipped. Another time she moved off when he was facing her rump.

Of these two mothers Flo was the most competent when initiating dorsal riding and it seems likely, since Flint was not her first offspring, that previous experience helped her to cope with the situation.

At first, as we have seen, the mother pushes or lifts her infant on to her back. When the infant is older she may try to 'scoop' it on to her back, from the ground – she puts her hand under its rump and pushes as it scrambles up her side. Later she may stand with slightly flexed knees and either look back over her shoulder, or reach back to touch or gesture towards the child. These serve as signals for the latter to climb on (Fig. 3a, b and c). Signals of a similar nature occur also in the baboon in this context (DeVore,

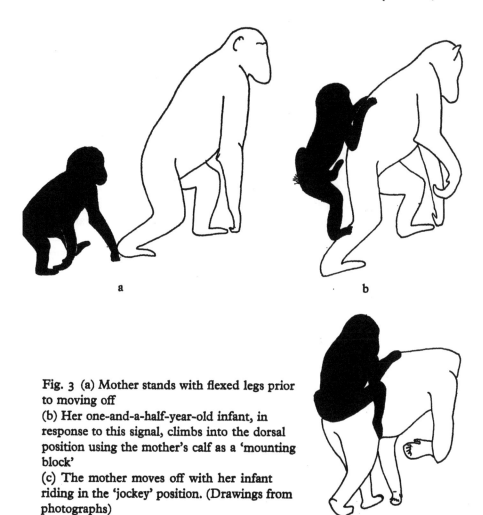

a

b

Fig. 3 (a) Mother stands with flexed legs prior to moving off
(b) Her one-and-a-half-year-old infant, in response to this signal, climbs into the dorsal position using the mother's calf as a 'mounting block'
(c) The mother moves off with her infant riding in the 'jockey' position. (Drawings from photographs)

c

1963). If the mother moves off a short way without the infant it usually hurries after her, grabs on to her rump, and pulls itself up while she continues to walk.

By the time the infant is two and a half years old it seldom rides on its mother. If there is a sudden alarm it may hurry to its mother and cling on, but this is for protection rather than transport: two juveniles were observed to climb briefly on to their mothers' backs for the same reason. Reynolds saw a female travelling with an infant ventral and another dorsal, but does not state whether the larger climbed on in alarm (Reynolds and Reynolds, 1964).

(iii) *Dangling and unusual positions.* Small infants, when they lost their foot holds, often hung momentarily from their hands whilst they alternatively flexed and extended their legs until they found new foot holds or until their mothers pressed them back into place. Both Flint and Goblin, at three months old, and Cindy at four months old, were frequently observed dangling under their mothers, sometimes gripping with both hands, sometimes with one only. This differed from the earlier brief hanging behaviour in that the infants made no efforts to grip on with their feet, nor did their mothers press them into place unless they were moving fast. This dangling persisted until dorsal riding became the normal means of transport.

When infants started to crawl around on their mothers or move a few feet away, they often gripped on in a variety of abnormal places when their mothers started off: thus one was dragged along the ground clinging to his mother's ankle; another grasped his mother's arm with one hand and her leg with the other. Both mothers stopped after a few steps and pulled their infants into the ventral position. Similar behaviour has been observed in rhesus infants (Hinde, Rowell and Spencer-Booth, 1964). Chimpanzee infants also clung on in unconventional places, particularly to one arm, when they slipped down from their mothers' backs at the onset of dorsal riding.

4. Suckling

Suckling continues throughout infancy. So far it has not been possible to obtain data on suckling behaviour during the night, and the information below refers to daytime observations only.

It is not easy to record the frequency of suckling since, during an observation period, the infant is frequently hidden by the mother's body, vegetation, etc. This means that whilst two feeds may be recorded as consecutive there is the likelihood, particularly with small infants which do not leave their mothers, that an interim feed was missed by the observer. However, the data for the infant Flint, on whom the most observation was possible, suggest that during his first year the longest interval between feeds was one hour and thirty minutes, and the shortest was ten minutes. This held good for all infants under one and a half years. Fig. 4, showing the frequency of suckling per calendar month for Flint, whilst it may

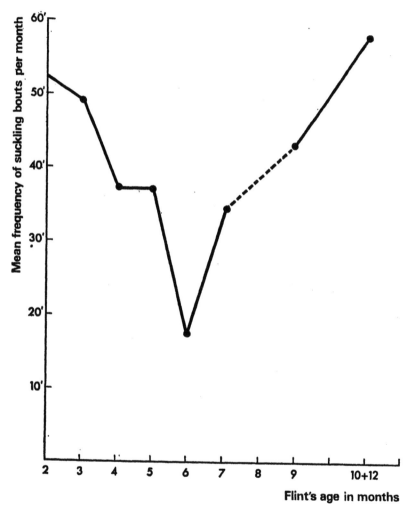

Fig. 4 Mean suckling frequency of the infant Flint during his first year. This figure cannot be regarded as completely accurate (see text for explanation)

be slightly inaccurate for the above reasons, shows that during his sixth month the frequency increased. This coincided with the time when he first began to explore around his mother and usually, after a short spell away, hurried back to her for a short feed before tottering off again.

Data for older infants are limited, but on the twenty occasions when it was possible to accurately time intervals between feeds for the infant Gilka during her third year, the longest was one hour fifty minutes and the shortest thirty-seven minutes, with an average mean of one hour.

The length of individual feeds remains fairly constant throughout infancy. Suckling bouts of more than seven minutes' duration were only recorded during the first month of any infant when they were sometimes as long as twelve minutes (the reason for these long feeds at this time is not yet known). Fig. 5 shows the mean length of feeds per month for three infants during their first year (or parts thereof). In this connection suckling bouts of less than half a minute were not taken into consideration since, in all observed cases, these represented feeds that were interrupted (nipple jerked out as mother made a sudden movement) or 'reassurance nipple contact' (when the infant was suddenly frightened or hurt, etc. (see discussion below)). Examination of this figure shows that the mean length of suckling per feed for the infant Flint decreased during the period when the frequency of his feeding increased (Fig. 4) so that, in fact, the actual milk intake did not decrease.

Data concerning length of suckling bouts for older infants is again limited. The mean length for a two-year-old infant was one minute fifty-three seconds and for a three-year-old one minute thirty seconds. (In both cases the range was half a minute to five minutes.)

Feeding is normally initiated by the infant. During its first few months it searches for the nipple by a combination of pulling itself upwards by its arms, pushing upwards with its legs, and turning its head from side to side whilst, at the same time, rubbing its face up and down against the body of the mother. During the first six to eight weeks such nuzzling behaviour was normally only successful when the infant concerned was being cradled in a position from which one or other breast was easily accessible. Frequently the infants seemed unable to reach either nipple unless assisted by their mothers. The infant Pom, however, was able to reach the breast unaided on several occasions during her first week even when she was being cradled low down on her mother's lap.

The normal response of a mother to nuzzling behaviour was to cradle the infant higher and in a position from which it could reach one of her breasts. On one occasion, however, Melissa pressed Goblin's head against her shoulder in response to his nuzzling: similar behaviour has been observed in mothers of young rhesus infants (Hinde, Rowell and Spencer-Booth, 1964).

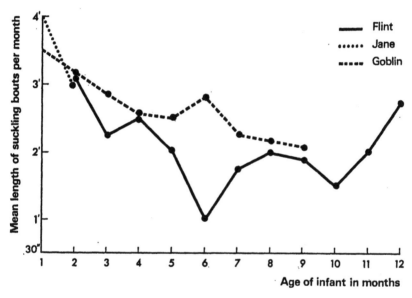

Fig. 5 Mean length of suckling bouts for three infants during their first year, or part thereof

For the first two to three months the infants, during a single feed, suckled from one breast only. After this, however, they normally suckled from each breast alternately, usually half the feeding time at each breast. This change over was initiated by the infants themselves.

Table V shows that, in some cases, infants showed a definite nipple preference. However, whilst Goblin showed a 76 per cent preference for the right nipple when he suckled from one breast only during a single feed, he was as likely to choose either nipple first when he began to suckle from both during a feed.

During the first ten weeks or so mothers normally held their infants to their breasts during suckling. Subsequently as the mothers gradually relaxed their cradling behaviour the infants usually maintained position themselves during feeds, even when mothers got up and moved off before they had finished. On several occasions, however, when Flint slipped during a feed at four months or more Flo pressed him back and clamped him into position with her knees or one hand during the remainder of the feed.

Usually when small infants had a sudden fright or when mother-infant contact was threatened, they immediately nuzzled and suckled, often only for a couple of seconds. Older infants, alarmed when moving about near their mothers, often hurried back to suckle briefly, after which they invariably looked round in the direction of the disturbance as though they had been reassured by nipple contact. Similar behaviour has been observed in rhesus infants (Hinde, Rowell and Spencer-Booth, 1964). 'Reassurance nipple contact' of this sort was frequently observed when Flint's sibling tried to pull him away from his mother (see Sect. II, 6) and this may be related to the fact that hair pulling, toe pinching and loss of support as well as other forms of intense stimulation may reinstate suckling in human infants (Jensen, 1932).

The exact cause which terminates suckling behaviour in individual infants has not yet been determined, but in the wild chimpanzee mothers do not appear to play an active part in weaning behaviour. The circumstances relating to the weaning of three infants will be discussed in Section II, 8.

5. *Grooming, Hygiene and Care of Injured Infant*

(a) *Grooming*. Infant chimpanzees, during the first few weeks of their lives, were seldom groomed, apart from a few quick movements on their heads or backs. These early groomings, moreover, often occurred in contexts which suggested that they were primarily displacement activities. Chimpanzees frequently groom themselves in conflict or anxiety situations and mothers often interspersed such grooming with quick grooming movements on the head or shoulders of their infants. Small babies often looked dirty since pieces of sticky food which fell on to them were seldom removed by their mothers. One infant had a hard pellet of dung stuck over his anus to which his mother paid no attention. In respect of grooming behaviour, chimpanzee mothers differ from many other primate species. Gorillas, rhesus, langurs and vervets all groom their newborn babies frequently and intently; similar intent grooming of a newborn baby has been recorded for captive female chimpanzees (Schaller, 1963; Hinde, Rowell and Spencer-Booth, 1964; Jay, 1964, pers. obs.; Smith and Budd, 1943). Gorilla mothers apparently pay particular attention to the anal region of their infants.

As the chimpanzee infant grows older its mother grooms it more frequently and more thoroughly. The development of this behaviour, since it plays, to a large extent, a social rôle, will be discussed in Section III, 2.

(b) *Hygiene.* The position in which an infant is normally cradled is such that its urine and faeces often either fall clear of the mother or touch only her feet. When urine did touch mothers they normally paid no attention, but on five occasions when soiled with their infants' dung, mothers picked leaves and wiped themselves clean. Only one mother, Melissa, was seen with the hair on her inner thighs in a dirty condition from her infant's excretions, but by the time Goblin was three weeks old this was clean again.

So far no mother has been seen to wipe dung off her infant, although Mandy, after first wiping her infant's faeces off her own leg, picked the infant up, peered at her bottom, and made a few flicking movements with one finger. On another occasion she watched carefully when Jane defecated, caught the hard pellet in her hand, sniffed it, and then dropped it beside her.

(c) *Care of injured infant.* Only one badly injured infant was observed during four years in the field. As a result of this injury Mandy's infant was dead within three days. The cause of the accident was not known, but it resulted in ripping off the flesh of her inner right forearm. Bones and tendons were clearly visible, and from the position of the bones it appeared that they were broken, or at least dislocated.

Mother and infant were seen soon after the accident: the wound looked fresh and Mandy was still dabbing at a wound on his own ear. They were under observation for approximately thirty minutes, during which time Mandy scarcely looked at the infant and made no attempt to touch or lick the wound. She was normally the most solicitous of the mothers and usually if her infant made the slightest sound she looked down and cradled the baby. On this occasion even when the baby screamed in pain Mandy merely pressed her more closely without looking down.

Two days later Mandy was seen carrying the infant who was dead. Mandy laid the body beside her whilst she groomed a companion, but she kept pausing to hit away the cloud of flies that surrounded the corpse. When she left the tree Mandy gathered up the body and held it to her in the ventral position with her hand under its back. The following evening Mandy was seen again, and by that time she was no longer carrying the body.

Schaller observed a gorilla mother carrying her dead baby for four days before abandoning the stinking carcass. She laid the body beside her whilst feeding, and then held it to her breast as she moved off (Schaller, 1963).

On two occasions infants were observed with small injuries. One had a cut on her finger causing it to swell considerably: one had a sore on his foot. In neither instance was the mother seen to touch or lick the wound. When Flint, at eleven weeks, was bitten by a large ant he whimpered and rubbed his face in his mother's hair for twenty minutes until the ant fell from his lip. During that time Flo repeatedly embraced him or settled him into more comfortable positions, but made no attempt to remove the cause of his distress – and was, almost certainly, unaware of it.

6. *Leaving the Mother*

As the locomotor abilities of the infant develop it begins to crawl about on the body of its mother. Later it tries to crawl away from her into the surrounding environment. All mothers restricted such early attempts at independence for several weeks. In the three cases for which accurate data are available, mother-infant contact was broken for the first time between three and a half and five and a half lunar months (apart from occasions when the infants accidentally fell or slipped away from their mothers).

(a) *Mother restricts early efforts.* When infants first started to pull themselves about actively on their mother's laps their movements were often curtailed. Flo, for example, frequently pushed Flint's head down into her groin, or behind her knee, when he was struggling to look around him.

Jane, Cindy and Goblin all tried to struggle over their mothers' thighs between seven and nine weeks, pushing off with their feet, but after a few moments were hauled back into position. Flo repeatedly pushed Flint back when, at ten weeks of age, he made efforts to turn from his back into a crawling position. Both Flint and Goblin during their eleventh weeks and Cindy during her fourteenth, tried to pull themselves from their mothers by means of twigs, grasses, etc., but were prevented and held tightly by their mothers.

(b) *Initial breaking of mother-infant contact.* Data as to when and how contact is broken are available for three infants only. Flint was pulled away from his mother by his sibling Fifi when he was fourteen weeks old (see below).

The first time that Flint was observed out of contact with either his mother or sibling occurred when he was sixteen weeks old. On that occasion he himself let go his grip on his mother as he sat beside her on the ground, but when she got up and moved a step forward he at once gave a soft 'hoo' whimper at which Flo gathered him up.

The first observed instance when Goblin was out of tactile contact with his mother occurred when he was twenty-two weeks old and was initiated by Melissa who carefully detached his hand as he sat beside her. After about one minute he reached out and held on to her again. Cindy was first seen out of contact with her mother at twenty-one weeks old; when she reached out to a little branch and clung to it with both hands and feet.

(d) *Breaking of mother-infant contact initiated by elder sibling.* Only two of the infants born during 1964–5 had siblings – the others were first-borns. One of these infants, Flint, was extremely attractive to the youngest of his siblings, Fifi, who was about four and a half years old when he was born. The other infant, Honey Bee, was only observed on a few occasions, but on those occasions the elder sibling, Little Bee, showed little interest in the new infant.

Observations on the infant Flint commenced when he was seven weeks old; during this week Fifi was only observed to touch him on four occasions. The following month, however, she touched him more and more frequently and, by the time he was thirteen weeks old she made repeated attempts to pull him from his mother. Initially Flo allowed Fifi to touch or groom the infant briefly, but when Fifi tried to play with Flint, or became rough, and later when she tried to pull him away, Flo frequently tried to protect the infant. She sometimes got up and moved away. At other times she either drew the infant away or gently pushed off Fifi's hand. When Fifi persisted Flo sometimes played with (Plate VII) or groomed the child: both these activities had the effect of temporarily distracting Fifi's attention from the infant. Fig. 6 shows that Flo was most active in trying to protect her infant from his sibling when Fifi first began to try and pull him away. Once Fifi had accomplished this, and thus broken mother-infant contact, she was allowed to touch or carry Flint away more and more frequently.

Fifi was first observed to pull Flint from his mother when he was fourteen weeks old. Flo made no attempt to take the infant back until, after about thirty seconds, he looked towards his mother with a pout face (lips pursed and pushed forward) and gave a soft 'hoo' whimper. At this Flo at once reached out and gathered him up. Fifi was observed to pull him away once more on the same day: once more the mother took the infant back in response to his pout face and call. The following day Fifi was observed to carry Flint some 100 yds along a track, supporting him with one hand as he clung ventrally. Flo followed behind and took Flint back when Fifi sat down.

During the following weeks Fifi took Flint from his mother on many occasions, either sitting with him near Flo or moving a few feet away to groom or play with him. Flo frequently reached him back almost immediately after Fifi had taken him or, if Fifi walked off with him, Flo followed and took him back when Fifi stopped. (On ten occasions during the next three months Fifi hurried away with Flint as Flo tried to take him back – the mother then chased her child until she caught up and then usually tried to get ahead of Fifi before reaching out to retrieve her infant. Only on two occasions was the mother observed to grab hold of Fifi in order to stop her and so 'rescue' Flint.)

On 58 per cent of the occasions when Fifi pulled him from his mother during his fourth month, Flint himself initiated the re-establishment of mother-infant contact by reaching towards Flo with a pout face and 'hoo' whimper (see Sect. III, 4). During the following month Flint initiated the re-establishment of contact with his mother on 70 per cent of the occasions when Fifi took him away. This increase was probably in part due to the fact that Flo seemed less anxious and seldom followed Fifi when she carried Flint away unless she went out of sight. Also Flint was able to crawl along the ground by himself, and usually soon after Fifi settled down with the infant he struggled away and returned to his mother. When Flint was only fifteen weeks old, Fifi climbed with him up an 80-ft palm tree. Flo looked up but did not move until the infant screamed (he was out of my sight), when she rushed up the tree to rescue him.

Observations on the relationship between the other infant,

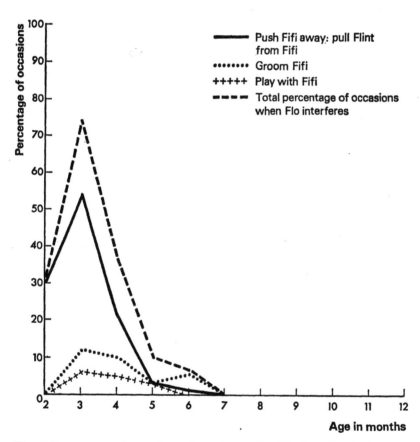

Fig. 6 Percentage of occasions when the mother Flo interfered when her juvenile daughter tried to play with or pull away the infant Flint. The types of maternal interference have been divided into three categories: (1) when Flo either pushed her daughter away from the infant, or pulled Flint away from his sibling; (2) when Flo groomed Fifi and thus distracted her attention from Flint; and (3) when Flo played with Fifi and thus distracted her

Honey Bee, and her juvenile sister were limited, during her first month, to about six hours. During that time the elder sibling, although she often sat close to her mother, was only seen touching the infant, briefly, twice. When Honey Bee was four months old, Miss Koning was able to make more regular observations, and although the infant grabbed out at her sister frequently, the latter paid little attention.[1]

(e) *Attitude of mother to infant's first movements away.* Data concerning the mothers' attitude to early infant independence relate mainly to Flo and Melissa, and I left the reserve soon after Goblin had taken his first steps. Flo reached out to hold on to Flint or pull him back every time he tried to move away until he was nineteen weeks old. After this he was gradually permitted to move further and further away, but his mother was quick to reach out and gather him to her embrace if he fell over when tottering near her. Melissa was not observed to restrict Goblin when he took his first steps, but the infant, apparently of his own accord, was never seen to move more than six inches away from her until his thirtieth week. Unlike Flo, Melissa was not observed to embrace Goblin when he tumbled over, but merely reached out one hand and laid it on the infant until he had righted himself.

So far no mother has been observed to 'encourage' her infant to walk or climb, behaviour which has been described in captive chimps (Yerkes, 1943) and wild gorillas (Schaller, 1963).

7. *Maternal Protection*

Maternal protection extends over the whole period of infancy and, indeed, to the juvenile and on rare occasions the adolescent. However, it is during the first two years that the infant is most in need of protection. During this time the mother protects her infant from unfavourable climatic conditions, and from dangerous objects of the environment. She protects it from falling, initially from her own body and subsequently from branches, rocks, etc., until its locomotor ability is well developed. Finally she protects it from

---

[1] One significant incident which suggests a possible adaptive value of a sibling bond has been reported by Miss Koning. When the mother of the two-year-old infant Merlin died – that is, she disappeared and was presumed dead – the infant was virtually 'adopted' by his juvenile sister, Miff, who sometimes carried him, and on various occasions showed 'maternal' protective behaviour towards him. His adolescent brother, Pepe, also moved around with him and protected him on occasions.

social encounters with other individuals until it is able to react correctly to the various expressive calls and movements of the adult.

(a) *Protection from the environment.* The very young infant is well protected and sheltered from rain and wind by its mother's body. During heavy rain the mother hunches over her infant which is often completely invisible. After such a storm infants looked dry, though they were often bedraggled after light rain when their mothers did not protect them to the same extent. After the first few months the mother does not necessarily hunch right over her infant during rain, though she continues to embrace it when it cuddles up to her for at least two years.

Mothers of small infants usually moved carefully through thick undergrowth. Once a twig hit a small infant: he screamed and his mother backed, held him close, and moved on in a slightly different direction. When infants are riding in the dorsal position their mothers often reach back and gesture them to move into the ventral position when they move from the open into undergrowth.

There is no serious predator to the chimpanzees in the Gombe Stream Reserve (with the possible exception of the leopard). One mother, however, embraced her infant as a hawk flew overhead. Mothers often threatened or chased off baboons when the latter tried to take bananas from their infant or juvenile offspring at the artificial feeding area. All mothers protected their infants, at first, from any contact with human observers, although eventually Flo allowed Flint (at nine months) to touch us or take pieces of banana from our hands.

(b) *Protection from falling and other bodily harm.* Some mothers moved particularly carefully when carrying small infants. Mandy, indeed, seldom made any movement without supporting her infant. She often climbed from the trunk of a tree to the lowest branch in a sloth-like position and then manoeuvred herself with great care on to the branch, supporting Jane with one hand.

When infants lose their footing or balance whilst clambering around on their mothers, the latter are quick to try and catch them. When Flint took his first tottering steps, Flo frequently placed her hand under his tummy and was thus able to catch him as he collapsed. She often placed her hand behind his back during his first attempts to climb.

When an infant was permitted to move away for short distances, its mother invariably kept a watchful eye on it and hurried to its rescue if it climbed into positions from which it could not get down. This will be further discussed in Section III, 4. Frequently during Flint's first year Flo followed behind when he went on exploratory expeditions, and was thus able to snatch him up if necessary.

By the time the infant is one and a half years of age it can normally move about easily on the ground and through most branches. Often, however, the mother waits at the lowest branch to carry her infant down a thick trunk, or waits to assist it over some difficult gap as it follows her from one tree to another. She is quick to reach out and catch hold of her infant if it should lose its balance whilst playing or feeding beside her in a tree: one mother of a three year old made a lightning grab to catch the infant as the branch it was playing on cracked.

(c) *Protection from other chimpanzees.* During the first few months of its life the infant is usually protected from almost all contact with other individuals. When mature animals approached most mothers embraced their infants closely. Melissa threatened a juvenile who peered at Goblin closely during his second day, and attacked a female who tried to touch him. Cerce, Melissa and Mandy all pushed away other infants who tried to touch their own babies (Plate 8) and Flo, as we have seen, repeatedly tried to stop Flint's sibling from pulling him away from her. Passion, however, permitted a juvenile and adolescent (both females) to touch and briefly groom her infant during its first two or three days but subsequently prevented such contacts. Sometimes quite small infants grabbed on to the hair of individuals sitting close to their mothers. On those occasions the mothers usually hastily detached the babies' hands and hugged them close.

As the infant grows older its mother gradually permits it more social contacts, though she is quick to seize it up at any sign of danger, particularly if one of the mature males shows signs of social excitement. Flint, from the age of ten months onward, invariably tottered towards any chimpanzee to arrive in the group. Sometimes Flo hurried after him and gathered him up before he reached his objective. At other times she permitted him to approach but often followed and began to groom the individual concerned. Olly, during her child's third year, often hurried to groom or lay her hand on males with whom Gilka was trying to play. Once she took the infant three times away from a group of mature individuals, leading her by the hand (Plate IX): each time Gilka pulled away from her mother and returned to the group.

When mothers carrying infants were attacked by males they usually remained in one place and crouched over their infants which, at the start of such an incident, usually slid rapidly into the ventral position (Plates Xa and Xb). Only on one occasion was a male seen to attack, very mildly, a female with an infant on her back, whereas twenty-five attacks were recorded on females with infants in the ventral position. It seems possible, therefore, that the sight of the infant on the mother's back may act as an inhibitory signal to the male. This signal may be enhanced by the white tail tuft of the infant which first becomes conspicuous between four and six months of age, just prior to the start of dorsal riding.

Sometimes mothers protected or went to the assistance of their older offspring. Fifi, as a juvenile, usually hurried to sit close to or behind her mother when she was threatened by a mature male. Equally, Fifi frequently threatened other individuals, including adult females and adolescent males, but normally only when they were subordinate to her mother. When Flo was not nearby, Fifi was not observed to threaten individuals older than herself.

On one occasion Figan, as a young adolescent, was attacked by an older male and ran, screaming, to his mother. Flo, a high-ranking female, immediately threatened and chased the aggressor, closely followed by the then courageous Figan. On two other occasions the same mother retraced her steps from a considerable distance when Figan screamed after being attacked by a mature male but, on those occasions, Flo made no attempt to 'help' her son. Another mother, Olly, a low-ranking and timid female, frequently appeared worried when her adolescent son, Evered, was attacked. She did not approach too closely but gave loud barks directed towards the aggressor. After such incidents the mother, on three occasions, approached the males concerned and made appeasing gestures. The mother Marina, on the other hand, was never seen to show anxiety when her juvenile daughter was attacked or threatened, but on one occasion went to the aid of her adolescent son who had been chased by two adult females. Mother and son together chased the others away. Similar protective behaviour to older offspring has been observed in rhesus monkeys (Sade, 1965).

8. *Maternal Tolerance*

In general it can be said that chimpanzee mothers are tolerant towards their offspring although, of course, individual mothers vary in this respect. When incompatibility does occur the mother normally plays a 'denying' rôle rather than an aggressive one – that is, she withholds an object or desired contact without threatening or attacking her offspring. Indeed, although mothers were seen to mildly threaten their older offspring fairly frequently, only nine instances of physical punishment were recorded, and these were mild: on seven occasions mothers seized and gently bit their children's hands; one mother hit her adolescent male offspring with the palm of her hand and, another time, briefly pounded on him with her feet. Moreover, the mothers concerned after biting the hands of their infant (five times) or juvenile (twice) offspring followed this up by laying their hands on the youngsters or embracing them and thus giving them immediate reassurance.

(a) *Tolerance of discomfort caused by clinging, etc.* Inevitably during the first few years of her infant's life the mother suffers some discomfort and restriction of movement as a result of its clinging to her and clambering over her. Infants often grasped hair in places where it probably hurt the mothers – such as on the inner thigh, breast, inner arms, or cheeks. Mandy was the most tolerant mother in this respect, and seldom detached Jane's hand – even when the infant grabbed her beard. Also, she often sat for up to fifteen minutes with her foot held off the ground in order to support her infant when she was using both hands for some other task (Plate II). Melissa was less tolerant: she invariably moved Goblin's hands if he gripped hair in the wrong place, and often sat in positions which caused Goblin to slide off her lap. She usually ignored his subsequent efforts to climb to a place where he could relax and rest.

Even large infants were normally allowed to ride on their mothers' backs or cling ventrally. Flo permitted Fifi to ride dorsally even when the child was over three years old, and sometimes Fifi actually clung on in the ventral position, her back brushing along the ground. One mother occasionally reached back and pushed her two and a half year old (Little Bee) from her back when she was standing still. In this instance the situation was not normal as the infant had a club foot and rode on her mother more frequently than most infants of her age.

(b) *Tolerance during suckling and weaning.* In most primate species mothers play an active and aggressive rôle in the weaning of their infants – wild baboon and langur infants, for instance, are repeatedly rejected when they try to suckle and may be hit or even bitten by their mothers (DeVore, 1963; Jay, 1963). Aggressive behaviour of mother to young during weaning has also been observed in captive chimpanzees (Yerkes, 1943). This, however, has not been observed in wild chimpanzee mothers and, from the scanty evidence available, it seems that the drying up of the mother's milk is the factor which normally terminates suckling behaviour.

Wild mothers were never seen with sexual swellings until their infants were between two and four years old. In each of the two cases where it was possible to observe an infant, at fairly regular intervals, before, during and after weaning, it seemed that the resumption of the menstrual cycle of the mother was in some way connected with the drying up of her milk.

One infant, Fifi (at approx. three and a half years), was seen to suckle three days before her mother's sexual swelling. (The first observed since Fifi's birth.) On the third day of her mother's swelling, however, she put her lips to each nipple in turn for a few seconds only and then sat close to her mother whimpering softly. She continued to try and suckle for the next few days and then, apparently, gave up. Flo was not seen to threaten or push away her child. The other infant, Sniff (at approx. three years) was observed during three oestrous periods of his mother but was not seen at all during the intervals between them. He suckled, apparently normally, during the first, but was observed to suck each nipple in turn for a few seconds only during the second (about a month later). During his mother's third period he approached, leaned forward as though to suckle, and then suddenly backed away screaming although no threat gesture on the part of his mother had been observed. He continued to scream until his mother reached out to embrace him; he then put his mouth near her nipples but made no attempt to suck. He was not observed to try and suckle at any time after this.

The behaviour of another infant, Gilka, is not understood. On almost every occasion, during eleven months, Gilka, after approaching to suckle, backed away from her mother with whimpers. During this period of time her mother showed one sexual swelling but, at the time of writing (July 1966) the infant had not been weaned. Again no sign of aggression was seen on the part of the mother who either approached and tickled the crying child and, thus, temporarily distracted her from her objective (see Sect. II, 8d) or drew the infant close. In the latter case Gilka then suckled, apparently normally (mean length one minute thirty seconds per feed). It is difficult to interpret this behaviour. Possibly the mother had threatened her child on some previous occasion.

(c) *Tolerance in feeding situations.* Small infants are often allowed to bite or grab at food from their mothers' mouths or hands: in the same way very young infants of some other primate species (such as gorillas (Schaller, 1963) and langurs (Jay, 1963)) are permitted to take scraps from their mothers.

Chimpanzee mothers, however, continue to share food, on occasions, with their older offspring. One child of about two years old frequently begged by pushing his pouted lips towards his mother's – she invariably responded by pushing a lipful of chewed food into his mouth. Sometimes a child put one or both hands around its mother's mouth until she spat out the food she was chewing – or moved away. At the artificial feeding area older infants, juveniles and adolescents sometimes begged by reaching their hands to touch their mothers' bananas: often they were allowed to take a fruit. Kortlandt (1962) observed infants begging from their mothers.

However, despite this general tolerance, many instances of mother-offspring incompatability were observed in connection with feeding. Mothers often snatched food from their infant, juvenile or adolescent female offspring. When a small infant reached for its mother's food, this was frequently jerked out of reach. When an older infant persistently tried to take its mother's food she often seized hold of the child, drew it on to her lap, and then pinned it into the ventral position with her elbows and knees while she finished her meal in peace. Only on rare occasions, however, did a mother threaten her infant or bite its hand if it persisted in trying to get her food. By contrast captive rhesus monkey infants behaving in this way were sometimes cuffed when they were only eight weeks old. (Hinde, Rowell and Spencer-Booth, 1964.)

There was, of course, a good deal of individual difference in the toleration of the various chimpanzee mothers. Thus whilst Miff (juvenile) and Sally (adolescent) were, apparently, too afraid to even try and beg from their mothers, Fifi, when over five years of age, not only continued to beg persistently, but went into temper tantrums when the food was withheld. On such occasions she was often allowed to take a fruit after her tantrum, and in two instances Flo actually held out fruits to the screaming child. Only once was Flo observed to bite Fifi's hand as a result of this behaviour, and that was when the child had already been allowed to take one banana which she was holding, untouched, in her hand.

(d) *'Distraction' of infant.* Two mothers, when their offspring persisted in trying to attain a desired objective, reacted, on many occasions, by tickling or grooming their children rather than by punishing them.

This behaviour was illustrated as we have seen when Fifi persisted in her efforts to try and pull away her sibling, the infant Flint, from the mother, Flo. Frequently Flo either began to groom her or to play with her, both of which had the effect of temporarily 'distracting' Fifi from her interest in Flint.

Similar behaviour was observed when Gilka approached her mother to suckle and then backed away, whimpering (see Sect. II, 8b above). On many occasions her mother then played with her, often continuing until Gilka 'laughed'. This again had the effect of causing the child to be temporarily distracted from her objective.

Behaviour of this category does not appear to be widespread amongst primates, but Jay (1963) observed something similar in some langur mothers when their infants were being weaned: 'If she grooms her infant before it becomes tense and aggressive this has a calming effect and at least postpones a weaning tantrum.'

(e) *Infant fails to respond to signals.* Occasionally an infant fails to respond when its mother gestures it to cling on, or to follow when she starts to move off. Mothers are tolerant in such instances.

A year-old infant may refuse to cling when pressed to the ventral position by its mother.[1] After several attempts have failed the mother often walks for a short distance carrying the infant with one hand under its tummy.

One mother sometimes had to climb a tree to fetch her infant when she was ready to go, and even then he often climbed just out of reach and patted towards her outstretched hand. When she

---

[1] Infants were not observed to behave in this way when their mothers were in a hurry or alarmed.

finally retrieved him his mother sometimes pushed him rather roughly into the ventral position but showed no other sign of aggressiveness.

SECTION III – SOCIAL INTERACTIONS BETWEEN MOTHER AND OFFSPRING

1. *Family Tie between Mother and Offspring*

The family tie between a mother and her offspring is such that the offspring may continue to associate closely with its mother long after it has ceased to rely on her for food, transport and protection. The degree to which this occurs probably depends on both the mother and the offspring concerned. Plate XI shows the mother Flo with four of her offspring in a compact, relaxed group.

Fig. 7 shows that the offspring, as it grows older, moves about less and less frequently with its mother.[1] Infants were never seen travelling about away from their mothers,[2] and juveniles, up to the age of five and six, were normally seen with their mothers. Fifi (at age of four to five) was observed away from her mother on four occasions for periods ranging from a couple of hours to a day, and once for four days in succession. Miff (at age five to six) was seen away from her mother on eight occasions, once for five days. Both juveniles showed great distress, initially, when they were unable to see their mothers (see Sect. III, 4) and when Fifi lost her mother, Flo, one evening she was still whimpering and crying the following morning.

The relationship between these two juveniles (Fifi and Miff) and their mothers were very different. Flo, as we have seen, was generally tolerant of her daughter, Fifi, and often went to her defence in social or other dangerous situations. Normally she waited for Fifi to follow when she moved from one place to

[1] The number of occasions that an offspring is said to 'move about' with its mother does not necessarily correspond to the number of observation periods when the two are seen together (Table I). Moving or travelling about with, refers to occasions when individuals proceed from one location to another as a group. (Members of a group usually move within a few yards of each other, although a member may lag behind and even temporarily, lose sight of the others.) Frequently it was possible to observe a mother and her elder offspring together even when they were not travelling about as a group – such as when mother and offspring arrived at the feeding area at different times, rested or groomed together and then went off in different directions at different times.

[2] The exception to this occured when the mother of the two-year-old Merlin disappeared and left the infant on his own.

another. The mother Marina, on the other hand, was often intolerant of Miff, particularly during feeding situations, nor did she usually give her daughter protection. When Miff lost her mother it was usually when the latter simply walked off without waiting.

No relationship between a mother and her adolescent daughter has yet been observed in detail. Such data as are available suggest that young females remain with their mothers fairly consistently

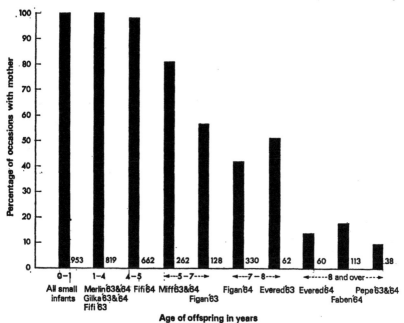

Fig. 7 The number of occasions when mother and offspring were seen travelling about together, as a percentage of the total number of occasions when either or both were seen. (The total number of occasions when the offspring were seen travelling with their mothers is shown at the foot of the relevant columns)

until their first period of oestrous. The adolescent female Sally was never seen with her mother when either of the two were in oestrous (a total of six times). Certainly this young female was always subordinate to her mother during feeding situations.

The juvenile male starts to leave his mother for several days at a time when he is five or six years old. As he reaches puberty he accompanies his mother less and less frequently: nevertheless, the mother-offspring relationship is clearly demonstrated in such things as grooming behaviour, tolerance during feeding, and the manner in which either may hasten to the defence or 'help' of the other. For instance, we have seen how a mother may defend her adolescent son (Sect. II, 7). In a similar manner, when Flo was attacked by a low-ranking mature male, her ten to thirteen-year-old son, Faben, hurried to her assistance and between them, mother and son chased the other male away.

As the male grows older, a gradual change occurs in his relationship with his mother. During his adolescence (from seven to eight to twelve or thirteen years) he slowly matures socially and, as he does so, his status changes. Thus some females who were socially superior to Evered in 1963 were normally dominated by him in 1964. Despite the gradual rise in status of adolescent males, their mothers, under normal conditions, continued to dominate them. Thus all four adolescent males usually allowed their mothers to take bananas in front of them, including Pepe and Faben when they were well over ten years old and would not tolerate such behaviour in other females.

However, despite the male's continued 'respect' for his mother, the data show that the mother does develop an increasing 'respect' for her son. When Figan gave frustration or branch waving displays (Goodall, 1965) during 1963 his mother ignored him. During the following year, however, she frequently screamed and rushed out of his way when he behaved thus. When Olly and her son Evered were handed bananas, Olly often held back, glancing repeatedly towards Evered, even though he was waiting for her to take the fruit offered to her.

On one occasion Flo hit Figan when he was about eight years old, and on another she stamped on him, but she was never observed to attack Faben, who was about three years older than Figan. Nor was Figan seen to attack his mother although, on one occasion, Faben jumped on Flo and stamped briefly on her back.

2. *Behaviour which probably Functions to Strengthen and Reinforce the Mother-offspring Relationship*

During infancy, affectionate behaviour, play and grooming are all probably helpful in establishing a strong social tie between a mother and her offspring. As the latter grows older, sessions of social grooming become increasingly longer and probably serve to re-inforce this tie when the offspring returns to its mother after leaving her for longer and longer periods.

(a) *Affectionate behaviour*. Mothers of small infants frequently stared down at them intently, often without touching them. On one occasion Flo put her index finger under Flint's chin and tilted up his face, and then stared at him for a full half minute (Plate IX). Mothers were also observed to 'caress' their infants, either gently stroking them with their fingers, or making a few idle grooming movements whilst gazing down at their babies. All mothers 'kissed' their infants by lightly putting their lips to some part of their bodies: Mandy and Flo both picked up their infants' hands and then 'kissed' the palms on several occasions. Similar behaviour has been observed in rhesus monkey mothers (Hinde, Rowell and Spencer-Booth, 1964) and langurs (Jay, 1963).

Mothers of small infants often lifted up one of their feet and then moved it up idly and down so that the knee and hip joints were alternately flexed and extended. Flo sometimes did this for at least half a minute at a time during Flint's first three months.

When an infant is over one year old the above types of behaviour are not normally observed. The mother of an older infant, however, may occasionally lay her hand on her child, or hold its hand, seem-ingly for social or 'affectionate' reasons only. Similarly an infant often leaves some game, for no apparent reason, to go to sit in contact with its mother. Schaller (1963) observed similar behaviour between a gorilla mother and her infant.

(b) *Play*. All chimpanzee mothers (with the exception of Mandy whose infant dies at three months) were observed to play with their infants, and one, Flo, was also seen to play with her juvenile and adolescent offspring. Fig. 8 suggests that the frequency of this behaviour depends on the age of the offspring concerned, and may vary between different mother-offspring pairs. The figure shows that both Flo and Melissa played most frequently with their infants when they were between four and six months old. This, since it coincided with the early struggles of the infants to pull away from their mothers, suggests that the mother's play behaviour was, at least in part, related to the 'distracting' behaviour described above (Sect. II, 8d).

Flo, Melissa and Circe initiated playing during the first few months of their infants lives (Table III) by tickling them with their

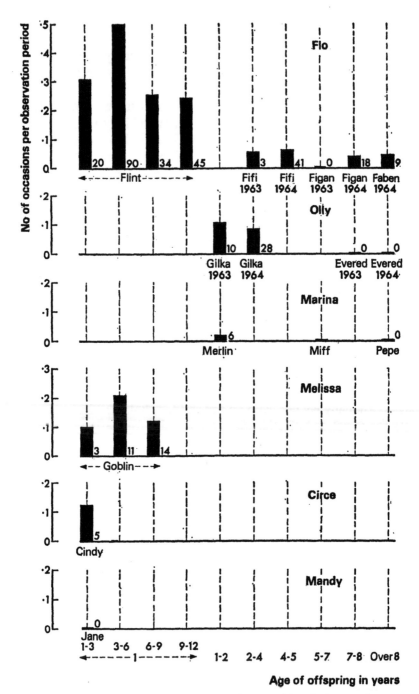

Fig. 8 Changes in frequency of maternal play with relation to the age of the offspring. (The total number of occasions when play was seen between a particular mother-offspring pair in each specified period is shown at the foot of the relevant column)

fingers, usually under the chin or in the groin. Flo frequently tickled Flint with small nibbling movements of her worn teeth. Also, on many occasions, she lay on her back, held Flint above her with one of his ankles or wrists clasped in her foot, and tickled him. Melissa was only observed to play with Goblin in this way on one occasion, and Circe not at all.

When Flint was about twelve weeks old Flo began to tickle or fiddle with his penis, either with her fingers or her lips and continued to do this frequently for several months. Flo also occasionally tickled Fifi's clitoris during a game, and twice tried to tickle Figan's penis when he was an adolescent. This behaviour was not observed in other mothers. Schaller observed one gorilla mother manipulate the genital area of her three-and-a-half-month-old offspring until he showed a penile erection which she then held several times with her lips (Schaller, 1963).

Flint was first observed to respond to tickling, with a play-face (see Plate XIII) when he was eleven weeks old: he 'laughed'[1] for the first time the following week. Goblin, who was less frequently played with by Melissa, was not seen with a play-face until he was eighteen weeks old, and had not been heard laughing at twenty-four weeks (when I left the reserve). Both infants, however, made play responses between eighteen and nineteen weeks of age, grabbing and patting at their mothers' hands during tickling. After this both infants frequently tried to initiate play, pulling their mothers' hands towards them with play-faces: Flo invariably responded, but Melissa often ignored Goblin.

Play behaviour of mothers with their one-year-old infants included rolling the infants over, mock biting and tickling them, gently sparring with them, or pushing them to and fro as they dangled from a branch. The infants responded by grabbing and biting the mothers' hands, flinging themselves on to the mothers with flailing arms, or tickling them.

As infants grow older, mothers normally play with them less frequently. Marina seldom played with Merlin after he was about one and a half years old unless he himself initiated the playing, pulling her hand towards his tummy for tickling. Olly played with Gilka fairly frequently but seldom intensively unless it was to distract the child from suckling (see Sect. II, 8d above).

---

[1] During play chimpanzees frequently make a series of low 'panting' sounds 'aach-e-aach' which are roughly equivalent to human laughter.

The play behaviour of Flo seemed unusual. During 1963, before the birth of Flint, she was observed to play with Fifi on three occasions only, and was not seen to play with her two sons. In 1964, however, she played intensively with all her offspring. It seems possible that this sudden increase in playful activity in the mother was a direct result of the birth of another infant. For one thing, as we have seen, she frequently tickled Fifi in order to divert her attention from the infant. For another, chimpanzees sometimes develop 'temporary habits' in their behaviour patterns, and possibly as a result of tickling Flint and Fifi, playing became such a 'habit' for Flo, stimulating her to playful activities with her older offspring. Fig. 9 shows how Flo's frequency of play with each of her offspring did, in fact, follow a similar pattern during the first year in Flint's life. (The sudden decrease which occurred during Flint's ninth and tenth months was between October and January when Flo spent a great deal of each day in feeding on termites.) However, the possibility cannot be overlooked that Flo was a particularly playful female but that during 1963 her protracted periods of oestrous interfered with normal behaviour.

Play between Flo and her adolescent offspring frequently commenced when they grasped each other's hands or feet; this was followed by tickling or mock biting, particularly of the fingers and ears. On two occasions Flo joined her three older offspring in a game which consisted of chasing each other round and round the base of a tree. The children sometimes played this game with Flo sitting in the middle (Plate XIV).

No mother was observed to play with youngsters other than her own offspring except occasionally with an infant or juvenile that was trying to touch or play with her own small infant.

(c) *Grooming*. Grooming behaviour, as a social interaction between a mother and her offspring, appears gradually to become more important as the offspring grows older.

Fig. 10 shows how the mother grooms her infant for longer at a time as it grows older. Mothers were only occasionally observed to groom infants of under three months for more than a few seconds at a time, and never for more than three minutes. Not until Goblin and Flint were twelve and thirteen weeks old respectively were their mothers seen to groom them for as long as five minutes at a time and, in fact, Flo was only once seen to groom Flint for longer than fifteen minutes during his first year. Gradually, however, the mother grooms her offspring for longer and longer at a time – thus the percentage of groomings for four mothers that were longer than

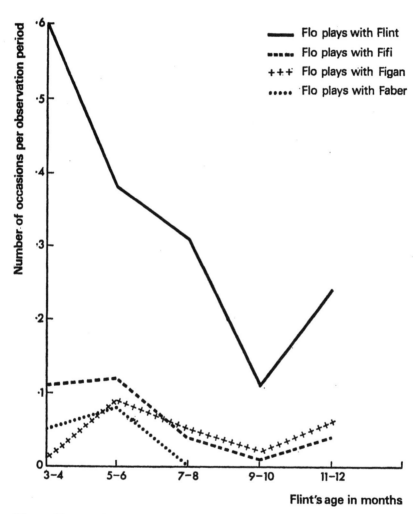

Fig. 9 Changes in frequency of the mother Flo's play with her four offspring during nine months of her infant Flint's life

five minutes increased from 1 per cent (for infants under six months) to 58 per cent (for adolescents of nine years or more).

As the infant grows older a gradual change in its attitude to grooming is apparent. At first Flint and Goblin were restive whilst being groomed, and not until Flint was nineteen weeks old was he observed to lie passively during grooming. When he was thirty weeks he appeared to present himself for grooming for the first time, climbing on to Flo's lap and stretching out, side uppermost. Subsequently, infants present for grooming frequently.

Fig. 10 also shows how the infant gradually grooms its mother for longer at a time as it grows older. Flint was twenty-six weeks old when he was first observed to make an attempted grooming movement on his mother's back. An adult chimpanzee parts the hair for grooming with the backs of its fingers or with the side of the thumb. Flint, however, seized hair between his flexed fingers, pulled it towards him, and poked at the exposed skin with the index finger of his other hand. He then put his lips to the skin in an adult manner. Similar behaviour was seen on four subsequent occasions, but he had not been observed to make adult grooming movements when I left the reserve: he was then one year old.

During its second year the infant gradually develops an adult grooming technique, but infants under two years were never observed to groom their mothers for more than one minute at a time.

Fig. 11 shows how the total number of grooming sessions between a mother and her offspring initially decreases as the child becomes increasingly independent and spends more time investigating the environment or playing. Gradually, however, as the child approaches adolescence, playful activities become less frequent, and grooming begins to play an important rôle in its social interactions – including those with its mother. The figure shows that the number of grooming sessions between a mother and her adolescent son becomes more frequent (relative, of course, to the amount of time he spends with her) as he grows older.[1] In addition it should be borne in mind, when examining Fig. 11, that only 6 per cent of the grooming times in the first two columns are of over five minutes' duration, whilst in the last column 46 per cent of the mother's grooming sessions and 59 per cent of the offspring's, are longer than fifteen minutes.

Finally, Table VI shows how the percentage of the total number of grooming sessions recorded between mothers and their offspring, when the behaviour was mutual, also increases as the

[1] It is probable that a similar increase in grooming behaviour occurs as an adolescent daughter grows older, but reliable data in this respect are not yet available.

youngster grows older. The fact that 97 per cent of the sessions of thirty minutes or more occurred when both individuals took part suggests that grooming is more rewarding to the animals concerned when the behaviour is mutual.

3. *Reaction of Infant to Interactions between its Mother and other Individuals*

(a) *Copulation.* Four known mothers with children of between two and a half and three and a half were observed when they came into oestrous: one was seen during three successive periods of oestrous, two during two, and one during one. The response of all the infants concerned to males copulating with their mothers was to hurry up and try to push the male off: this was invariably tolerated by the males. Of these infants, three were females and one a male. Another infant female was observed when a large group of males displayed at, and copulated with, her mother. She screamed, as if frightened, during six copulations. Then, when most males had left the tree, she approached and watched closely during a seventh copulation, touching the place where the penis was inserted into the vagina. When her mother screamed and tried to escape from another male who persistently rushed after her, the infant screamed at him and hit him. He mildly threatened her and then gave up the chase (Goodall, 1965).

Fifi was about three and a half years of age when Flo came into oestrous for a total period of nearly five weeks.[1] For the first few days Fifi hurried up at each successive copulation, often jumping on to Flo's back to push the male harder. In between copulations she often sat close to Flo, sometimes with one hand on her mother's back. The pushing-away behaviour persisted throughout Flo's period of oestrus although, after the first few days, Fifi's efforts were less intensive and she sometimes missed copulations.

Occasionally, infants or juveniles pushed at males copulating with females who were not their mothers: in such instances the males tolerated the infants but hit out at the juveniles. Kortlandt observed a number of copulations in wild chimpanzees but although he saw infants watching closely he does not describe pushing behaviour (Kortlandt, 1962). Langur infants, however, sometimes hit at males copulating with their mothers (Jay, 1963)

[1] There was one period of detumescence lasting for a few days during a five-week period of conspicuous anogenital swelling.

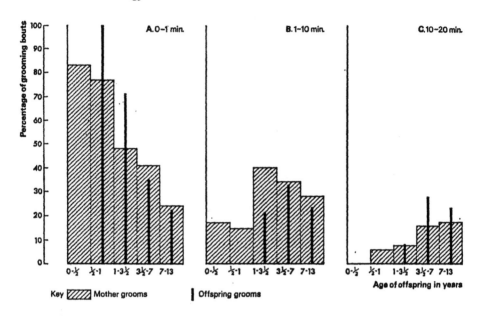

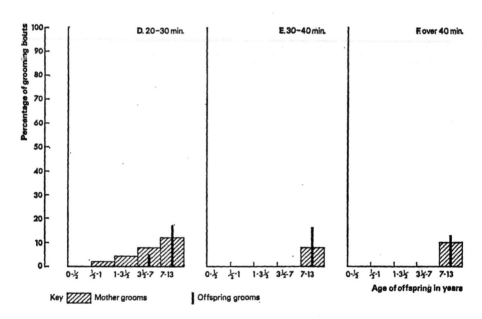

Fig. 10 Distribution of grooming sessions of different lengths between mothers and their growing offspring

	0-½ year	½-1 year	1-3½ years	3½-7 years	7-13 years
1. 0-1'	Flint (56) Goblin (29) Cindy (18) Jane (11)	Flint (36) 5 Goblin (10)	Fifi (7) 2 Gilka (20) 13 Merlin (14) 2	Fifi (30) 14 Figan (1) Miff (2)	Figan (22) 6 Evered (4) 1 Faben (5) Pepe (4) 1
3. 1-10'	Flint (15) Goblin (7) Cindy (1)	Flint (7) Goblin (4)	Fifi (8) 3 Gilka (19) 2 Merlin (7)	Fifi (19) 10 Figan (6) 1 Miff (3) 2	Figan (27) 10 Evered (3) 3 Faben (8) 7 Pepe (3) 2
2. 10-20'		Flint (3)	Fifi (3) Gilka (3) 2	Fifi (9) 10 Figan (3) Miff (1) 1	Figan (10) 7 Evered (8) 7 Faben (5) 5 Pepe (3) 3
3. 20-30'		Flint (1)	Fifi (1) Gilka (3)	Fifi (5) Miff (2) 2	Figan (6) 4 Evered (3) 3 Faben (2) 2 Pepe (7) 7
1. 30-40'					Figan (5) 5 Evered (3) 5 Faben (2) 2 Pepe (3) 3
7. OVER 40'					Figan (1) 1 Evered (7) 4 Faben (3) 3 Pepe (4) 4

Figure *in* brackets shows number of occasions when mother grooms offspring.
Figure *after* brackets shows number of occasions when offspring grooms mother.

and Gartlan records similar behaviour in vervet monkeys (pers. comm.).

(b) *Grooming.* On many occasions when Flo groomed another individual Fifi hurried up and pushed herself between the two, herself presenting to Flo for grooming. Flo invariably responded by grooming the child, at least for a few minutes. Figan and Fifi sometimes pushed each other out of the way time and time again,

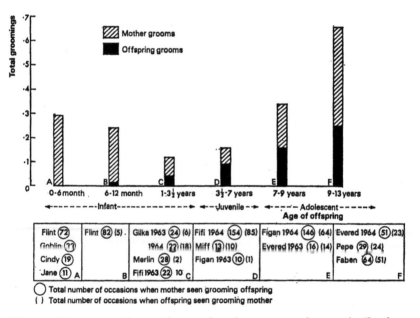

Fig. 11 Frequency of grooming sessions between mothers and offspring of different ages

and Figan twice lifted Flo's hand away from Fifi and pulled it towards himself. Small infants were often observed to fling themselves between or on top of groups where their mothers were involved in grooming activities. On such occasions the mothers frequently paused to play briefly with their infants.

4. *Mother-Infant Co-ordination*
In order that the infant can be transported from place to place, fed and protected it is vital that close physcial contact between mother and infant be maintained. As we have seen, the infant is by no

means helpless: it is able to cling tightly to the mother for long periods of time, after the first week or so, and to initiate suckling by nuzzling and pulling itself towards the nipple. However, there are times when the infant loses its grip on the mother, and times when

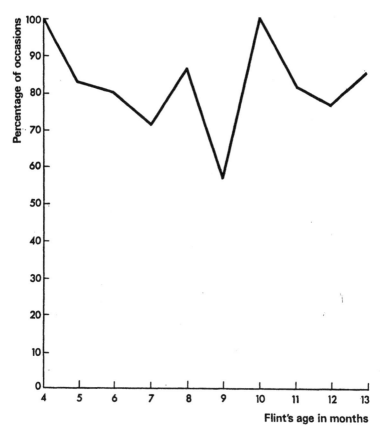

Fig. 12 Percentage of occasions when the mother Flo responded to the 'hoo' whimper of her infant during his first thirteen months

it is unable to reach her nipples unaided, and in these respects, as well as in many others, co-ordination between the efforts of the infant and the assistance of the mother is necessary if maternal care is to function efficiently.

There are a number of movements, sounds and gestures, some made by the infant and others by the mother, which apparently serve as communicatory signals and help to make co-ordination possible. The signals that will be discussed in this chapter relate specifically to mother-infant co-ordination: many of them are non-ritualized and form no part of the adult communicatory system.[1]

(a) *Tactile signals.* During the first few months of a chimpanzee infant's life tactile signals play a major role in mother-infant co-ordination. Indeed, Harlow suggests that contact clinging behaviour, at least in rhesus monkey infants, plays an important rôle in eliciting maternal behaviour (Harlow, 1963).

In the chimpanzee, the pressure of the infant's body against that of the mother and the intensity of 'hair pull' where the infant grasps her hair with its hands and feet, together with changes of intensity in either of these stimulations, function as signals. Thus, as we have seen, a mother does not necessarily support her infant when she moves off, provided it is already clinging tightly (Sect. II, 3). If the infant loses a hand or foot hold as the mother is walking, there is an immediate alteration in the intensity of 'hair pull': the mother responds by pressing her infant back into place. The mother normally reacts also to any sudden change in the body position of the infant in relation to her own: thus when a sleeping infant slips from her lap she gathers it up again.

Similar changes in 'hair pull' and body position are also important signals in effecting successful suckling, for the infant not only nuzzles but, when it cannot reach the nipple, tries to pull and push itself towards the mother's breast. Again the mother reacts by cradling her infant: this normally enables it to reach one or other of her nipples.

Many signals of the mother to her infant are also tactile. As we have seen, she pushes her small infant in order that it should adopt a ventral or a dorsal position on her body (Sect. II, 3). As it grows older it learns to respond to much abbreviated forms of the same signals and will usually cling on, climb on her back, or slide into the ventral position in response to a light touch – the course adopted, of course, depending on the situation at the time.

Mother chimpanzees, as we have seen, are not normally aggressive to their infants, and tactile signals which serve to rebuff

[1] The gradual development of social gestures in the infant chimpanzee and its adjustment to adult society are all, of course, reflected in mother-offspring communication: this aspect will be analysed and presented in a later paper.

a child are normally limited to pushing away. Biting and hitting in this context are rarely observed – if they are the mother usually follows this up by laying her hand on the child and thus giving it immediate reassurance. The significance of this sparcity of pain eliciting signals will be discussed below (Sect. IV).

(b) *Auditory signals.* Three clear-cut vocal sounds are functional in mother-infant co-ordination, the 'hoo' whimper, the scream and the 'soft bark'.[1] These sounds all form part of the adult communicatory system, although the scream of a small baby differs from that of the older infant and may be referred to as the 'infantile scream'.[2]

During its first few months the infant normally utters the 'hoo' whimper when the mother fails to respond to changes in 'hair pull' or body position as the infant, for instance, tries to reach a nipple. The mother usually responds to this sound immediately, cradling the infant close and often looking down at it with a similar 'hoo' herself. When the infant is older and begins to move from its mother, it invariably utters this sound if it gets into any difficulty and cannot quickly return to her. Until the infant's locomotor patterns are fairly well developed the mother normally responds by going to fetch it at once, and Fig. 12 shows the percentage of occasions when Flo retrieved and embraced her infant after he had uttered a 'hoo' whimper.

The same sound is used by the mother when she reaches to remove her infant from some potentially dangerous situation or even, on occasions, as she gestures it to cling on when she is ready to go. The 'hoo' whimper, therefore, serves as a fairly specific signal in re-establishing mother-infant contact.

If a small infant has a fright, such as when it falls or nearly falls from its mother's body, or if there is a sudden loud sound, it emits a loud scream. In the former case mothers appeared to react to the sudden drastic change in 'hair pull' and grab up their infants almost before the latter screamed. The scream itself, however, invariably elicits intense cradling behaviour. This embracing may well function to minimize the possible traumatic effects of the

[1] This is a soft single syllable coughing sound which may be repeated and which is normally accompanied by a slight upward and backward jerk of the head.
[2] This is a fairly high-pitched, short single syllable sound. It was heard in a two-day-old infant. After the first few weeks this sound gives place to the typical scream which only differs slightly from the scream of a frightened adult.

fright since a laboratory experiment showed that young chimpanzees that were being held by, and were clinging to, a human experimenter, had a higher pain threshold (measured in terms of vocal response to painful shock stimuli) than when they were lying alone on bare boards (Mason and Berkson, 1962). In the present study only one instance was recorded when a mother did not respond to infantile screaming with intense cradling behaviour, and this was in abnormal circumstances when an infant was wounded and screaming in pain (Sect. II, 5).

On several occasions infants screamed loudly when their mothers started to move away without them. Each time the mothers immediately turned back and retrieved their infants. Indeed, throughout infancy, screaming normally results in the mother hurrying to rescue her child. When a juvenile screams the mother, if she cannot actually see the child, often hurries towards the sound and may then actively defend her offspring if occasion demands. Similar maternal reaction to screaming in a juvenile is recorded for baboons (DeVore, 1963). As we have seen, one chimpanzee mother retraced her steps when her adolescent son screamed (Sect. II, 7 above).

The 'soft bark' is occasionally used by the mother as a mild threat to her offspring – if it begs too persistently, for instance. It is seldom directed towards infants, but more frequently towards juveniles and adolescents. No data are as yet available as to when a mother first threatens her infant in the typical adult manner, or whether the infant immediately responds with submissive behaviour.

(c) *Visual signals*. Visual signals did not appear to play an important rôle in mother-infant co-ordination during the first few months, although an experiment by Cross and Harlow suggests that the sight of a newborn monkey may be an incentive to its mother and may play some part in eliciting maternal affection (Cross and Harlow, 1963).

After the chimpanzee infant has started to move away from its mother visual signals play an increasingly important rôle. There are various facial and postural attitudes of the infant which serve as clear-cut signals. Thus the 'pout face' is always made by an infant seeking to re-establish physical contact with its mother (when it is normally accompanied by the 'hoo' whimper). In addition the infant may reach one or both arms towards the mother.

She normally reacts to this expression and gesture by going to collect her infant. If the infant bounces up to the mother with a playface she normally responds by tickling or sparring with the child. When the infant sits with its back to her, head bent forward, she normally responds by grooming it.

When the infant is first permitted to move away from the mother she keeps a sharp eye on its activities and there are a variety of non-ritualized postural attitudes of the infant which serve as cues to the mother. Thus if the infant is simply tottering peacefully nearby she does not normally interfere. If, however, it suddenly tumbles over, or if it starts to walk determindly away from the mother, she often hurries to gather it up. If her infant moves from her sight she usually reacts by hurrying to find and retrieve it.

Finally, certain aspects of the environmental situation also serve as cues to the watchful mother. Thus when Flint climbed so high up a thin twig that it started to bend Flo stood up to reach him down, although the infant had shown no signs of distress himself. Again, if a mature individual approaches too closely, or if a fight or other violent social activity takes place nearby, the mother of a small infant normally hurries to gather it up. This is particularly necessary until the infant is about ten months old since prior to this it does not react to many of the ritualized sounds and gestures made by adults. As the infant grows older the mother gradually relaxes her vigilance, but by the time the child is about one year old it has begun to respond to adult communicatory signals, and hurries back to its mother of its own accord in potentially dangerous situations. Schaller (1963) observed that the wild gorilla infants failed to react to some ritualized signals during their first year.

During the year following its first five or six months the infant gradually responds to more and more visual signals of its mother. Thus if she reaches her hand towards it, it will normally return to her. If she walks towards it and pauses, standing almost over it, the infant normally clings on. If she stands with slightly bent knees and looks back at the infant, it usually climbs on to her back, and if she walks away from the infant and looks back it normally follows her. In addition, the child apparently begins to respond to an increasing number of subtle postural and facial expression variations in its mother, each of which indicates her mood and her probable subsequent behaviour.

(d) *Occasional inadequacy of mother-offspring communicatory*

333

signals. The various tactile, auditory and visual signals described above permit co-ordination between a mother and her offspring under normal circumstances. There is evidence, however, which suggests that in some cases maternal responses are too stereotyped to adapt to abnormal circumstances.

Two examples of this have been described. Thus Flint, when bitten by an ant, whimpered and rubbed his face on Flo for twenty minutes (Sect. II, 5) but elicited merely the cradling response from his mother. When the infant Jane was seen with her wounded arm it was obvious that any sudden movement of her mother caused her great pain so that she screamed: Mandy's response was to clasp the infant tighter. This increased the pressure on the arm and caused the infant to scream even louder.

There is another type of situation which illustrates lack of co-ordination, and this is when an infant or a juvenile accidentally becomes separated from its mother.

On one occasion a two-year-old infant moved some 100 yd and out of sight of her mother whilst playing. The mother, when she noticed her child was not there, became anxious and began peering around and uttering the 'hoo' whimper. This sound, as we have seen, may serve as a specific signal in the re-establishment of mother-infant contact – however, since it can only be heard if the recipient is within a 15 to 20 yd radius it was totally inadequate to the situation. Finally the mother saw her child and hurried towards her.

On several occasions juveniles were observed who had accidentally lost their mothers. In each instance, after peering round from various trees, whimpering and crying as they did so, they hurried off – often in the wrong direction. On three occasions I was able to observe the reaction of the mother and every time, although she set off in the direction of her offspring's crying, she herself made no sound to indicate her whereabouts. In one instance the juvenile stopped crying before her mother found her, and this resulted in a separation of several hours.

SECTION IV – DISCUSSION

One fact which emerges from this study is the strength and long-term duration of the tie between a mother and her offspring. This is a fact which has also become apparent in two other long-term studies of free-ranging primates – namely on Japanese macaques and on rhesus macaques on Cayo Santiago island. In both these species it has been noted that the bond between a mother and her infant may persist into adult life (Imanishi, 1960; Yamada, 1963; Koford, 1963; Sade, 1965). However, even though the mother-offspring tie may persist in many primate species, most monkey infants experience rejection by the mother during weaning. In some instances this may be quite severe as, for instance, in the baboon (DeVore, 1963) and langurs (Jay, 1963). In both these species the infant spends more and more time away from the mother after weaning, and joins other youngsters in a play group. It is, however, within the troop – it may at any time be defended by its mother and, as a troop member, enjoys the protection of the adult males. In chimpanzee society, however, there is no closed troop. Instead, every adult is, basically, an independent unit, and the growing child, in order to benefit from the most reliable adult protection (that of its mother), must remain close to her until it is old enough to become, itself, an independent unit.

In two respects maternal behaviour, as described above, can be regarded as adaptive to this specialized social situation. Firstly, the infant constantly receives a variety of 'pleasant' social stimulations from its mother, such as affectionate caressing, play and grooming. Mason tested a group of young captive chimpanzees with a number of social stimulations, including petting, play and grooming, and concluded that such interactions had definite rewarding effects and might be regarded as relevant factors in the development and maintenance of companionship preferences (Mason, Hollis and Sharpe, 1962). Secondly, not only is there no apparent rejection of the infant during weaning, but the child is seldom punished physically by its mother, and on the rare occasions when it is the mother is quick to follow this up with a reassuring touch. It seems likely that these two factors both contribute to rendering the mother socially attractive to her growing child and this, in turn, probably helps to ensure that the youngster does not stray too far from maternal protection.

Another point which emerges from the study is the possibility that, in some instances, the mother-infant relationship may be inadequate and that this may contribute to the high infant mortality rate (Sect. I). Firstly, in two cases at any rate, mothers failed to either eat or detach the placenta from their neonates during the first day or so. This might well prove fatal to an infant if, for instance, the cord became tangled whilst the mother was moving fast, particularly through the trees. The newborn infant of a captive rhesus monkey was jerked from its mother when the placenta, which was attached, caught on a branch (Hinde, Rowell and Spencer-Booth, 1964).

Secondly, in striking contrast to most other primate species, the small chimpanzee infant often appears unable to remain securely attached to its mother if she makes a sudden movement. For several months after the birth of her infant the mother may have to support it, thus hindering her movements, and Plate III shows that when moving through the trees the mother may be severely handicapped when carrying a small infant. Female chimpanzees, particularly the younger ones, are quick to leap out of the way during branch-waving displays of the males or other violent social activities, and it seems not unlikely that one of the major causes of infant mortality may be from injuries caused by falling from the mother.

Finally, I should like to emphasize that this field study is by no means completed. The data presented here concern only a very small number of individuals and it may well prove that further information will drastically alter some of the views expressed in this chapter. It becomes increasingly apparent that, in chimpanzee society, individuality plays a highly significant rôle in many aspects of behaviour. Only a more detailed study of the individuals described in the foregoing pages, and further studies on other females and their infants, can fill in some of the many gaps in our understanding of mother-offspring relationships.

ACKNOWLEDGMENTS

I should like to express my thanks to all those who made this field study possible: to Dr L. S. B. Leakey, Honorary Director of the National Museum Center for Prehistory and Palaeontology, who initiated the expedition; to the Wilkie Foundation, which provided funds to launch the project; to the National Geographic Society which has subsequently financed my research; to the Tanzania Government officials in Kigoma and the Tanzania Game Department for their co-operation and practical assistance. My gratitude is due to Baron Hugo van Lawick, my husband, for his valuable photographic and film record and for his practical assistance in all aspects of the research; to my assistants Miss Koning and Miss Ivey for their help; to my African staff and helpers; and to the mothers and infants, particularly Flo and Flint, without whose co-operation this chapter could not have been written. My thanks are also due to my supervisor, Professor Robert Hinde of the Sub-Department of Animal Behaviour, Cambridge University, for his help and advice in the preparation of this manuscript.

TABLES

CAPTIONS FOR TABLES

Table I. Listing the various mothers with their offspring which visited the artificial feeding area. This table also lists the known or probable dates of birth of the offspring concerned and the number of observation periods (see p. 288 in text) during which it was possible to observe the relationship between a mother and any one particular offspring.

Table II.—Showing the age in weeks of four wild born chimpanzee infants when a particular behaviour pattern was first *observed*. A few comparative ages are given for wild gorilla infants (from Schaller, 1963) and for wild baboon infants (from De Vore, 1963).

Table III.—Listing the number of hours per week during which it was possible to observe four infants during their first six months.

Table IV.—Listing stages in development of behaviour in wild born chimpanzees.

Table V.—Percentage of occasions when six infants showed a preference for right or left breast during feeds.

Table VI.—Change in proportion of mother-offspring grooming sessions, in which grooming was mutual, with age of offspring.

TABLE I

MOTHER	OFFSPRING	DATE OF BIRTH	NO. OF OBSERVATION PERIODS	
			1963	1964–65
Flo	♂Flint	1/3/64		666
	♀Fifi	1960[1]	141	658
	♂Figan	1957[1]	119	430
	♂Faben	1953[1]		188
Olly[2]	♀Gilka	2–4/1962	87	309
	♂Evered	1956[1]	45	110
Marina	♂Merlin	5–6/1963	10	272
	♀Miff	1958[1]	10	263
	♂Pepe	1953[1]	5	81
Melissa	♂Goblin	7/9/64		201
Mandy	♀Jane	20–24/11/64–14/2/65		37
Circe	♀Cindy	22–24/1/65		49
Madam Bee[2]	♀Honey Bee	12/1964–1/1965		5
	♀Little Bee	1–6/1962	5	20
Sophie	♂Sniff	1962	12	20
	♀Sally	1954[1]	10	10
Passion[3]	♀Pom	13/7/65		

[1] Probable year of birth.
[2] Seen with an infant in 1961 which presumably died.
[3] Seen with an infant in 1964 which presumably died.

TABLE II

Age in Weeks

	4 WILD BORN CHIMPANZEE INFANTS				WILD GORILLA	WILD BABOONS
	FLINT	GOBLIN	JANE	CINDY		
ook round and seems to take interest	—	3	3	3	—	—
uck or bite on thumb	by 7	5	6	7	—	—
lother tickles infant	11	10	—	5	—	—
cratches Self	9	8	7	8	—	—
each towards object, looking at it	7	10	7	6	—	—
rasp object	12	11	—	11	—	—
rawl on mother's body	14	17	—	14	10–12	—
hange breasts during single feed	12	11	9	—	—	—
truggle to pull or push from mother	12	11–13	7	13	—	—
eet on ground push up-right holding mother	12	14	—	15	—	—
angle below mother from two or one hands	12	13	—	19	—	—
ooth eruption:						
Upper incisors (1st)	14	—	—	—	—	—
Lower incisors (1st)	15	16	—	12	—	—
Upper incisors (2nd)	15	—	—	15	—	—
Lower incisors (2nd)	—	—	—	14	—	—
lay face	13	19	—	—	—	—
each to other chimpanzee	14	10	—	15	—	—
espond to social play, grabbing and play-biting	14	19	—	—	—	—
Laughing' first heard	13	—	—	—	—	—
lother-infant contact broken	16[1]	22	—	21	10–12	4–8
ick or suck at food objects	14	19	—	—	—	—
hew and swallow solid food	15	26	—	25	10–12	16–20
tands tripedally, holding mother with one hand	16	19	—	15	—	—
lother push infant towards back	15	21	—	—	—	—
lother puts over her shoulder on to back	16	21	—	—	6–7	—
nfant rides dorsal 10 yards or more	17	21	—	—	—	5
st quadrupedal steps	17	24	—	25	16–18	—
limbs small branch	20	24	—	23	16–18	—

[1] This was the first time the infant was seen completely on his own, in contact with neither his mother nor his sibling (see Sect. II–6(c) for explanation).

TABLE III

Number of Hours of Observation per Week

AGE IN WEEKS	FLINT[1]	GOBLIN	JANE[2]	CINDY[3]
1		4.50	3.50	0.30
2		3.35	0.30	0.50
3		2.00	1.00	0.55
4		0.10	1.10	2.50
5		9.10	4.30	3.45
6		1.00	4.20	2.30
7	11.15	—	5.50	2.50
8	19.15	—	5.40	2.55
9	16.35	1.30	3.50	2.35
10	17.30	3.35	1.15	
11	20.25	11.20		
12	17.45	6.25		
13	9.50	8.50		
14	28.30	5.20		
15	18.15	2.55		
16	21.30	3.40		
17	26.25	0.35		
18	24.40	1.10		
19	14.20	7.45		
20	11.50	3.00		
21	11.20	1.40		
22	15.20	3.35		
23	19.55	11.20		
24	8.10	5.55		
	312.50	99.20	31.55	19.40

[1] I did not arrive at the Reserve until Flint was seven weeks old.
[2] The infant died during her eleventh week.
[3] I left the Reserve when Cindy was nine weeks old.

TABLE IV

1. *The Infant*

The behavioural criteria used to define an infant include: suckling, being transported by the mother either continuously or occasionally, sleeping in the same nest with the mother at night.

(a) *0–6 months* (Infant 1)[1]

1st week	Supported almost continually by the mother during first few days. Little voluntary movement apart from head movements when searching for nipple and some attempt to pull upwards towards the breast. One infant was able to pull its head towards mother's body when this lolled backwards and away from her. Eyes do not appear to focus. White tail tuft represented by some hairs of a few millimetres in length.
2–6 weeks	Able to grip fairly well by end of second week. Is more active, and makes random movements, stretching out to side, grabbing at mother or own head with hands. Pulling and pushing movements stronger. During third week seemed able to focus eyes. Looked round at sudden sounds but not usually in right direction. Mother grooms infant more frequently than during first week but seldom for more than a few seconds at a time. She shows 'affectionate' behaviour, gazing down at her infant (see Section III).
7–10 weeks	Movements less random. Begin to make apparently purposeful reaching movements towards objects such as mother's face or leaves. First co-ordinated scratching movements observed between seventh and ninth weeks.
11–15 weeks	Pulling and pushing movements much more vigorous and infants able to push themselves into bipedal position, feet on ground, holding mothers' hairs with hands. Infants begin to crawl about on mothers, grasping hair with hands and feet. Try to push themselves away, or to pull away by grasping grasses, twigs, etc. Mother restricts these efforts. Infant Flint repeatedly

[1] The period of infancy can be roughly divided into three categories, each of which is characterized by certain behavioural characteristics. These are useful guides when the exact age of an infant is not known (Goodall, 1965).

pulled from his mother by his sibling during this period. Mothers play frequently with infants. Infants may reach towards other chimpanzees when they approach. Tooth eruption commences during this period. White tail tuft now conspicuous. Both male infants first seen with penile erections during week 14.

16–24 weeks   Most infants break physical contact with mother for first time. Begin to lick or suck at solid foods. Mothers initiate dorsal riding but this is not normal manner of transport during this period. Infants start to take their first quadrupedal steps and to climb small branches.

(b) *6 months–2 years*   (Infant 2)

6 months–1 year   Locomotor patterns rapidly improve. Invariably rides in dorsal position from 7th or 8th month. Small amount of solids eaten. During this stage one infant showed first attempts at nest making. Mother frequently plays with and grooms infant. Social interactions with other individuals frequent – infant may be patted or groomed by other adults, and played with, groomed and carried by older infants, juveniles and adolescents. Infant frequently approaches individuals joining its group to 'greet' them. White tail tuft fully developed.

1–2 years   Continues to ride about on mother's back when she travels for long distances, but may walk for short distances. Locomotor abilities almost fully developed. Towards end of second year proportion of solids eaten increases. Mother continues to play with infant. Other individuals continue to show tolerance. During this stage most, if not all, adult ritualized gestures and postures appear.

(c) *2–3/3½ years*   (Infant 3)

During early months of this period continues to ride frequently on mother's back, but gradually this occurs less often. Proportion of solids eaten increases greatly and, during most of this stage, diet differs little from that of adult. Older individuals still generally tolerant but infant receives a number of gentle rebuffs and behaves with increasing caution. Still protected by its mother.

2. *The juvenile*

3/3½–6/7 years

No longer dependent on mother for food or transport. Makes own nest at night. The mother still protects her offspring on occasions, and the juvenile continues to move around with her for most of the time. Rebuffs from older individuals become increasingly severe. White tail tuft gradually becomes less conspicuous.

3. *The adolescent*

6/7–11/13 years

In the male chimpanzee the period of adolescence is fairly well defined and can be said to commence with puberty (about seven years) and continue until the individual is socially mature and becomes integrated into the adult male hierarchy. In the female adolescence is not easy to define, but probably commences between six and seven years. During the first year or so of adolescence the female may show a very slight swelling of the ano-genital region, but this in no way approximates to a normal sexual swelling and does not arouse attention from the males. At a later period of adolescence the female shows the normal external features of the oestrus cycle, although small females, judged to be adolescent, were not seen with swellings as large as many seen in adult individuals.

(a) *Adolescent male.* Moves about with mother less and less frequently. Gradually becomes dominant over young mature females and then older females. Is cautious during interactions with mature males, and may be severely threatened or attacked by them on occasions. Shows, for the first time, complete branch-waving and dragging displays.

(b) *Adolescent female.* Often seen with mother but seldom when either have sexual swellings. She is normally extremely timid in social interactions with superiors.

4. *The adult*

The age at which a chimpanzee becomes socially mature is not certain. As it grows older its actions (except when socially excited, frightened, etc.) become increasingly deliberate and, when moving in trees, cautious. Even the oldest of the known individuals still play on occasions, particularly with infants or juveniles. The age to which chimpanzees may live in the wild is not known, but may be in the region of 30–50 years.

TABLE V

Name	Age	Infant suckles from one breast only during feed			Infant suckles from both breasts during feed		
		LEFT	RIGHT	TOTAL OBS	LEFT 1ST	RIGHT 1ST	TOTAL OBS
Flint	0–6 months	55%	—	195	69%	—	16
	6–12 months	45%	—	36	57%	—	38
Goblin	0–9 months	—	76%	59	—	52%	25
Jane	0–3 months	—	62%	24	—	—	—
Merlin	1–2 years	54%	—	11	45%	—	11
Gilka	1½–3 years	83%	—	12	83%	—	66

TABLE VI

AGE OF OFFSPRING	% OF TOTAL NO. OF GROOMING BOUTS WHEN BEHAVIOUR WAS MUTUAL
0–1 year	—
2 and 3 years	9·5
4–6 years	23
7 and 8 years	32
9–13 years	50

REFERENCES

BUDD, A. and L. G. SMITH (1943). 'On the birth and upbringing of the female chimpanzee "Jacqueline" ', *Proc. Zool. Soc. Lond.* **113**, p. 1.

CROSS, H. A. and H. F. HARLOW (1963). 'Observations of infant monkeys by female monkeys', *Percept. mot. Skills*, **16**, p. 11.

DEVORE, I. (1963). 'Mother-infant relations in free-ranging baboons', *Maternal behavior in Mammals*, ed. H. L. Rheingold, New York and London, John Wiley and Sons.

GOODALL, J. (now VAN LAWICK) (1962). 'Nest-building behavior in the free-ranging chimpanzee', *Ann. N.Y. Acad. Sci.* **102**, p. 219.

—— (1963). 'Feeding behaviour of wild chimpanzees: A preliminary report', *Symp. Zool. Soc. Lond.* **10**, p. 39.

—— (1965). 'Chimpanzees of the Gombe Stream Reserve', *Primate behavior*, ed. I. DeVore, New York, Holt, Rinehart and Winston.

HARLOW, H. F., M. K. HARLOW and E. W. HANSEN (1963). 'The maternal affectional system of rhesus monkeys', *Maternal behavior in mammals*, ed. H. L. Rheingold, New York and London, John Wiley and Sons.

HINDE, R. A., T. E. ROWELL and Y. SPENCER-BOOTH (1964). 'Behaviour of socially living rhesus monkeys in their first six months', *Proc. Zool. Soc. Lond.* **143**, p. 4.

IMANISHI, K. (1960). 'Social organization of subhuman primates in their natural habitat', *Current Anthropology*, **1**, p. 393.

JAY, P. C. (1963). 'Mother-infant relations in langurs', *Maternal behavior in mammals*, ed. H. L. Rheingold, New York and London, John Wiley and Sons.

—— (1965). 'The common langur of North India', *Primate behavior*, ed. I. DeVore, New York, Holt, Rinehart and Winston.

JENSEN, K. (1932). 'Differential reactions to taste and temperature stimuli in newborn infants', *Genet. Psychol. Monogr.* **12**, p. 361, cited by W. A. Mason in 'The social development of monkeys and apes', *Primate behavior*, ed. I. DeVore, New York, Holt, Rinehart and Winston.

KORTLANDT, A. (1962). 'Chimpanzees in the wild', *Scient. Am.* **206** (5), p. 128.

MASON, W. A. and G. BERKSON (1962). 'Conditions influencing vocal responsiveness of infant chimpanzees', *Science*, **137**, p. 127.

MASON, W. A., J. H. HOLLIS and L. G. SHARPE (1962). 'Differential responses of chimpanzees to social stimulation', *Journal of Comp. and Physiol. Psych.* **55**, 6, p. 1105.

REYNOLDS, V. and F. REYNOLDS (1965). 'Chimpanzees of the Budongo Forest', *Primate behavior*, ed. I. DeVore, New York, Holt, Rinehart and Winston.

SADE, D. S. (1965). 'Some aspects of parent-offspring and sibling relations in a group of rhesus monkeys, with a discussion of grooming', *Am. J. Phys. Anthrop.* **23**, p. 1.

SCHALLER, G. (1963). *The mountain gorilla: ecology and behavior*, Chicago, University of Chicago Press.

VAN DEN BERGHE (1959). 'Naissance d'un gorille de montagne a la station de zoologie experimentale de Tshibati', *Folia Scientifica Africae Centralis* **4**, 81 p., cited by G. Schaller in *The mountain gorilla: ecology and behavior*, Chicago, University of Chicago Press.

YAMADA, M. (1963). 'A study of blood-relationship in the natural society of the Japanese macaque', *Primates* **4**, p. 43.

YERKES, R. M. (1943). *Chimpanzees, a laboratory colony*, New Haven, Conn., Yale University Press.

Chapter Ten

An Ethological Study of Some Aspects of Social Behaviour of Children in Nursery School

N. G. BLURTON JONES

INTRODUCTION

ONE of the research programmes in the Department of Growth and Development at the Institute of Child Health, University of London, is an attempt to apply ethological methods of observation and interpretation to the behaviour of normal children. Our aim is to extend this into a longitudinal study of development, after doing the groundwork on suitable age cross-sections.

Between November 1963 and May 1964 I carried out a pilot study on four- to five-year-old children in a London nursery school. Previously I had observed three- to five-year-old children in a variety of nursery schools in two other cities.

It became obvious that one can study human behaviour in just the same way as Tinbergen (1953 and 1959), and Moynihan (1955) and others have studied gulls, and van Hooff (1962), Andrew (1963, who also gives comparative data from a child), and others have studied non-human primates.

This paper describes the crude preliminary observations which gave rise to this conclusion. It is a provisional, descriptive account, much like a report on the first season's ethological field work on any new species. It is subject to similar criticisms. Fortunately, we will shortly have the facilities needed to take these criticisms into account in our future studies.

In places I speculate on comparison with other primates, and on possible implications for human psychology and education. I do this mainly to show that this kind of study can engender such speculations and can suggest ways of verifying them, not because I believe the speculations are necessarily correct.

METHODOLOGY

My approach is best described by reference to ethological studies of bird behaviour, and by comparison with Darwin's (1872) study of facial expressions and vocalizations in 'The Expression of the Emotions in Man and Animals'.

In ethology the methods and concepts used in studying the motivation of behaviour have gradually been made more rigorous. When an early ethologist might have said 'this piece of behaviour is aggressive', he now says that (a) 'it is causally related to attack, or shares common causal factors with attack', and (b) 'its effect is to make other animals move away'. To the first statement must be added some definition of attack, usually certain fairly specific motor patterns most commonly directed at a conspecific. The evidence for causal relationship between two movements is usually circumstantial (see Tinbergen, 1959), but the relationship is described in a way that makes it susceptible to experimental testing. In some cases, e.g. Tugendhat (1960) and Blurton Jones (1960, 1964), such statements about causation have been checked by experimental manipulation of causal factors.

The circumstantial evidence about causation was described by Tinbergen (1959) as being of three kinds: (1) comparison of the form of movement; sometimes a movement of unknown causation is seen to be made up of components of other movements, e.g. the Upright display of the Herring Gull can be analysed into a mixture of components of actual attack and fleeing, suggesting that causal factors for both these activities are operating. (2) Analysis of temporal association of movements; causally related activities should tend to occur together, at roughly the same time; since when the causal factors for one movement are present they are also present for any other movements they influence (although other factors may favour the occurrence of one rather than the other). (3) Examination of situations in which the behaviour occurs; which sometimes gives useful insight into their causation.

Because of these advances in the ethological approach, my work differs from Darwin's mainly in the study of the 'meaning' of the various facial expressions and other movements. In description of facial expression particularly I have added nothing to Darwin's descriptions, and any improvement will have to wait until our film facilities are set up. But where, for example, Darwin describes puckered eyebrows as signifying grief, I have to ask what 'grief' means in terms of observables. I have on one side to find out whether puckered eyebrows elicit certain responses in others, perhaps the same responses as weeping elicits but different from responses elicited by a stare and a shout, or a smile. On the other side, and this is the side I have paid most attention to, I have to find out what factors influence the occurrence of this expression and whether these same factors also influence other responses, either those which we label by popular tradition as 'grief', or others which perhaps we label differently.

In this study I have attempted to apply the circumstantial or interpretative methods described above to the social behaviour of children. This depends on first describing motor patterns which are relatively constant in form, which look the same each time one sees them and occur again and again in one individual (one could make ethograms of individuals) or in many individuals. As a layman one may feel that human behaviour should be infinitely varied and plastic, in which case this approach would be doomed. But it evidently is not, at least for children up to the age of five years, and no doubt long after this.

THE OBSERVATION SITUATION

My observations have all been made on nursery school children, aged three to five, in a variety of schools but chiefly at one school in London during 1963–4. The aspects of behaviour I deal with in this paper vary little from school to school, despite differences in social background. Many aspects of behaviour do vary but I saw all the fixed action patterns described here in all the schools that I visited.

I used no special observation techniques, but simply visited the school repeatedly, and sat on a chair in a corner with a notebook. The children gradually reacted less and less to me, as I tried to be as unresponsive as possible without antagonizing them. Of course I can never say that my presence does not affect their behaviour but their initial responses to me do certainly disappear. They always have at least one teacher there and visitors and students are common. Nursery school teachers seem to have a policy of non-interference broken only by giving limited guidance and encouragement when asked, or direly needed. This makes nursery school an ideal situation for studying the unrestrained behaviour of three to five year olds.

In nursery schools there are no formal lessons, though in many there are organized music sessions, stories, milk distribution, at certain times. There are abundant facilities for painting, modelling, climbing, sliding, etc., and the children spend most of their time using these.

Much of this time is spent in investigative-manipulative-creative occupations with the materials provided. They paint, draw, mould clay or plasticine, nail together bits of wood. This is the most 'human' side of their behaviour and, because of the plasticity of the motor patterns used and the emphasis on the objects made rather than on the behaviour performed, it is the hardest part of the child's life for the ethologist to investigate.

The ethologist has more scope in studying the interactions of the children with each other and with adults. Indeed, I feel he really has the only proper (though sometimes impracticable) approach to this side of the child's behaviour. There are many fixed action patterns in the social behaviour and I describe these below and comment on their inter-relationships and possible causal organizations. The headings of the sections which follow should not be taken to be anything more than names and divisions for convenience in writing.

AGGREGATIVE BEHAVIOUR AND SOCIAL ORGANIZATION

Nursery school children are often based on a specific room in the school according to age but are able to go into other rooms or outdoors and mingle with children of other ages in the school. Outside they spend most time with children of their own age, exceptions centring around siblings in other classes.

Within their own class there are differences in the amount of time individuals spend with each other. The class studied in 1963–4 included two small groups of two to four each, but these had remarkably little interaction with each other as groups, and certainly no defended territory. All the members of a class know each other's names and meet and interact at milk-time, stories, music. Most of the children had at least one commonest associate but three were very solitary, though two of these often talked to the teacher or to me. Boys seemed more social than girls but neither showed any preference as to which sex they associated with. There was no tendency to attack members of other groups except when they took objects in use by the first group and then usually only one group member was involved.

No doubt one could stretch the concept of a dominance hierarchy to fit this organization but I suspect that there is a better case for applying at least the orthodox concept of dominance only to groups of older children. 'Dominance' says nothing useful or instructive about the social organization of the class of three- to five-year-olds I observed or of the groups within it. Certainly some individuals regularly won fights but these were not all also leaders or peace keepers or given priority access to objects. However, dominance or submission might say something useful about the behaviour of some timorous individuals who always give up objects, lose fights if they get in them, often leave a game in fright, and are seldom followed. These 'submissive' individuals show a syndrome which looks as if their behaviour is organized on the lines of 'win or lose at everything'. But this does not apply to winners.

Members of groups talk to each other a lot and use words to ensure group cohesion ('come on'). When an individual joins his friends after being away from them for a short time he greets them saying 'Hello' and smiling and stops to see what they are doing. Friends also tend to go and sit together at milk-time and for stories. They play together in chasing games, at manipulative constructive play, including painting, at imaginative games like firemen, shopkeepers, office-workers, etc.

In later sections I describe some of the behaviour seen in the interactions between the children, and the way the different pieces of behaviour cluster into groups (e.g. Table 1), as a first step in the causal analysis of the social behaviour. There are also noticeable constancies in the occurrence of behaviour in response to the child's mother when she fetches it at the end of the day, and in the responses to strangers visiting the school.

RESPONSES TO ADULTS: THE TEACHER

Most of the children spend little time directly interacting with the teacher although they respond rapidly to any request or suggestion which she makes. Many go to her or are led to her (by another

Table 1. The number of times some of the behaviour patterns occur with each other in a small series of time samples. There were ten observation periods of five minutes on each of ten children, a hundred in all. Values expected if all the patterns were associated by chance alone were calculated on a separate 2 × 2 table for each pair of movements. Although many of the expected values are very small I follow Cochran's (1954) recommendation that χ^2 can be used when $N > 40$, with the Yates correction for continuity.

	No. of observation periods in which recorded:	Paint	Work with paint, sand, clay, wood, paper	Laugh	Run	Jumps	Open beat	Wrestle	Beat	Low frown	Fixate	Red face	Pucker brows
Paint	5												
Work with paint, sand, clay, wood, paper	24	5											
Laugh	38	2 / 1·9	9 / 9·1										
Run	50	3 / 2·5	9 / 12·0	31 / 19·0 ***									
Jumps	30	0 / 1·5	3 / 7·2	19 / 11·4 ***	27 / 15·0 ***								
Open beat	13	0 / 0·6	3·1	9 / 4·95 *	11 / 6·5 **	8 / 3·9 **							
Wrestle	14	1 / 0·7	4 / 3·4	12 / 5·3 ***	14 / 7·0 ***	8 / 4·2 *	5 / 1·8 **						
Beat	7	0 / 0·3	1 / 1·7	3 / 2·7	5 / 3·5	3 / 2·1	0 / 0·9	3 / 1·0					
Low frown	11	0 / 0·5	1 / 2·6	5 / 4·2	7 / 5·5	6 / 3·3	2 / 1·4	2 / 1·5	4 / 0·8 ***				
Fixate	9	0 / 0·4	2 / 2·2	2 / 3·4	6 / 4·5	4 / 2·7	1 / 1·2	2 / 1·2	3 / 0·6 **	7 / 1·0 ***			
Red face	15	0 / 0·7	6 / 3·6	6 / 5·7	5 / 7·5	3 / 4·5	2 / 1·9	1 / 2·1	2 / 1·5	2 / 1·6	1 / 1·3		
Pucker brows	7	1 / 0·3	2 / 1·7	2 / 2·6	2 / 3·5	1 / 2·1	0 / 0·9	2 / 1·0	2 / 0·5	0 / 0·8	0 / 0·6	5 / 1·0 ***	
Cry	2	0 / 0·1	1 / 0·5	0 / 0·8	0 / 0·1	0 / 0·6	0 / 0·3	0 / 0·3	1 / 0·1	0 / 0·2	0 / 0·2	1 / 0·3	2 / 0·1 ***

$N = 100$ Observation periods   * $P < ·050$ ** $P < ·025$ *** $P < ·001$

child which walks alongside almost always with its open hand on the back of the other, at the bottom of the shoulder blades; a conspicuous gesture only seen in this situation) after any mishap and they are all successfully comforted by her. If hurt or frightened they may cling to the teacher and sometimes she picks them up. New children spend much time by her, making sorties out to the other children and retreating to the teacher. Many children show her their paintings or clay or wood creations. One or two spent a lot of time talking to the teacher or just standing or sitting near her looking at books.

RESPONSES TO ADULTS: STRANGERS

The commonest response to me on my first visit, and to people making rare visits to the nursery school, is initially to stop and stare with no marked expression at the stranger. I find that if I look back at a staring child or make any approach to it, it is likely to look away or go away. But if I make no response the child stops staring and often then brings some object to me and holds it out towards me at about the level of its waist. Sometimes I seem to be expected to take the object, sometimes just to look at it. Subsequently the child may ask me to help it or join its constructive play, or rarely may try to initiate rough-and-tumble or chasing play, or it may simply return to what it was previously doing.

RESPONSES TO ADULTS: PARENTS

The responses to the child's mother when she comes to fetch it at the end of the day seems to fall into two patterns. One is when the child looks up, smiles, walks, or more often runs, to the parent and embraces her or takes her hand. Touching the mother and smiling usually go together $\chi^2 = 5\cdot0$ $P < 5$ per cent for fifty-three meetings of nearly fifty-three different children) and tend to go with a fast approach. In the other pattern the child usually approaches by walking, does not smile or touch its mother, but gives her or shows her an object ($\chi^2 < 5\cdot6$ $P = 2\cdot5$ per cent for touching rarely with showing), usually a painting or something it has made during the day, or points to something. Often a child appears quite to ignore its parent and continues whatever it was doing until she insists that it comes home.

The separation into two clusters of responses seem to be statistically valid but is not absolute and I do not know how consistent it is within an individual. Whatever the reason for the child not touching its mother it is interesting that, at the same time, the likelihood of the child smiling is also reduced. And, in contrast with other primates, there is no evidence of smiling having any causal relation to fleeing, as in many species. Smiling goes with the most overtly aggregative behaviour in children of this age and is less common when aggregative behaviour is mixed with other behaviour.

One can produce a slender argument that one reason why children show things rather than touching and smiling is that they are partly motivated to flee from the parent. The 'not touching but showing' response is a common response to strangers (but also to the teacher, though not in a greeting situation). Nursery school children do not greet strangers by touching them and most only smile when they know the stranger quite well (I understand that institution and day nursery children do touch and smile at strangers at the first meeting). Pointing to objects leads to going away to get them. I have only twice seen a child step back after approaching its parents but both times the child then pointed away to some object.

It is hard to say whether in addition or alternatively the child's 'pride in its work' causes it to show things to its mother. No doubt a child learns that its mother is pleased by being shown or given things, but is there really any *a priori* reason why a successful achievement should be demonstrated to others? Why should the child need to please its mother, and where does this 'pride in its work' come from? Why should showing her things tend to exclude touching her and smiling at her?

AGONISTIC BEHAVIOUR

Among three to five-year-old children in nursery school, fights occur over property and little else. One child pulls at another's toy and the owner pulls back, then one of them kicks, pushes, bites or pulls the hair of the other. The beating movement is the commonest and is rather consistent in form. It is an overarm blow with the palm side of the lightly clenched fist. The arm is sharply bent at the elbow and raised to a vertical position then brought down with great force on the opponent, hitting any part of him that gets in the way. Biting seems to be more commonly done by girls than by boys. These attacks are often preceded and accompanied by fixating the opponent and by what looks like a frown with lowering of the eyebrows and rather little vertical furrowing of the brow ('low frown') and no conspicuous modification of the mouth expression. Often the child shouts 'no', or 'let go', with a characteristic tone, low pitch and hard explosive quality being evident. Andrew (1963) published a spectrograph of this sound. Usually in these property fights little locomotion is involved, they end quickly with one child gaining possession of the object and one or the other walking a short way off. The beating movement, and the preparatory position with bent arm held high and clenched fist, is confined to property fights; it occurs with low frowns and fixation and in situations where the opponents end up separated.

When there are signs, such as stepping back from the opponent, that something is inhibiting the attack, the mouth expression changes and the child shows a 'fierce' expression with lower teeth bared and the corner of the mouth drawn down. I haven't seen this often but it has an interesting place in 'folklore'. It is the expression illustrated by Netter (1958) for rage in man (whatever rage may be), but Netter adds an eye component, with brows up and eyelids wide open, which is characteristic of a child fleeing from an unusually violent opponent!

Occasionally a robbed child would scream long and loud, a low-ish-pitched scream or roar, and, red in the face, beat at, with closed fist, but not hit, and not approach its opponent. This performance sometimes looks rather like what has been called a temper tantrum and the child gives the appearance (because of his beating movements and orientation towards the opponent) of being highly stimulated to attack but for some reason not actually doing so and in addition showing features (scream and red face) of a defeated child.

The child who gets hit, or whose property is grabbed often gives a brief high-pitched scream. This is rarely repeated, subsequent behaviour being either a verbal call for help, or retaliation, or let-ting go often followed by weeping. Weeping is associated with, and usually preceded by puckering the brows (see also Darwin, 1872) and reddening of the face. Often the child then stays immobile for a minute or more, frequently sitting down, and may suck its thumb, hold on to a lock of its own hair, and even rock back and forward. The last two are not common but occur mostly after crying. Sucking often also occurs when the children are sitting listening to a story, and so sometimes do hair-holding and rocking. Thumb-sucking is also common as a child goes to sleep, so that perhaps it is most closely associated with 'inactivity' or low arousal rather than with 'distress', or is going to sleep a lonely, distressing situation? (Do children who do not suck never complain about going to bed, and go to sleep quicker than those who do, or does a child suck more on those occasions when he complains about going to bed?) Weeping and the associated patterns seem to be most commonly elicited by social factors, although of course a child does often cry after it falls over. But a child pushed over by another in rough-and-tumble play rarely cries, whereas one pushed over in a fight frequently does so. Also it seems that crying, puckered brow, red face, inactivity, differentiate sharply from actual escape or fleeing behaviour. If a child runs away from another it doesn't cry or go red or pucker its brows, it screams in briefly interrupted bursts and has raises eyebrows and wide-open eyelids. The raised eyebrows can precede actual fleeing, as when a child suddenly meets another who commonly attacks and pursues others (there was one such rare individual in my study group).

One other conspicuous but rare response of a child in fights is to raise the arm over or in front of the head, with forearm horizontal providing a protection against the opponent's beating. Sometimes the eyebrows are raised as well. This seems to be an alternative to fleeing, happening when the child is cornered, and once I saw it made by a child being beaten by one it spent a lot of time with and from which it would be unlikely to go away.

The more precise description of attacking, and threatening and tantrums and weeping, and fleeing should clarify the relationship between them and between ill-defined categories such as frustration, distress, fear and aggression. I would speculate that a temper tantrum is the behaviour shown when attack is stimulated but the opponent is overwhelmingly powerful and/or a friend (e.g. a parent; in my observations it has been a child opponent, but a playmate and one who usually wins his fights). It must somehow relate to weeping and the inactivity of the robbed or beaten child, but this again is quite separate from fleeing and the components accompanying fleeing.

To judge from the child behaviour, popular ideas of human threat gestures seem to be quite erroneous. Mostly they refer to more intimidated displays. The cold direct stare is the real danger sign (and the one people react to most readily). It is interesting that punching, wrestling, use of weapons or of more sophisticated methods of attack are not in the child's battle repertoire at this age. Wrestling occurs, not with any of the behaviours described above but with alternating chases and laughing and falling over, between children who stay together for a long time and neither of whom is holding a toy. It falls into a group almost identical with 'rough-and-tumble play' in the Harlow laboratory's monkey studies.

ROUGH AND TUMBLE PLAY

Harlow's and his colleagues' (Harlow and Harlow, 1962; Hansen, 1962; Rosenblum, 1961) recent experiments have emphasized the importance of playmates in the development of social and sexual behaviour in rhesus monkeys, and stressed the part occupied by the chasing and fighting play between individuals. It is therefore interesting, though obvious, that almost identical patterns of play occur, and are clearly definable, in human children. It is important to define this kind of play firstly because it surely must not be confused with all the other things which we call 'play' just because they are done by children, and secondly because in investigating its distinctiveness from other behaviour we may get some idea how it differs from other behaviour in its effects and function and in its casual organization.

The human 'rough-and-tumble play', as I shall call it (Harlow, *et al.* use this term for one out of three or more kinds of social play) consists of seven movement patterns which tend to occur at the same time as each other and not to occur with other movements (see Table 1). These are running, chasing and fleeing; wrestling; jumping up and down with both feet together ('jumps'); beating at each other with an open hand without actually hitting ('open beat'); beating at each other with an object but not hitting; laughing. In addition, falling seems to be a regular part of this behaviour, and if there is anything soft to land on children spend much time throwing themselves and each other on to it.

There seems to be a common facial expression in this play besides the smile-like expression involved in laughing. This is seen

357

when a child is about to be chased by another and stands slightly crouched, side-on to the chaser and looking at it with this 'mischievous' expression, an open-mouthed smile with the teeth covered, which morphologically resembles the 'play-face' of *Macaca* and *Pan* (van Hooff, 1962).

Rough-and-tumble play subsequently seems to develop rather sharply into formalized games like 'tag' and 'cowboys and indians'. There are the same motor patterns but rules and verbal explanations have been added.

Most of the rough-and-tumble play consists of behaviour which on the surface looks very hostile: violent pursuit, assault and fast evasive retreat. However, the rôles of the participants rapidly alternate and the behaviour does not lead to spacing out or capture of objects; the participants stay together even after the chasing ends. Also the movements involved are quite different from those involved in fights over property. The facial expressions and vocalizations, and the motor patterns involved separate out into two quite different clusters. Thus beating with clenched fist occurs with fixating, frowning, shouting, and not with laughing and jumping. Wrestling and open-handed beats occur with jumping and laughing and not with frown, fixate and closed beat. So although rough and tumble looks like hostile behaviour it is quite separate from behaviour which I call hostile because of its effects, i.e. involving property ownership and separation of individuals.

Not only does rough and tumble include patterns like wrestling or beating sticks at each other (or 'shooting' at each other) which adults might think and often do think are aggressive; in some circumstances children react to them as dangerous and conceivably aggressive. Children new to school do not join in these games straight away. Those I have seen have watched the games and followed them around but always ran to the teacher or some refuge (a wall or corner or seat) if the players happened to move towards them. If they got caught up in a game they were more likely to cry than established children.

Some children seem permanently unable to join in and take the rough and tumble 'in fun'. What are these individuals like as adults? They are not all 'only children', but is there a critical period for developing the ability to rough and tumble, or are these children unusual in some way other than deprivation of the chance to play at the right age? Since the rough-and-tumble motor patterns and expressions appear quite as early as eighteen months old and maybe even younger, the nursery school starting age of three years could be too late for those who had no playmates of the right age in or near home.

Sometimes the fleeing involved seems to 'turn real'. A child fleeing for a long time without chasing back, going faster and faster, may raise its eyebrows and stop smiling and its laugh changes and becomes a more continuous vocalization, a tremulous scream. I have heard the same noise in response to an insect running about on the ground, the child stamping rapidly with alternate feet and looking at it and running away, returning, running away again etc.

In this age group the attack-like behaviour involved in rough and tumble does not seem to turn into real attacks, but in older rhesus, to judge from Hansen (1962), (and in rats too) it gets more difficult to distinguish play-fighting from real fighting; there is for instance more actual biting. Possibly human children follow the same development; I have not observed older children enough to say.

Rough-and-tumble play relates to real hostile behaviour in that:

1. It looks like it to adults, and quite often one sees adults responding as if a play pattern (e.g. open-handed beat plus play-face) were really hostile.

2. Some children respond as if it were hostile, e.g. they flee from play attack movements.

3. Sometimes play fleeing becomes real fleeing.

4. Some motor patterns are similar, e.g. orientation of loco-motion (though there is little locomotion in property fights), and a possible similarity in form between laughing and screaming with intermediates between them, and the arm position and movement of beating, in both rough and tumble and hostile behaviour.

Most of the time, despite these similarities, the players neither respond as if their playmates were hostile nor show any indication. of their own motivation being hostile (i.e. of the causes of rough and tumble being at all related to the causes of fighting). Short-term effects of this play are eventual exhaustion, continuing to stay with the playmates, seeking them out another time to play with. If anything, its short-term effect is to gain friends rather than to lose them.

Atypically some children initiate rough-and-tumble play with newly arrived strange adults. Some do this by running up and beating at the adult, or jumping and making beating movements with an object. These are the children who get called 'cheeky'. They evoke hostility in adults and can only be persuaded to stop this behaviour, and subsequent wrestling, by the adult genuinely giving way to his hostile feelings. Making a fixating-threat face with no trace of a smile is the least one can get away with. Other children use a less objectionable approach, inviting the adult to chase them both verbally and by adopting the posture and play-face that elicits chasing and precedes their play fleeing with other children. These two performances, in a greeting situation, are not typical of the situations in which rough-and-tumble play occurs (mostly with friends and not decreasing with familiarity), but they are suggestive of some remote connection between rough and tumble and more aggregative and hostile behaviour.

No doubt some would argue that the laughing and jumping up and down which are characteristic of rough-and-tumble play are just signs that the children are 'excited' or 'enjoying themselves'. It is hard to tell what this would mean in terms of observables. Perhaps it would mean that laughing and jumping occur in any situation where the child was in a state of high general arousal (and this still does not mean anything in terms of observable behaviour) and was in sight of positive reinforcement. In any case it would suggest a rather wide occurrence of the behaviour, in many situations and accompanying a variety of other behaviours. But the evidence at present is that laughing has mainly to do with chasing and wrestling and is not linked to a wider range of active behaviour. A child running to greet its mother is presumably aroused, but I have never seen them laugh or jump up and down. A child painting is surely enjoying it (they keep doing this!) and must have an aroused EEG but the child does not laugh and jump. I think that laughing and jumping are both very specific signals (in causation and function) indicating the friendly meaning of the hostile-looking behaviour involved in rough-and-tumble play.

DISCUSSION: EVALUATION

I have shown above how one can treat the behaviour of three to five-year-old human beings exactly as ethologists have treated the behaviour of many different animals in field studies. The next question is whether this is a good way to investigate the behaviour of any animal. Such ethological soul-searching mainly belongs elsewhere than in this paper, except to say that I think one cannot do an ethological study of human behaviour without becoming even more concerned about methodology. Secondly, one must ask whether this is a useful way to study *human* behaviour, and what the differences are between ethology and various branches of psychology.

Correlations and other statistical investigations of fixed action patterns may give rise to a more objective and accurate picture of the organization of behaviour than the arbitrary categories of motivation in use in everyday life and in the more armchair branches of psychology. But perhaps the ethologist is merely pushing the arbitrariness one step further down. Thus just as factor analysis studies of motivation or personality ultimately depend on the selection of questionnaire items, so an ethological study relies on the identification of fixed action patterns. In the present paper some of my items of behaviour are not even all at the same level of description, e.g. 'beating' and 'painting'. Obviously one has ideally to examine the fixed action patterns carefully to show just how constant they are, and just how one differs from another, right down to description of the particular combinations and sequences of components or even the muscle contractions and skeletal movements involved. This, it seems to me, is something that any further ethological study of human behaviour should go into quite deeply.

In addition there is the probability that this approach stems from much too simple a view of the mechanisms of behaviour. This shows up most when one tries to think about the creative activities of the children and about some of their more complex, verbally organized games. However, I believe that the approach illustrated in this paper (at its most primitive and preliminary level) can provide a lot of new and useful information on social and emotional behaviour, even quite complex aspects of it, and may eventually go further than one can at present see in analysing verbal and intellectual behaviour. I also feel that to criticize direct observation as a method, on the grounds that it is incapable of elucidating some of the more complex (and obscure) aspects of motivation is wrong in two ways. One is that it may reflect a simple lack of talent and experience as an observer; the other is that the concepts about motivation may themselves be at fault.

Many of these obscure and highly verbalized ideas about motivation or organization of behaviour are attempts to describe very complex and barely understood relationships between environment and behaviour. One should aim to make these relationships explicit and phrase them in terms of observables, thereby making them available for scientific study. Sometimes the ideas will turn out to be meaningless but sometimes not. But in addition, studies based on observation of normal uncontrolled behaviour show up previously neglected important aspects of behaviour. The extremely obvious rough-and-tumble play has always been regarded as unimportant by psychologists. Although its importance was shown up by Harlow's experimental studies, it is such conspicuous behaviour and so interesting in its relationship to hostile behaviour that one could not have long disregarded it and its effects in an ethological study of children.

A more specific common criticism of the ethological approach is one that is made by laymen of psychology, and, surprisingly, by psychologists of ethology. This is that the wide range of individual differences makes such a study nearly impossible. I do not think it does. There seems to be less individual difference in the measures I used than in many, because I concentrate on movements that I find to occur in most individuals. While the frequency of occurrence of the movements may vary greatly the association between movements usually persists, i.e. if one scores individuals separately the correlations I report here would be found in most individuals. For example, although some may rarely chase others they also more rarely laugh and jump up and down (Fig. 1). The behaviour of one child in my 1963-4 study group cut across some of the correlations, and I would think this provided a useful characterization of his behaviour, which was conspicuously abnormal to me, the teachers, and the children (each in our different terms!). Even at this age there are individual mannerisms and even occasional complex individual stereotypes. In such a case one could still usefully analyse the place of that individual fixed action-pattern in the behaviour of that individual. I think its occurrence would be very predictable.

In using the term 'action pattern' I have intended to stress the relative constancy of some observable, complex motor activities, and not necessarily to imply species specificity or universality, nor imply anything about the ontogeny of the behaviour. Development is a topic to investigate rather than to classify out of existence with terms like 'innate' and 'learned'. But there is an interesting possibility concerning rough-and-tumble play. There appears (as in Fig. 1) to be a sex difference in the amount of rough-and-tumble

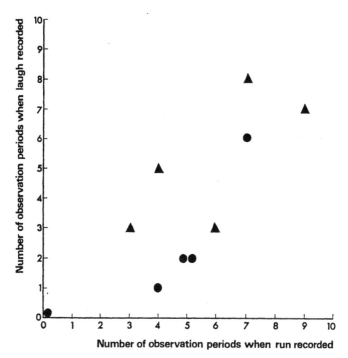

Fig. 1. Graph showing for each individual the number of observation periods in which laughing occurred against the number in which running occurred. It appears that individuals who run little also laugh little, and that girls do less of both these rough and tumble play patterns than do boys. Triangles represent boys, solid circles represent girls.

play by children, males playing more often than females. The same difference is well established in rhesus monkeys. Young, Goy and Phoenix (1964) report nearly male frequencies of rough-and-tumble play in young female rhesus whose mother was injected with testosterone during pregnancy. It looks as if this may be one sex difference in human behaviour which is not culturally determined, although it could result from differences in physique rather than direct CNS effects of hormones.

DISCUSSION: COMPARATIVE ASPECTS

An advantage of this kind of study for comparative primatology is that it allows comparisons of the same kind of data about man and other primates, rather than comparisons of our everyday inside knowledge of man with a more empirical knowledge of other species. Thus we could compare counts of occurrence of smiles before, after or with attacks, avoidances, approaches and other behaviour, from both man and other species. Although it is premature to attempt extensive comparison, one or two points are worth mentioning here.

In the children smiling indeed comes out as closely related to approaching and embracing or staying with, as opposed to the closer relationship to avoiding which occurs in macaques and chimpanzees. But one sometimes feels that human smiles are also partly 'fear' motivated. These 'frightened smiles' may look different only because they are often momentarily interrupted if there is a strong tendency to flee, although they still result from a simultaneous strong stimulus to approach and stay with the other person. Also, the flickering timid smiles often occur with respiration reminiscent of laughing and I wonder if one might not find they resembled laughing more closely than smiles. I have already mentioned that laughter and screams during fleeing can intergrade in form. Andrew (1963) uses his concept of 'stimulus contrast' to relate these various expressions and vocalizations in a more uni-dimensional scheme.

While fixation in attack, with relatively inconspicuous facial changes (lowering brows), raised eyebrows in fleeing, and the play face are described in other species, I have not identified in the children any equivalent of grooming or lip-smacking, or the appeasement presenting.

In the age group studied, rough-and-tumble play seems to be more clearly differentiated from hostile behaviour than it is in any other species. Besides showing frequent alternation of chaser and chased, lack of threat postures, persistence, and lack of spacing effect and lack of injury, as in most animals, rough and tumble has at least four characteristic motor patterns which do not occur in hostile behaviour. These are wrestling, jumping up and down with both feet together, laughing, and play face. Perhaps this species difference is only apparent because there are so few published accounts in which an attempt has been made to document the difference between subjectively identified play fighting and real fighting. But the comparison suggests that rough-and-tumble play has a very important function in man if there are so many signals to indicate its difference from serious hostility.

More frivolous comparisons can be made with behaviour of adult humans in various cultures. Is it just coincidence that one revolutionary movement has a salute which is identical to the child's raised arm position in beating? Is the tradition of early explorers taking gifts with which to greet the savages nothing to do with nursery school children giving things to strangers? A different kind of comparison is with children in some societies who perform or attempt useful work. Children in nursery school spend much time simulating adult occupations in their play ('Firemen', 'Policemen', 'Shops', 'Offices', 'Tea Parties', etc.). Is it because they are deprived of the chance to work?

DISCUSSION: SPECULATIONS

I have already commented on the possible significance of rough-and-tumble play and raised the question of whether there is a critical period for developing this behaviour and whether the critical period might not end before three years old. This would have obvious educational implications where an only or oldest child is concerned. One could only produce answers to these suggestions by longitudinal study of a rather large number of children, but they would be good answers. More indirect evidence can be gathered in short-term studies, for instance on the differences in the other social behaviour of children who do or do not do rough-and-tumble play. Are those who do not play usually extreme in their aggression or fearfulness?

Only two of the children in my study group spent much time looking at books. These two spent little time making things or painting. They were also often alone, seldom moving about alone or with others, almost never joining in rough-and-tumble play. They talked a lot and very well, to themselves, to the teachers, one of them to me, and to any passing child. The other children spend a lot of time making things or painting, move about a lot, don't talk to me or to teachers much, do not hold long monologues, do join in rough-and-tumble play now and again. Is this the beginning of the 'two cultures', or just a chance occurrence? If a division of 'doers' and verbalists'' is a general division at this age, it corresponds remarkably closely with Bibby's (1964) modified delineation of the 'Two Cultures' (which themselves need proper documentation) to which Tanner (1964) drew my attention after I commented on these 'verbalist' children. The division is surprising in the way it goes so widely through other aspects of the child's behaviour. Is one of the clues to the cause of a child's taking to one 'culture' or another to be found in these other aspects of the behaviour, for instance its early sociability?

These are speculations but speculation is justified when one can so easily see what further evidence to gather. I felt it worth indicating some of those implications of ethological study of human behaviour which potentially ethology can investigate rather thoroughly.

SUMMARY

The behaviour of children in nursery school when greeting adults, fighting and fleeing, and in certain parts of their 'play', is readily analysable into rather constant motor patterns. Certain of these patterns occur most commonly with certain others, clustering into several different groups.

One such group, called 'rough-and-tumble play', is clearly distinct from hostile behaviour even though it superficially looks like fighting. This behaviour is discussed at some length because Harlow's experiments with rhesus monkeys suggest that it has an important rôle in development.

Some speculations about the importance of this and other aspects of their behaviour are included to indicate the kind of result one might expect and the kind of further evidence one would need to gather. Some of the areas in which the precision and objectivity of the study will be improved are also mentioned.

ACKNOWLEDGMENTS

I wish to thank the following for their varied and valued help in starting this work and providing facilities:

Mrs C. M. Lee, Miss Eileen Molony, Miss M. Brearley, Miss

K. Thompson, Miss F. M. Parsons, Miss M. Taylor, Prof.
K. R. L. Hall and Dr J. M. Tanner and the Nuffield Foundation.
Mr H. Goldstein advised me on some of the statistics. Dr Tanner
and many ethologists have entered into encouraging and helpful
discussions, including Dr G. M. Manley whose unpublished work
on play in lower primates has been a very direct help in thinking
about rough-and-tumble play in the children.

REFERENCES

ANDREW, R. J. (1963). 'The origin and evolution of the calls and facial
expressions of the primates', *Behaviour* **20**, pp. 1–109.
BIBBY, C. (1964). 'Science as an instrument of culture', *Nature* **202**,
pp. 331–3.
BLURTON JONES, N. G. (1960). 'Experiments on the causation of the
threat postures of Canada Geese', *Wildfowl Trust* 11th *Annual
Report* 1958–9, pp. 46–52.
—— (1964). 'Motivation studies with birds; with special reference to
conflict and thwarting', unpublished thesis, Oxford.
COCHRAN, W. G. (1954). 'Some methods for strengthening the common
X² tests', *Biometrics* **10**, pp. 417–51.
DARWIN, C. R. (1872). *The expression of emotions in man and animals*,
London, J. Murray.
HANSEN, E. W. (1962). 'The development of maternal and infant
behavior in the rhesus monkey', unpublished thesis, Wisconsin.
HARLOW, H. F. (1962). 'The Heterosexual affection system in monkeys',
American Psychologist **17**, pp. 1–9.
HARLOW, H. F. and M. K. HARLOW (1962). 'Social deprivation in
monkeys', *Scientific American* **207** (5), p. 136.
VAN HOOFF, J. A. R. (1962). 'Facial expressions in higher primates',
*Evolutionary aspects of animal communication, Symp. Zool. Soc.
Lond.* **8**, pp. 97–125.
MOYNIHAN, M. (1955). 'Some aspects of reproductive behaviour in the
Black-headed Gull (*Larus ridibundus L.*) and related species',
Behaviour Supplement **4**.
NETTER, F. H. (1958). 'Nervous System', *Ciba Collection of Medical
Illustrations Volume* 1. Ciba.
ROSENBLUM, L. A. (1961). 'The development of social behavior in the
rhesus monkey', unpublished thesis, Wisconsin.

TANNER, J. M. (1964). 'Human biology in general university education', *Teaching and research in human biology* 7, pp. 23–37.

TINBERGEN, N. (1953). *The herring gull's world*, London, Methuen.

—— (1959). 'Comparative studies of the behaviour of gulls (*Laridal*): a progress report', *Behaviour* 15, pp. 1–70.

YOUNG, W. C., R. W. GOY and C. H. PHOENIX (1964). 'Hormones and sexual behaviour', *Science* 143, pp. 212–18.

Index

Alba, J. de and S. A. Asdell, 110

Allogrooming, 148–75, 238–44; as cleaning behaviour, 163–5; description of, 151–7; distribution of, 149–50; relationships within troops, 157–63

Altman, J., 221, 228–9

Altmann, S. A., 65, 82, 86, 109–10, 116, 123, 154, 158, 161, 168, 181, 259–60

Andrew, R. J., 7–8, 11, 13–15, 22, 24, 27–30, 34, 36, 41, 43, 48–9, 50–2, 53–62, 156, 242–3, 245, 250, 256 f, 347, 355, 364

Andrew, R. J. and J. Buettner-Janusch, 150, 153, 156

Anthropomorphism, 1–2, 12

Antonius, O., 10, 55, 110

Appelman, F. J., 91

Armitage, K. B., 110, 150

Arnhem Zoo, 16, 29

Backhouse, K. M., 116

Bally, G., 180

Barnett, S. A., 150, 201

Bartelmez, G. W., 74

Baumann, H., 128, 132

Beach, A. F., 176–7, 180, 186–7, 201

Berg, I. A., 113, 115

Berkson, G., W. A. Mason and S. V. Saxon, 197

Berlyne, D. E., 193

Bernstein, I. S. and W. A. Mason, 77, 106, 231

Bibby, C., 366

Bierans de Haan, J. A., 176, 186, 191

Bindra, D., 187

Bingham, H., 4, 109, 208

Birch, H. G., 202–4

Birch, H. G. and G. Clark, 78

Bishop, P. M. F., 125, 150, 156

Blin, P. C. and J. A. Favreau, 80, 109

Blurton-Jones, N. G., 6, 348

Bolwig, N., 7–8, 10–13, 18–19, 23, 24, 27, 32, 35, 37, 40, 46–7, 49–51, 58, 79–80, 114, 158–9, 161, 183, 186, 205, 222

Booth, A. H., 89, 93–4, 303

Bopp, P., 78–9, 104

Brearley, M., 366

Breder, C. M., 176

Brownlee, A., 181, 185

Budd, A. and L. G. Smith, 292, 303

Budd, A., L. G. Smith and F. W. Shelley, 190, 200

Büttikofer, J., 88

Cabrera, A. and J. Yepes, 259

Cannon, Laura, 171

Carpenter, C. R., 4, 7, 78, 80, 86, 109–10, 112, 133, 150, 155, 157–9, 161, 163–4, 167, 188, 191, 201, 206–8, 210, 221, 237–8, 243, 261, 263

Chance, M. R. A., 80, 82, 86, 106, 109, 158–9, 166

Chance, M. R. A. and A. P. Mead, 132

Cole, J., 262

Collings, M. R., 86

Communication, comparative aspects of, 236–66; allogrooming, 238–44; facial expressions, 250–2; olfactory signals, 244–50; pilo-erection, 252–5; sounds, 255–7

Conaway, C. H. and C. B. Koford, 77–8, 101

Crawford, M. P., 47, 80, 109, 161

Cross, H. A. and H. F. Harlow, 332

Cullen, J. M., 148

Dandelot, P., 70, 89

Darwin, Charles, 3, 7, 21, 23, 27, 51, 56, 79, 119, 348–9, 355

Dathe, H., 114–15

Davenport, R. K. Jr. and E. W. Menzel Jr., 196

Davis, D. D. and H. E. Story, 104

De Beaux, 88

DeVore, I., 65, 133–4, 153, 164, 204–5, 313, 332, 335

DeVore, I. and S. L. Washburn, 83, 228

Ditmars, R. L., 9

Dobzhansky, 72

D'Osbonville, 209

Draper, W. A. and I. S. Bernstein, 197

Duffy, E., 187

Eckstein, P. and S. Zuckerman, 74, 86, 89

Eibl-Eibesfeldt, I., 7, 9, 54, 56, 113, 132

Eisenberg, J., 166

Elliot, D. G., 73, 88

Epsmark, Y., 113

Erikson, G. E., 246

'Ethograms', 5

Ethology, beginnings of, 4–6

Evolution, 263

Ewing, H. E., 153, 163

Facial displays, 7–68, 250–2; conditions influencing appearance of, 8–11; description of compound expressions, 17–64; methods and terminology, 10–16

Falk, J. L., 163

Fiedler, W., 86, 88, 93, 114, 236

Finch, 28

Fischer, E., 125

Fisher, 79

Fitzgerald, A., 150, 152, 207, 238

'Fixed motor pattern', 5

Folia Primatologica, 263

Fooden, J., 237

Frankfurt Zoological Garden, 86, 91

Freedman, L., 70–1

Frijda, N. H., 57

Furuya, Y., 158, 161, 167–8

Gajdusek, D. C., 134

Galt, W. E., 80, 109

Garrod, A. H., 91–2

Gartlan, J. S., 221, 224

Gessner, K., 128

Gilbert, C. and J. Gillman, 74

Glickman, S. E. and R. W. Sroges, 193, 201

Goldstein, H., 367

Gombe Stream Reserve (Tanzania), 201, 204, 287

Graham-Jones, O. and W. C. O. Hill, 97

Grant, E. C. and J. H. Mackintosh, 150

Gregory, W. K., 9

Grimm, 124

Groos, K., 184, 186

Gusinde, M., 125

Haber, 14

Haddow, A. J., 157

Hall, Fae, 171

Hall, K. R. L., 18, 22, 32, 34, 40, 62, 70, 77, 110, 118, 157, 159, 171, 226, 228, 367

Hall, K. R. L., R. C. Boelkins and M. J. Goswell, 80, 91, 110, 113, 118, 152–4, 167, 260 f

Hall, K. R. L. and I. DeVore, 80, 109, 260 f
Hall, K. R. L. and J. S. Gartlan, 157
Hamilton, 80, 109
Hansen, E. W., 33, 50, 357, 359
Harlow, H. F. and M. K., 163, 165, 185, 211, 267, 286, 330, 357, 366
Harlow, H. F. and M. K., and E. W. Hansen, 292
Harms, J. W., 77, 90, 91
Harrison, C. J. O., 149, 152, 188, 192
Harrison-Matthews, L., 171
Hartman, C., 86, 97
Haselrud, G. M., 197
Hävernick, 128
Hebb, D. O., 187, 189
Hediger, H., 113, 149, 195, 201
Hediger, H. and F. Zweifel, 79, 97, 112
Herschkovitz, P., 69, 236, 262
Hess, E. H., 211
Hewes, G. W., 129
Hidden Persuaders, The, 122
Hill, W. C. O., 74, 86, 89, 91, 93–4, 112–13, 119, 237, 240, 245
Hill, W. C. O., A. Porter and M. D. Southwick, 96
Hinde, R. A., 15, 337
Hinde, R. A. and T. E. Rowell, 6–8, 18–19, 22–5, 30, 32, 35, 39, 47–8, 53, 61–2, 109–10, 262, 267
Hinde, R. A., T. E. Rowell and Y. Spencer-Booth, 268–9, 292, 294–5, 298, 301–3, 314, 319, 336
Hines, M., 159
Histoire naturelle des Mammifères (Cuvier), 85
Hokororo, A. M., 127
Holst, von, 13
Holzapfel, M. M., 195
Hooton, E., 27, 47, 124
Huber, E., 9, 11, 56
Hudson, Anita, 171
Hutchinson, G. E., 106
Huxley, 78–9

Ilse, D. R., 113
Imanishi, K., 88, 101, 335
Inhelder, E., 80, 109
Itani, J., 133

Jacobsen, C. and M., and J. Yoshioka, 188–9
Jay, P., 77, 80, 93, 109, 133–4, 153, 159, 169, 206, 221, 260 f, 294, 303, 313–15, 319, 325, 335
Jensen, G. D. and R. A. Babbitt, 222
Jensen, K., 320
Jolly, A., 224
Jürgens, H. W., 134

Kacher, H., 72
Kaufman, C., 227
Kaufmann, J. H. and A., 70
Kawamuru, S., 223
Kellogg, R. and E. A. Goldman, 237 f
Kennedy, J. S., 14
King, Christine, 171
Kirchshofer, R., 110, 113–15
Kitzler, G., 9
Knottnerus-Mayer, T., 79, 163
Koford, C. B., 133, 209, 335
Köhler, W., 3, 27, 48, 51, 61, 202
Kohts, Nadie, 4, 7, 27, 51
Koning, Edna, 288, 308 f, 337
Kortlandt, A., 104, 108, 290, 314, 325
Kroeber, 170
Kühme, W., 72, 102, 115
Kummer, H., 19, 23–4, 27, 30, 36, 50, 71, 77, 79, 82, 109, 133, 155, 158, 219
Kummer, H. and F. Kurt, 82, 230
Kunkel, P. and L., 114–15, 157

Lancaster, J. B. and R. B. Lee, 101
Lashley, K. S. and J. B. Watson, 36, 230
Lawick, Baron H. van, 336
Lawson and Marx, 28
Leakey, L. S. B., 336
Lee, C. M., 366

Leonhard, K., 128, 134
Leyhausen, P., 10, 104
Lightoller, G., 9
Loizos, C., 6
London Zoo, 4, 16, 151–2, 155–6, 171, 182–3, 205, 208, 214, 230, 292
Lorenz, K., 5–7, 58, 72, 124, 134, 149, 178–80, 186–7, 195 f, 204
Lowther, F. de L., 207
Lullies, 131.

Maclean, P. D., 111, 115, 128, 132
Malinowski, 170
Manley, G., 150, 171, 208–9, 214, 367
Manyara, Lake (Tanganyika), 85
Marler, P., 7, 65, 158, 161, 169, 229, 241, 246, 260 f
Martin, R. D., 72
Maslow, A. H., 107, 161
Mason, W. A., 184, 187, 206, 213, 256, 258
Mason, W. A. and G. Berkson, 332
Mason, W. A., G. Berkson, J. H. Hollis and L. G. Sharpe, 335
Mason, W. A., P. C. Green and C. J. Posepanko, 107
Matschie, 88
Matthews, L. H., 91–2, 104
McClelland, D. C., 14
Mead, M., 134–5
Men and Apes, 3
Menzel, E. W. Jr., 193, 196–7, 199, 201
Menzel, E. W. Jr., R. K. Davenport Jr. and C. M. Rogers, 196, 198, 201
Meyer, 88
Michael, R. P. and J. Herbert, 162
Miller, R. A., 73
Miller, R. E. and J. H. Banks, 110
Mitchell, C. P., 184
Molony, Eileen, 366
Monkey Centre (Inuyama, Japan), 4
Morris, D., 6, 9, 37, 114, 151–2, 154, 171, 179–80, 186–8, 191–2, 196, 202, 214

Morris, R. and D., 3
Mother-infant relations, in free-ranging chimpanzees, 287–346; in rhesus monkeys, 267–86
Moynihan, M., 6, 8, 13–15, 152, 201, 236, 244, 245 f, 257, 259, 347
Müller-Using, D., 110
Murie, J., 73

Nairobi National Park, 85, 229
Napier, J., 246
National Geographic Society, 336
Nelson, K., 123
Nesturkh, M. F., 74
Netter, F. H., 355
Nicolai, J., 134
Nissen, Henry, 4, 221
Nolte, A., 109, 150, 158, 164, 237
Nowlis, V., 78

Orange Park (Florida), 4
Ottow, B., 115
Owen, Richard, 3

Packard, V., 122
Parkes, A. S., 74
Parson, F. M., 367
Petter, J. J., 101, 237, 263
Petter-Rousseaux, A., 96
Play behaviour, 175–218; causation of, 186–7; function of, 184–6; nature of, 177–84; ontogeny of, 188–91; social, 204–14; solitary, 191–204
Ploog, D. W., 112, 134
Ploog, D. W., J. Blitz and F. Ploog, 111–12, 238, 241 f
Ploog, D. W. and P. D. Maclean, 243
Pöch, R., 125
Pocock, R. J., 74, 77, 86, 89, 91, 93, 100
Polyak, St, 10
Pycraft, W. P., 186

Rand, R. P., A. C. Burton and T. Ing, 72

Reitzenstein, Frhr. F. v., 127
Reynolds, V., 78, 82, 86, 107, 109, 122, 133, 151–3, 156, 158–9, 161, 163, 166–7, 188, 192, 205
Reynolds, V. and F., 158–9, 161, 166–7, 171, 209, 290, 298
Ripley, S., 221
Romanes, G. J., 208
Rosenblum, L. A., 227, 357
Rotterdam Zoo, 16
Rowell, C. H. Fraser, 165
Rowell, T. E., 78, 82, 86, 97, 221, 228, 260 f
Rowell, T. E. and R. A. Hinde, 164, 260 f, 262
Rowell, T. E., R. A. Hinde and Y. Spencer-Booth, 133, 162, 164, 167, 267, 282
Ruge, G., 9, 11
Russell, W. M. S., 170

Sade, D. S., 116, 229, 311, 335
Saint-Paul, von, 13
Saller, K., 128
Sanderson, I. T., 112, 237, 243, 246
Schaller, G. B., 20, 26, 47, 51, 61, 151–2, 155–9, 161, 164, 188, 206–8, 295, 303–4, 308, 314, 319–21, 333
Schenkel, R., 10, 82, 110
Schiller, P. H., 181, 202–3
Schlegel, 88
Schloeth, R., 34, 80, 109, 114
Schlosser, K., 128
Schmidt, H. D., 114
Schneider, K. M., 44–5
Schultz, A. H., 115
Schultze, L., 125
Schwarz, E., 70
Scott, J. P. and E. Fredericson, 24, 150
Serengeti National Park, 89
Shadle, A., R. Smelzer and M. Metz, 115
Sibley, C. S., 134
Simonds, P. E., 80, 87, 109, 164, 167, 169, 205, 208–9

Simons, E. L., 243
Social behaviour of children, ethological study of, 347–68
Social organization of primates, 219–35; environmental factors, 226–33; illusions affecting observations, 222–6
'Social releaser', 12
Socio-sexual signals, 69–147; comparison of Hamadryas with other catarrhines, 84–98; genital display, threat and territory marking, 113–20; Hamadryas' hindquarters problem, 72–84; interference with rank order, 77–8; intraspecific signal-imitation, 98–100; male sexual activities, 108–13; motivational aspect, 106–8; parallels with human social behaviour, 124–35; presenting as a 'greeting ceremony', 79–84; socio-sexual link, 100–6
Sorenson and Conoway, 263
Southern, H. N., 114
Southwick, C. H., M. A. Beg and M. R. Siddiqi, 161–2, 204, 206–8, 226
Sparks, J. H., 6, 149, 152, 157–9, 163, 166–7
Spiegel, A., 77, 88, 97
Spindler, P., 134
Spiro, M. E., 124
Spitz, R. A., 133
Stieve, H., 116
'Stimulus contrast', 14–16
Stirton, R. A., 257
Struhsaker, 221
Sugiyama, Y., 221, 224
Swynnerton, 89

Tanganyika National Parks, 72
Tanner, J. M., 366–7
Taylor, M., 367
Tembrock, G., 133, 180
Thompson, K., 367
Thorington, R. W. Jr., 237
Tinbergen, N., 5–7, 12–13, 152, 347–8

Tinklepaugh, O. L., 121, 186
Tokuda, K., 88
Tollan, E. C., 186
Tombs, Valerie, 171
Torday, E. and T. A. Joyce, 125
Trumler, E., 10, 55
Tsavo National Park (Kenya), 85
Tugendhat, 348

Ullrich, W., 76

Valen, van, 263
'Valency', 14
Van den Berghe, 292
Vandenbergh, J. G., 242
Van Hooff, J. A. R. A. M., 6–8, 11, 16, 19, 23, 31, 33–4, 37, 40–1, 45, 53, 54, 62, 153, 205, 250, 347, 358
Veit-Stoeckel, 138

Wallis, D. I., 149
Walls, G. L., 10
Washburn, S. L., 69, 73
Washburn, S. L. and I. DeVore, 77–8, 101, 158–9, 161, 164, 168, 221

Washburn, S. L. and D. A. Hamburg, 206
Watson, J. B., 150, 164
Weber, H., 9, 88
Welker, W. I., 177, 190, 193, 199–201, 213
Whipsnade Zoo, 16, 82
Wickler, W., 6, 84, 92, 100, 102–4
Winkelsträter, K. H., 107, 110

Yamada, M., 335
Yerkes, R. M., 4, 27, 78, 82, 155, 159, 164, 169, 221, 290–2, 308, 313
Yerkes, R. M. and A., 48
Yerkes, R. M. and J. H. Elder, 77, 82
Young, W. C., R. W. Goy and C. H. Phoenix, 363
Young, W. C. and W. D. Orbison, 82

Zeeb, K., 10, 54, 55
Zuckerman, S., 4, 27, 40, 77, 100–2, 107, 150, 153, 157–9, 161–2, 164, 166, 168, 221–2, 230